Lecture Notes in Mathematics

Edited by J.-M. Morel, F. Takens and B. Teissier

Editorial Policy
for the publication of monographs

1. Lecture Notes aim to report new developments in all areas of mathematics and their applications- quickly, informally and at a high level. Mathematical texts analysing new developments in modelling and numerical simulation are welcome.

   Monograph manuscripts should be reasonably self-contained and rounded off. Thus they may, and often will, present not only results of the author but also related work by other people. They may be based on specialised lecture courses. Furthermore, the manuscripts should provide sufficient motivation, examples and applications. This clearly distinguishes Lecture Notes from journal articles or technical reports which normally are very concise. Articles intended for a journal but too long to be accepted by most journals, usually do not have this "lecture notes" character. For similar reasons it is unusual for doctoral theses to be accepted for the Lecture Notes series, though habilitation theses may be appropriate.

2. Manuscripts should be submitted (preferably in duplicate)) either to Springer's mathematics editorial in Heidelberg, or to one of the series editors (with a copy to Springer). In general, manuscripts will be sent out to 2 external referees for evaluation. If a decision cannot yet be reached on the basis of the first 2 reports, further referees may be contacted: The author will be informed of this. A final decision to publish can be made only on the basis of the complete manuscript, however a refereeing process leading to a preliminary decision can be based on a pre-final or incomplete manuscript. The strict minimum amount of material that will be considered should include a detailed outline describing the planned contents of each chapter, a bibliography and several sample chapters.

   Authors should be aware that incomplete or insufficiently close to final manuscripts almost always result in longer refereeing times and nevertheless unclear referees' recommendations, making further refereeing of a final draft necessary.

   Authors should also be aware that parallel submission of their manuscript to another publisher while under consideration for LNM will in general lead to immediate rejection.

3. Manuscripts should in general be submitted in English. Final manuscripts should contain at least 100 pages of mathematical text and should always include
   - a table of contents;
   - an informative introduction, with adequate motivation and perhaps some historical remarks: it should be accessible to a reader not intimately familiar with the topic treated;
   - a subject index: as a rule this is genuinely helpful for the reader.

Continued on inside back-cover

Lecture Notes in Mathematics 1829

Editors:
J.-M. Morel, Cachan
F. Takens, Groningen
B. Teissier, Paris

Springer
Berlin
Heidelberg
New York
Hong Kong
London
Milan
Paris
Tokyo

Eitan Altman
Bruno Gaujal
Arie Hordijk

Discrete-Event Control
of Stochastic Networks:
Multimodularity
and Regularity

Springer

Authors

Eitan Altman
INRIA
2004 Route des Lucioles
06902 Sophia-Antipolis Cedex, France
e-mail: altman@sophia.inria.fr

Bruno Gaujal
ENS Lyon, LIP
46 Allée d'Italie
69364 Lyon Cedex 07, France
e-mail: Bruno.Gaujal@ens-lyon.fr

Arie Hordijk
Mathematical Institute
Leiden University
P.O. Box 9512
2300 RA Leiden, The Netherlands
e-mail: hordijk@math.leidenuniv.nl

Cataloging-in-Publication Data applied for
Bibliographic information published by Die Deutsche Bibliothek

Die Deutsche Bibliothek lists this publication in the Deutsche Nationalbibliografie;
detailed bibliographic data is available in the Internet at http://dnb.ddb.de

Mathematics Subject Classification (2000): 60-XX, 60C05, 49-XX, 93-XX

ISSN 0075-8434
ISBN 3-540-20358-3 Springer-Verlag Berlin Heidelberg New York

Springer-Verlag is a part of Springer Science+Business Media

springeronline.com

© Springer-Verlag Berlin Heidelberg 2003
Printed in Germany

Typesetting: Camera-ready TeX output by the authors

SPIN: 10964620   41/3142/du - 543210 - Printed on acid-free paper

To our families

Preface

This work has been made possible largely thanks to the kind support of the Van Gogh project N. 98001 "Multimodularity and Control" (French-Dutch scientific cooperation project) and of INRIA *Action de Recherche Coopérative Maddes*. The work of Arie Hordijk on this book was initiated while he was on sabbatical leave at INRIA, Sophia-Antipolis; partially supported by the Ministère Français de l'Éducation Nationale et de l'Enseignement Supérieur et de la Recherche.

We wish to thank colleagues with whom we have had stimulating discussions and who helped us with different theoretical points presented here. In particular, we wish to thank François Baccelli, Sandjay Bhulai, Jerome Galtier, Alex Heinis, Emmanuel Hyon, Alain Jean-Marie, Ger Koole, Zhen Liu, Rob Tijdeman and Dinard van der Laan.

This book summarizes several years of research work of its authors. Part of the material in the book was obtained jointly with other researchers: Sandjay Bhulai, Emmanuel Hyon, Ger Koole and Dinard van der Laan.

Leiden, Lyon, Sophia-Antipolis
November, 2001

Eitan Altman
Bruno Gaujal
Arie Hordijk

Table of Contents

Introduction

A dynamical system which changes state at discrete points in time is called a discrete event dynamic system. The theory of discrete event dynamic systems is mainly used in the study of manufacturing systems, telecommunication networks and transportation systems.

The development of this theory has undergone a dramatic change of perspective recently thanks to the introduction of general principles which are useful to a wide range of application domains. Such principles include the use of the (max,plus) algebra [23] and more generally topical functions [57] to model the synchronizations present in the system as well as the network calculus [39, 37]. Another example is the work of [52, 53] where rather elementary monotone structures are used to derive powerful results.

The approach adopted here goes in the same direction. This book will introduce several general principles useful in the control of discrete event systems.

The aim of this monograph is not to offer a complete theory of discrete event control of stochastic networks, but to derive a theory and applications based on multimodularity and regularity. The main objective is to show that for a large class of stochastic discrete event systems and under rather natural assumptions on the behavior of the system as well as on the stochastic processes driving its evolution, the smoother the input process, the better the performances of the system.

Of course, the notions of smoothness and the performance criteria have to be made precise. This requires several technicalities using several notions from convex analysis, stochastic processes and word combinatorics. However, this must not hide the general underlying goal of the whole work.

This book is focused on a wide class of control (or of optimization) problems over sequences of integer numbers. We know that the theory of convex functions plays a key role in the theory of optimization over convex spaces. An important objective is to construct a counterpart for our setting in which the optimization is not done over a convex set any more. A natural candidate to replace a convex function over some set Z^n of n-dimensional vectors of integers is an integer-convex function, i.e. a real valued function f that satisfies the standard convexity condition

$$f(\alpha x + (1-\alpha)y) \leq \alpha f(x) + (1-\alpha)f(y),$$

for all $x, y \in Z^n$ and for all $\alpha \in (0, 1)$ for which $\alpha x + (1 - \alpha)y \in Z^n$. Unfortunately, this intuitive counterpart of convexity turns out to be too restrictive for our purpose. In particular, it does not even guarantee that a local minimum is a global minimum!

The natural counterpart of convex functions over integer sets turns out to be the so called *multimodular functions*, introduced in [59], and for such functions we have indeed the property that local minima are global minima. The property of multimodularity turns out to be useful in a much more general context: the control of discrete event systems. As we shall illustrate throughout this monograph, the natural performance measures in many problems in queuing, in telecommunications and in other areas of applications turn out to be multimodular. This includes many admission control problems to networks, routing control in networks, service assignment problems and control of vacations. Typical performance measures that are multimodular in these problems are expected waiting times, sojourn times and queue lengths. For all these problems, it turns out that the multimodularity of the costs induces, in many case, a particular form of optimal policies. They turn out to be very "regular", and can be described by the well known "bracket" sequences. Thus the study of multimodular functions is strongly related to the study of bracket policies.

The definition of multimodularity goes back to the seminal paper by Hajek [59], who introduced this term in order to study a problem of optimal admission control into a single queue under no queue information. The precise problem was to admit customers to a single queue, under the constraint that the long run fraction of customers admitted be at least p. The optimality of a policy based on a bracket sequence of admission actions was obtained in [59] for the number of customers in a one-server queue with exponential service and a renewal arrival process.

Another application of multimodular functions is in the control of queues with full state information. Weber and Stidham [114] and Glasserman and Yao [52, 53] obtained monotone properties of the optimal control policies as a function of the state, in a variety of queueing control and related problems. The methodology was strongly based on the multimodularity properties of the immediate costs and the cost-to-go functions.

In this book, we develop mainly the tools for control problems with no state information. This is done both in a deterministic setting as well as in a very general stochastic framework. Two classes of problems are handled. In the first one, the control sequence is one dimensional; it covers admission control, service assignment and vacation control problems. Although the control is one dimensional, the systems to which it is applied to may be quite general and complex. We focus on general discrete event models which can be described as linear in the max-plus algebra. This type of problem is fully solved, and an optimal policy is identified. A second type of problems we handle is the one in which the control is multi-dimensional. It covers routing as well as polling problems. Optimal policies are obtained only in special cases: the case of symmetrical models, systems with dimension two and many others. The restriction is due to the fact that the regularity properties that characterize optimal policies in dimension one cannot be generalized to dimension greater than two, except for very special cases. The identification of all cases in which regularity is possible in dimension larger than two has been a challenging open problem since several decades since it was formulated in the well known Fraenkel conjecture [46], and it prompted much research [55, 91, 107, 115, 93, 46, 90]. From the cases of low-dimension (up to six) which have been fully solved [108] we know that sequences that are "regular" in all components are very rare. We therefore need other tools to handle higher dimensions. This motivates us to consider the question of regular ordering between policies, which basically aims at identifying orders between policies such that if a policy is greater than another in that ordering then it yields a better cost. We also derive bounds on the performance measures by using a new combinatorial notion, called unbalance of a multidimensional control sequence.

Although our main concern in the book is the handling of control problems with no information, we also investigate the problem of closed-loop control in which the control has full or partial state information. We develop a general framework for handling multimodular cost functions, and we establish the optimality of optimal monotone policies: these are of threshold type for the one dimensional case, and of a switching-curve type for the two-dimensional case.

This book is structured in four parts.

Part I: Theoretical foundations, presents the theoretical foundations which consists of three notions: multimodularity, balanced words, bracket sequences and stochastic Petri nets. The material in Part I is used throughout the monograph. The other chapters can be studied independently.

In Chapter 1 we present the basic definition of multimodularity and investigate the properties of multimodular functions. Using these properties, we obtain general optimization results which are in particular useful for average cost criteria in open-loop control. We show in that context that the expected average cost problem is optimized by using bracket sequences.

Chapter 2 introduces the balanced sequences, the Sturmian words and bracket sequences and shows the close relations existing between them. It also details several ways to construct such sequences as well as some of their properties, useful in this context. More details on the combinatorial properties of this sequences can be found in [84].

In order to apply the tools developed in Chapters 1 and 2 to networks, and more generally to discrete event dynamic systems, we present in Chapter 3 the formalism of Petri nets, and more precisely of the so called stochastic event graphs; These are systems whose dynamics follows a linear vectorial evolution equation in the so-called (max,+) algebra, and it covers many queuing networks.

Part II: Admission and routing control, covers the application of the optimisation theorems for optimal admission and routing control in discrete event stochastic systems.

Using the above formalism we consider in Chapter 4 the first application of the general theory to problems of admission control into networks, where we show the main theorems. We show that open loop admission control in a stochastic network with (max,plus) dynamics is optimal for the travelling time of customers when the routing policy is Sturmian.

Chapter 5 discusses the relevance of the assumptions made so far for networks of queues, and in particular, the issue of cross traffic.

The objective pursued in Chapter 6 is to generalize the admission control problems to the case of routing control. The first part addresses the following combinatorial problem: is it possible to construct an infinite sequence over K letters where each letter is distributed as "evenly" as possible and appears with a given rate? The second objective of the Chapter is to use this construction in the framework of optimal routing in queuing networks. We show under rather general assumptions that the optimal deterministic routing in stochastic event graphs is such a sequence.

While Chapter 6 says that a bracket squence is optimal for routing into two parallel systems, it does not give any hint on the actual optimal policy. Chapter 7 gives the computation of the optimal policy in the simple case of deterministic queues. It is rather surprising that this computation uses the decomposition in continued fractions of the parameters of the system. It may also be surprising that greedy policies such as "join the shortest queue" are not always optimal here.

Part III: Several extensions, shows how the previous results can be useful in other cases.

In Chapter 8 we consider the problem of optimal routing of arriving packets into K servers having no waiting room. Packets that are routed to a busy server are lost. We consider two problems where the objective is to maximize the expected throughput (or equivalently, minimize the loss rate). We assume that the controller has no information on the state of the server. We establish the optimality of the bracket sequences, for exponential service times and general stationary arrival processes, which include, in particular, the interrupted Poisson process, Markov modulated Poisson Process (MMPP) and Markov arrival process (MAP). Based on this solution, we solve the dual problem of optimal assignment of a single server to several single server queues to which packets arrive according to Poisson processes. A first step for the solution of this problem is to establish the multimodularity properties of the costs, which together with some other conditions that we establish, imply the regularity of optimal policies. We go however beyond this characterization of optimal policies and show that (i) there exist optimal *periodic* policies; this is established using some tools from Markov decision processes; (ii) Policies that are regular in a weaker sense (related to the majorization ordering) are also optimal; this is done using the Schur convexity ordering. We finally apply our framework to the problem of robot scheduling for web search engines.

In Chapter 9 we consider the optimal open-loop control of vacations in queuing systems. The controller has to take actions without state information. We first consider the case of a single queue, in which the question is when should vacations be taken so as to minimize, in some general sense, workloads and waiting times. We then consider the case of several queues, in which service of one queue constitutes a vacation for others. This is the optimal polling problem. We solve both problems using concepts from multimodularity.

Chapter 10 shows how multimodularity can be used for closed loop control. Using this approach, we show that the optimal policy has monotonicity properties for a rather general class of problems.

Part IV: Comparisons, derives the theory of routing to three queues or more. Since the structure of the optimal policy is unknown in this case, the analysis is focused on getting lower and upper bounds for the performance measures, and on the comparison of the systems for different admisssion sequences.

Chapter 11 shows how the travelling times in a FIFO-stochastic event graph are compared in increasing convex ordering. Two comparison lemmas are proved. As application of the first one, we derive that:

- Independent sources perform better than coupled sources.
- Fixed batch sizes are better than random batch sizes.
- Fluid scaling improves the performance.

The second comparison lemma implies a lower bound on the travelling time.

In Chapter 12, the notion of multimodularity is generalized to simplex convexity. Whereas multimodular functions are defined through the L-triangulation, simplex convexity is defined on any triangulation. It turns out that the theory of Chapter 1 can be generalized to simplex convexity and the notion of cone ordering is introduced. It is related to the "distance" of a given sequence to the sequence minimizing a given simplex convex function. An application is made for periodic admission sequences in queues.

Chapter 13 introduces several other orders (graph order, unbalance) which are compared with the cone ordering. These new ways to compare sequences are used to derive bounds on the average waiting time in queues for several arrival processes.

Finally, Chapter 14 shows the link existing between regularity (in the multimodular sense) and majorization (in the Schur convex sense) and how this can be used in queues. The link between the two notion is given by the notion of regular perserving functions (which is stronger than multomodularity). It can be used to compare the maximum waiting time in stationary systems rather than the average waiting time, as for multimodularity.

Finally, note that the notation is uniformized wherever possible. However, the chapters in Parts II,III and IV are rather independent from each other and notations may sometimes vary slightly between different chapter. Most chapters are based on material which is published in journal papers ([6, 4, 5, 3]). We are grateful to the publishers of these journals for their permission to include adaptations of these texts in this monograph.

Part I

Theoretical foundations

The first part of the book is made of three chapters, dedicated respectively to the notions of multimodularity, balanced sequences (also called Sturmian words) and Petri nets.

In the following, we will mainly deal with discrete event systems modeled as *Petri net* with an input process controlled by *Sturmian words* with a cost function which is *multimodular*.

The three notions are used together in two general theorems: Theorem 7 which states that Cesaro limits of multimodular functions are minimized by Sturmian sequences and 18 which shows that the average traveling time of tokens in a Petri net is a multimodular function of the input process.

1 Multimodularity, Convexity and Optimization

1.1 Introduction

This chapter presents the theoretical framework useful in the following chapters dedicated to the control of the input in networks. It is mainly focused on the study of multimodular functions which can be seen as a discrete analog to convex functions.

We provide elementary proofs for properties already established by Hajek, who showed that the lower convex envelope of a multimodular function f is the piecewise linear interpolation on a specific triangulation of the space by simplices called atoms. In this chapter, we show directly that this linear interpolation is convex if and only if f is multimodular. A more general version of this result will also be presented in Chapter 12. This allows us to restrict the study of multimodular functions to convex subsets of $\mathbb{Z}^m$ which are convex unions of atoms (or faces of atoms). Additional interesting properties of multimodular functions are also presented.

In the second part of this chapter, we develop basic optimization tools for average costs. We establish lower bounds for average costs using Abel-type asymptotic techniques. We also show that the lower bounds are achieved by bracket sequences. Such costs depend on a sequence of multimodular functions, rather than on a single multimodular function. This is a nice feature of our approach since the optimization results can be applied directly to average costs as long as the assumptions used in Theorems 6 or 7 are satisfied. This is not the case when a single cost function is used, as in [59], where for any specific application, additional analysis has to be done before one may apply the general minimization results to average costs problems.

We illustrate the usefulness of this theory in admission control into a very simple queue and we provide a detailed analysis of the D/D/1 queue with fixed batch arrivals, with no state information. We show that the policy which is defined through a balanced sequence minimizes the average queue length for the case of an infinite queue, but not for the case of a finite buffer. However, when further restricting to those policies for which no losses occur, we obtain again the optimality of balanced policies. To conclude that example, we study also the case where it is possible to admit a part of an arriving batch.

The more general cases such as the G/G/1 queue and networks of queues are treated in Chapter 4.

1.1.1 Organization of the chapter

In the first part of the chapter, we develop basic optimization tools for average costs. We establish lower bounds for average costs using Abel-type asymptotic techniques.

In follow-up chapters, we shall make use of all the theoretical results of this chapter in order to study more general admission and service control problems in dynamic systems that can be described using the max-plus algebra, with general stationary inter-arrival and service times.

1.2 Properties of multimodular functions

We present in this section a short overview of Hajek's theory of multimodular functions. Some additional results are also established. We begin by presenting the definition of multimodularity, and some general properties (Subsection 1.2.1) which are interesting by their own and that will be used in subsequent work. We then present in Subsection 1.2.2 the relation between multimodularity and convexity. The properties presented in Subsection 1.2.2 are those needed in the following sections and subsequent chapters on optimization and control.

In the following, $\mathbb{Z}$ will denote the set of integer numbers and $\mathbb{N}$ the set of non-negative integer numbers, $\mathbb{N} = \{0, 1, \cdots\}$.

Let $e_i \in \mathbb{N}^m$, $i = 1, \cdots, m$ denote the vector having all entries zero except for a 1 in its ith entry. Define $s_i = e_{i-1} - e_i$, $i = 2, \cdots, m$ (for an integer i taking values between 1 and m, we understand throughout $i - 1 = m$ for $i = 1$).

Let $\mathcal{F} = \{-e_1, s_2, \cdots, s_m, e_m\}$, where

$$
\begin{aligned}
-e_1 &= (\ -1\ \ 0\ \ 0\ \cdots\ 0\ \ 0\ ), \\
s_2 &= (\ \ 1\ -1\ \ 0\ \cdots\ 0\ \ 0\ ), \\
s_3 &= (\ \ 0\ \ \ 1\ -1\ \cdots\ 0\ \ 0\ ), \\
&\quad\ \ \vdots \\
s_m &= (\ \ 0\ \ \ 0\ \ \ 0\ \cdots\ 1\ -1\ ), \\
e_m &= (\ \ 0\ \ \ 0\ \ \ 0\ \cdots\ 0\ \ 1\ ).
\end{aligned}
$$

$\mathcal{F}$ will be called a multimodular base of $\mathbb{Z}^m$.

In a typical application, an element $x \in \mathbb{N}^m$ will be used to denote a control sequence; in admission control problems x_k would then have the meaning of the number of customers accepted to a system at time slot k. The vector $x + e_m$ then has the meaning of admitting one more customer at the last epoch (as compared to the acceptance pattern of a vector x), whereas $x + s_j$ has the meaning of "shifting" a customer to the "left", i.e. accepting one customer less at the jth slot and one more at the $(j-1)$st slot (with respect to the acceptance pattern given by x).

Define $\hat{\mathcal{F}} = \{e_i, -e_i, s_i, -s_i, \ i = 1, \cdots, m\}$, and $\overline{\mathcal{F}} = \{-u : u \in \mathcal{F}\}$.
Let $\mathbb{R}$ be the set of all real numbers.

Definition 1 (Hajek). *A real-valued function $f : \mathbb{Z}^m \longrightarrow \mathbb{R}$ is multimodular with respect to $\mathcal{F}$ if for all $x \in \mathbb{Z}^m$, v and w in $\mathcal{F}$, $v \neq w$, the following holds:*

$$f(x+v) + f(x+w) \geq f(x) + f(x+v+w). \tag{1.1}$$

Remark 1. A function f on $\mathbb{Z}^m$ is multimodular with respect to $\mathcal{F}$ if and only if f is multimodular with respect to $\overline{\mathcal{F}}$. This can be checked easily using $y = x + v + w$ instead of x in Equation 1.1.

Unless otherwise stated, we shall say that f is multimodular if it is multimodular with respect to $\mathcal{F}$. A more general definition of multimodularity wll be given in Chapter 12, where the base may not be $\mathcal{F}$ or $\overline{\mathcal{F}}$.

1.2.1 General properties

For a function g defined on $\mathbb{Z}^m$, define

$$\Delta_i g(x) = \Delta_{e_i} g(x) = g(x + e_i) - g(x) \qquad and \qquad \Delta_{s_i} g = \Delta_{i-1} g - \Delta_i g.$$

We further define $\Delta_{-e_i} g = g(x - e_i) - g(x)$. Note that $\Delta_{s_i} g(x) = g(x + e_i + s_i) - g(x + e_i)$.
It is easy to check that

Lemma 1. *Δ_v is a linear operator for any $v \in \hat{\mathcal{F}}$ (i.e. $\Delta_v(\mu g + \lambda f) = \mu \Delta_v g + \lambda \Delta_v f$). For all $v, w \in \hat{\mathcal{F}}$, $\Delta_v \Delta_w g = \Delta_w \Delta_v g$.*

Lemma 2. *(a) f is multimodular if and only if*

$$\Delta_v \Delta_w f \leq 0 \tag{1.2}$$

for all $v, w \in \mathcal{F}$, $w \neq v$.
 (b) If f is multimodular then

− (b.i) For any $w \in \mathcal{F}$ with $w \neq e_1$.

$$\Delta_{e_1} \Delta_w f \geq 0. \tag{1.3}$$

− (b.ii) For all i, j,

$$\Delta_i \Delta_j f \geq 0. \tag{1.4}$$

− (b.iii) For all i, j,

$$\Delta_j \Delta_j f \geq \Delta_i \Delta_j f. \tag{1.5}$$

− (b.iv) $\Delta_{e_i} \Delta_{s_i} f \geq 0$.

– *(b.v)* $\Delta_{e_i}\Delta_{s_j}f \leq 0$, $j < i$ and $\Delta_{e_i}\Delta_{s_j}f \geq 0$, $j > i$.

– *(b.vi)* $\Delta_{s_1}\Delta_{s_i}f \leq 0$, $i \neq 1$.

– *(b.vii)* $\Delta_{s_i}\Delta_{s_i}f \geq 0$.

(c) Consider the 2-dimensional case: $\mathcal{F} = \{-e_1, s_2, e_2\}$. Assume that (1.4) and (1.5) hold. Then f is multimodular.

The proof of the lemma is technical and tedious. It is given in an appendix (Section 1.7).

Note that Equations (b.ii) and (b.iii) can be seen as a discrete counterpart of the characterization of convexity using second derivatives in the continuous domain. Equations (b.iv), (b.v), (b.vi) and (b.vii) are useful when dealing with functions of multimodular functions, like projections. Indeed, checking multimodularity of projections or restrictions of multimodular functions is easier using this approach.

For example, given a point X in $\mathbb{Z}^m$, a set I of indices $i_1 \leq \cdots \leq i_k$ with $k \leq n$ and the function $p_{X,I} : \mathbb{Z}^k \to \mathbb{Z}^m$ defined by $p_{X,I}(Y) = Z$, with $Z_i = X_i$ if $i \notin I$ and $Z_{i_j} = Y_j$ otherwise, we have the following property.

Lemma 3. *If $f : \mathbb{Z}^m \to \mathbb{R}$ is multimodular then for all I and X, the function $g : \mathbb{Z}^k \to \mathbb{R}$ defined by $g(Y) = f(p_{X,I}(Y))$ is multimodular.*

Proof. The proof holds by checking that relations (1.2) holds for g. We will denote by $\mathcal{F}^k = \{-e_1^k, s_2^k, \cdots s_k^k, e_k^k\}$ the multimodular base of $\mathbb{Z}^k$. The base of $\mathbb{Z}^m$ is $\mathcal{F}$. Using the previous notations, we have to check (1.2) for g with respect to $\mathcal{F}^k$.

$$\Delta_{-e_1^k}\Delta_{s_j^k}g(Y) = \Delta_{-e_{i_1}}\Delta_{\sum_{t=i_{j-1}+1}^{i_j}s_t}f(Z)$$

$$= \sum_{t=i_{j-1}+1}^{i_j} \Delta_{-e_{i_1}}\Delta_{s_t}f(Z)$$

$$\leq 0,$$

by Equation (b.v).

The other cases, $\Delta_{s_i^k}\Delta_{s_j^k}g(Y)$ and $\Delta_{s_i^k}\Delta_{e_k^k}g(Y)$ are checked similarly. ☐

1.2.2 Multimodularity and convexity

In the first part of this section, we present some details on the construction of a simplicial decomposition of $\mathbb{R}^m$ adapted to the base of multimodularity. This construction was given by Hajek. Theorem 4.3 in [59] proves that the lower convex envelope $\tilde{f}$ of a multimodular function f is the piecewise linear interpolation of f on this triangulation. In this monograph we define $\tilde{f}$ directly as the piecewise linear interpolation of f. The second part of the section is then devoted to Theorem 1, that shows that f is multimodular if and only if $\tilde{f}$ is convex, and some of its consequences which are useful to derive optimality results (see Section 5).

We first introduce the notion of *atoms*, which was used by Hajek . In the space $\mathbb{R}^m$, the convex hull of $m+1$ linearly independent points in $\mathbb{Z}^m$ forms a simplex . A simplex defined on the set of points $\{x^0, \cdots, x^m\}$ of $\mathbb{Z}^m$ is called an *atom* (defined in [59] § 3) if and only if for some ordering of the set and for some permutation $(i_0, \cdots, i_m)$ of $(0, 1, \cdots, m)$,

$$
\begin{aligned}
x^1 &= x^0 + g_{i_1} \\
x^2 &= x^1 + g_{i_2} \\
&\;\cdot \\
&\;\cdot \\
&\;\cdot \\
x^m &= x^{m-1} + g_{i_m} \\
x^0 &= x^m + g_{i_0}
\end{aligned}
\tag{1.6}
$$

where $g_{i_0}, \cdots, g_{i_m}$ are all the elements of $\mathcal{F}$.

Next we present a characterization of an atom (see [59]), which is essential for the optimization result that we obtain in the following sections. Denote by $\lfloor x \rfloor$ the largest integer smaller than or equal to x. Then the following trivially holds

$$
\int_0^1 \lfloor x + \theta \rfloor d\theta = x.
\tag{1.7}
$$

Given $z \in \mathbb{R}^m$, $\theta \in \mathbb{R}$, define the vector $u^z(\theta)$ in $\mathbb{Z}^m$:

$$
\begin{aligned}
u_1^z(\theta) &= \lfloor \theta + z_1 \rfloor - \lfloor \theta \rfloor, \\
u_i^z(\theta) &= \lfloor \theta + z_1 + \cdots + z_i \rfloor - \lfloor \theta + z_1 + \cdots + z_{i-1} \rfloor, \quad i = 2, \cdots, m.
\end{aligned}
$$

Then by (1.7),

$$
\int_0^1 u^z(\theta) d\theta = z.
\tag{1.8}
$$

$u^z(\theta)$ is periodic in θ with period 1, and piecewise constant with at most $m + 1$ jumps per period. Thus, the set $\{u^z(\theta) : 0 \le \theta \le 1\}$ contains at most $m + 1$ vectors, all integer valued. The next Lemma follows from [59]:

Lemma 4. *A point z is contained in an atom, say $S(z)$, if and only if the extreme points of $S(z)$ contain $\{u^z(\theta) : 0 \le \theta \le 1\}$. A point z is in the interior of atom $S(z)$ if and only if the extreme points of $S(z)$ equal $\{u^z(\theta) : 0 \le \theta \le 1\}$.*

Note that z may be a point in the intersection of two or more atoms. But each point $z \in \mathbb{R}^m$ is contained in some atom, say $S(z)$, and it can thus be expressed as the convex combination given by (1.8) of the extreme points of $S(z)$.

For any function f on $\mathbb{Z}^m$, we define the corresponding function $\tilde{f}$ on $\mathbb{R}^m$ as follows. It agrees with f on $\mathbb{Z}^m$, and its value on an arbitrary point in $z \in \mathbb{R}^m$ is obtained as the corresponding linear interpolation of the values of f on the extreme points of the atom $S(z)$. Note that $\tilde{f}$ is uniquely defined, if z belongs to atoms S_1 and S_2 then the points $u^z(\theta)$ belong to the extreme points of $S_1 \cap S_2$.

The following theorem establishes the equivalence between multimodularity of the discrete function and convexity of its continuous extension. In [59], the "only if" part was proved (this is the hard part), while the "if" part was omitted. In [27], the equivalence is established in the more general context of general multimodular triangulations. However, in both cases, the proofs are rather involved and use separating hyperplane techniques. The proof presented here is elementary and only uses a discrete counterpart of the second derivative argument. Furthermore, this proof technique enables us to prove this equivalence on special convex subsets of $\mathbb{Z}^m$, as shown in Corollary 10. A more general presentation is done in Chapter 12.

Theorem 1. f *is multimodular if and only if* $\tilde{f}$ *is convex.*

Proof. "only if": The function $\tilde{f}$ is continuous by definition. Moreover, along any direction d, it has only a discrete number of isolated points where it is not differentiable. By using the characterization of convexity given in [96], we will check convexity at a point z by showing that for point z, and any direction d, the right derivative is greater than or equal to the left derivative. It obviously suffices to check at points that are on the boundary of an atom, since, by definition, $\tilde{f}$ is linear in the interior of atoms. Hence, we first assume that the point z is on the interior of a face (of dimension $m-1$) which is common between two adjacent atoms. Without loss of generality, assume that the atoms (defined below by their extreme points) are

$$A = A(x_0, x_1, \cdots, x_m) \qquad \text{and} \qquad \overline{A} = A(x_0, x_1^*, \cdots, x_m).$$

where x_i satisfy (1.6) and

$$x_1^* = x_0 + g_{i_2}, \qquad x_2 = x_1^* + g_{i_1}.$$

<u>Case 1:</u> (see Fig. 1.1) $g_{i_1} = -e_1 = (-1, 0, \cdots, 0)$, $g_{i_2} = e_m = (0, 0, \cdots, 1)$.

Decompose direction d in its projection ∂_2 in the common face between the two atoms and in the component ∂_1 along the direction $(x_1^* - x_1)$. In the direction ∂_2, the left and right derivatives are equal. In the direction ∂_1, the right derivative is a constant c, depending on the length of ∂_1, times $\tilde{f}(x_1^*) - \tilde{f}(z)$. The left derivative is $c(\tilde{f}(z) - \tilde{f}(x_1))$. Omitting the constant c,

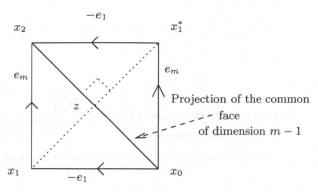

Fig. 1.1. Checking convexity at an interior point z of the common face, Case 1.

and using point $z = \frac{1}{2}(x_0 + x_2)$ hence $2\widetilde{f}(z) = f(x_0) + f(x_2)$, we get for the difference

$$(\widetilde{f}(x_1^*) - \widetilde{f}(z)) - (\widetilde{f}(z) - \widetilde{f}(x_1))$$
$$= (f(x_1^*) - f(x_0)) - (f(x_2) - f(x_1)) \qquad (1.9)$$

The fact that (1.9) is nonnegative follows by applying (1.1) with $x = x_0$, and

$$x_1^* = x_0 + e_m$$
$$x_2 = x_0 - e_1 + e_m$$
$$x_1 = x_0 - e_1.$$

<u>Case 2:</u> $g_{i_1} = e_m$ and $g_{i_2} = -e_1$. It is handled as Case 1.
<u>Case 3:</u> (see Fig. 1.2) $g_{i_1} = s_2 = (1, -1, 0, \cdots, 0)$, $g_{i_2} = -e_1$. We set $x = x_0$, and $x_1^* = x_0 - e_1, x_2 = x_0 - e_1 + s_2, x_1 = x_0 + s_2$. We decompose d along $-e_1$ on its projection ∂_2 in the common face, and in the projection ∂_1 along $(x_1^* - x_0)$. As in Case 1, it suffices to consider the direction ∂_1. The right derivative in this direction is $f(x_1^*) - f(x_0)$, and the left derivative is $f(x_2) - f(x_1)$ (both up to a multiplicative constant).

The difference between the right and left derivatives is indeed nonnegative: $f(x_1^*) - f(x_0) - (f(x_2) - f(x_1)) \geq 0$. This is obtained again by applying (1.1).
<u>Case 4:</u> $g_{i_1} = s_2 = (1, -1, 0, \cdots, 0)$, $g_{i_2} = s_3$. In this case, we project d along $(x_1^* - x_1)$, and the analysis is as for Case 1.

All other cases, in which z is in the interior of a face (of dimension $m-1$), common to two adjacent atoms, are similar to one of those considered above. It now remains to consider the case where the direction d in point z crosses from atom A to atom $\overline{A}$, and $A \cap \overline{A}$ is of dimension at most $m-2$. In that case, we consider a cylinder C in direction d containing point z and with an arbitrarily small diameter. This cylinder intersects atoms A and $\overline{A}$ and is covered by atoms. We consider the projection P_C of C along direction d. This has dimension $m-1$. The intersection (say of dimension k) of C with an

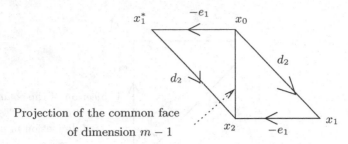

Projection of the common face

of dimension $m - 1$

Fig. 1.2. Checking convexity at a point on the common face, Case 3

atom is projected on the intersection of the projections and has dimension at most k. Therefore, P_C is almost everywhere (in Lebesgue measure) covered with projections of dimension $m - 1$. We can find a line L that belongs to C and intersects A and $\overline{A}$, with direction d, and which projection is a point in P_C not belonging to intersections of dimension smaller than $m-1$. Therefore, we can claim that L only intersects faces of atoms of dimension $m - 1$. The convexity in point z and direction d now follows from the convexity in points z_i corresponding to the intersections of line L with all the intermediate atoms between A and $\overline{A}$ and a continuity argument.

"if": Consider an arbitrary point x_0 and any two distinct elements g_i, g_j in $\mathcal{F}$. We have to show that

$$f(x_0) + f(x_2) - f(x_1) - f(x_1^*) \leq 0, \tag{1.10}$$

where $x_1 \overset{\text{def}}{=} x_0 + g_i$, $x_1^* \overset{\triangle}{=} x_0 + g_j$, $x_2 \overset{\triangle}{=} x_1 + g_j = x_1^* + g_i$, where $\overset{\text{def}}{=}$ is used to define the new variable x_1 to be equal to $x_0 + g_i$.

Define $z \overset{\triangle}{=} \frac{1}{2}(x_1 + x_1^*) = \frac{1}{2}(x_0 + x_2)$ and consider the line segment $x_1 \to z \to x_1^*$. The left derivative (l.d.) and right derivative (r.d.) in z are given by

$$l.d. = \widetilde{f}(\frac{1}{2}(x_1 + x_1^*)) - \widetilde{f}(x_1) = \frac{1}{2}f(x_0) + \frac{1}{2}f(x_2) - f(x_1),$$

$$r.d. = \widetilde{f}(x_1^*) - \widetilde{f}(\frac{1}{2}(x_0 + x_2)) = f(x_1^*) - \frac{1}{2}f(x_0) - \frac{1}{2}f(x_2).$$

Since $\widetilde{f}$ is convex, $r.d. - l.d.$ is non-negative, and hence (1.10) holds. $\qquad \square$

We can restrict the notion of multimodularity to some convex sets of $\mathbb{Z}^m$. Let A be a convex set which is a union of a set of atoms (or of faces of atoms). We restrict the definition of multimodularity to directions that only lead to points in A. More precisely, we say that f is multimodular in A if the following holds.

If $x_0, x_0 + g_i, x_0 + g_j, x_0 + g_i + g_j$ are all elements of A then

$$f(x_0) + f(x_0 + g_i + g_j) - f(x_0 + g_i) - f(x_0 + g_j) \leq 0.$$

Corollary 1. *The function f is multimodular in A if and only if $\tilde{f}$ is convex on A.*

Proof. It should be clear from the proof of Theorem 1 that the equivalence of the multimodularity of f and the convexity of $\tilde{f}$ still holds if we restrict the function f to A. □

The restriction to $A = \mathbb{N}^m$ turns out to be essential for the application of Theorem 1 presented in Chapter 4. Indeed, the function considered is only defined on non-negative coordinates. Also, in Section 1.5.1, an even smaller set A is used in order to consider a case of constrained optimization (see Lemma 9).

A second corollary of Theorem 1 concerns the minimization of multimodular functions. For a function defined on A, we call x a local minimum on A if $f(x) \leq f(x \pm e_i)$ for all i such that $x \pm e_i$ is in A.

Corollary 2. *Let the function f be multimodular in A. Then a local minimum is a global minimum on A.*

Proof. If f is multimodular in A, then $\tilde{f}$ is convex in A, and is linear on the (faces of) atoms forming A. The graph of $\tilde{f}$ (*i.e.* $\{x : \exists y \text{ s.t. } x \geq \tilde{f}(y)\}$) is a convex polytope. Therefore, all the local minima are global minima and are extreme points of atoms. □

Next we consider the integer convexity properties of a function f. A function f is said to be integer convex if the following holds. For vectors x and d in $\mathbb{Z}^m$, we have

$$f(x + d) - f(x) \geq f(x) - f(x - d).$$

Theorem 2. *Let f be multimodular. Then it is integer convex.*

Proof. Define $\partial_h^+(x) :=$ the right derivative of $\tilde{f}$ at x in the direction h and $\partial_h^-(x) :=$ the left derivative of $\tilde{f}$ at x in the direction h. Since $\tilde{f}$ is convex (Theorem 1) then

$$\partial_h^+(x) \geq \partial_h^-(x). \tag{1.11}$$

Since $\tilde{f}(y) = f(y)$ at the integer points, and since $\tilde{f}$ is convex, we have

$$\partial_h^+(x) \leq \frac{f(x + h) - f(x)}{|h|}, \qquad \partial_h^-(x) \geq \frac{f(x) - f(x - h)}{|h|},$$

where $|h|$ is the L_2 norm of h. This, together with (1.11) imply the integer convexity of f. □

The converse of the above theorem is not true:

Counter-example 3 Consider the convex function $f : \mathbb{N}^m \to \mathbb{R}$ given by $f(x) = \max_{i=1,\cdots,m} x_i$. It is integer convex since it is the maximum of convex (linear) functions. However, it is not multimodular. Indeed, consider $m = 2$, $x = (i+1, i)$ for some integer i. Then

$$2i + 2 = f(x - e_1 + e_2) + f(x) > f(x - e_1) + f(x + e_2) = 2i + 1.$$

Hence f is not multimodular.

For integer convex functions we need not have the useful property that a local minimum is a global minimum. This is illustrated in the next counter-example.

Counter-example 4 Consider the space $(\{0, 1, 2\})^2$. Define the function g such that

$$\begin{aligned}
g(0,2) &= -1, & g(1,2) &= 2, & g(2,2) &= 5, \\
g(0,1) &= 2, & g(1,1) &= 1, & g(2,1) &= 0, \\
g(0,0) &= 5, & g(1,0) &= 4, & g(2,0) &= 3.
\end{aligned}$$

One can easily check that g is an integer convex function, but not multimodular since $g\big((1,2)+b_0\big)+g\big((1,2)+b_2\big) = 0 < 4 = g\big((1,2)\big)+g\big((1,2)+b_0+b_2\big)$. Starting the local search algorithm at coordinate $(2,1)$ shows that all neighbors have values, which are greater than 0. However, the global minimum is $g\big((0,2)\big) = -1$.

This shows that when dealing with discrete functions, the counterpart of convexity is not integer convexity but rather multimodularity, which insures, for example, that a local minimum is a global minimum point (see Corollary 2).

The following result will be very useful when cost functions involve arbitrary convex functions of the quantities of interest, especially to prove optimality in the convex increasing order.

Theorem 5. *The following statements are equivalent,*
(i) f is multimodular and for all $v, w \in \mathcal{F}$, $\max(f(x + v), f(x + w)) \geq \max(f(x), f(x + v + w))$.
(ii) For all $h : \mathbb{R} \to \mathbb{R}$ convex increasing, $h(f)$ is multimodular.

Proof. We will use the notation introduced in [88]. Condition (i) is equivalent to the fact that the two-dimensional vector $(f(x), f(x + v + w))$ is weakly submajorized by $(f(x+v), f(x+w))$, denoted $(f(x), f(x+v+w)) \prec_w (f(x+v), f(x+w))$. By using proposition 4.C.1.b in [88] then this implies that for any convex increasing function h, $h(f(x))+h(f(x+v+w)) \leq h((f(x+v))+h(f(x+w))$. This is exactly the multimodularity of the function $h(f)$.

As for the converse, note that (ii) implies in particular that $h(f)$ is multimodular for all h continuous, convex and increasing. Using Proposition 4.B.2 in [88], this implies that $(f(x), f(x + v + w)) \prec_w (f(x + v), f(x + w))$. By definition of weak majorization, that is the same as statement (i). $\qquad\square$

1.3 The optimality of bracket policies for a single criterion

In this section, we will present a rather general framework under which multimodularity can be used in order to optimize a cost function based on a sequence of functions which will represent a quantity of interest in a given model. The problems that can be solved using this framework include the minimization of the average workload in a queue under general stationary assumptions, as well as many other similar problems (see for example Chapter 4). § 1.5.1 that presents a precise instance of such a problem.

The sequence of functions considered can be interpreted as cost functions, which are defined on a common sequence $(a_1, a_2, \cdots)$ of integers which we call a *control sequence*. For example, a_i can be the number of admitted jobs at the ith arriving epoch for an admission control problem in a queue. The control sequences will all belong to a set A which is a convex union of atoms in $\mathbb{Z}^k$. Our objective is to study optimization properties of Cesaro averages of the cost functions over the class of control sequences.

Consider a sequence of functions $f_k : \mathbb{N}_+^k \to \mathbb{R}_+ \cup \{\infty\}$ that satisfy the following assumptions:

- $< 1 >$ f_k is multimodular on A.
- $< 2 >$ $f_k(a_1, \cdots, a_k) \geq f_{k-1}(a_2, \cdots, a_k)$, $\forall k > 1$;

For a given sequence $\{a_k\}$, we define the cost $g(a)$ as

$$g(a) = \overline{\lim_{N \to \infty}} \frac{1}{N} \sum_{n=1}^{N} f_n(a_1, \cdots, a_n). \tag{1.12}$$

Definition 2. *Let p and θ be two positive reals. We define the bracket sequence $\{a_k^p(\theta)\}_{k \in \mathbb{N}}$ with rate p and initial phase θ as,*

$$a_k^p(\theta) = \lfloor kp + \theta \rfloor - \lfloor (k-1)p + \theta \rfloor, \tag{1.13}$$

where $\lfloor x \rfloor$ denotes the largest integer no larger than x.

Note that when p and k are fixed, the set $\{a_k^p(\theta), 0 \leq \theta < 1\}$ are extreme points of an atom containing the point $(p, p, \cdots, p)$.

The aim of this section is to prove that this sequence minimizes the function g, provided that some conditions (including $< 1 >$ and $< 2 >$) above hold. The sequence (1.13) was used by Hajek in [59], and we use several properties of the bracket sequence established in [59]. To establish the main optimization results, we need the following technical Lemma.

Lemma 5. *If f_k satisfies assumptions $< 1 >, < 2 >$, then the function $\widetilde{f_k}$ satisfies assumption $< 2 >$ for positive real numbers.*

The proof of this lemma is given in Appendix (section 1.8).

Lemma 6. *Under assumptions* $< 1 >$ *and* $< 2 >$, *let* Θ *be a random variable, uniformly distributed in* $[0, 1)$, *and denote the expectation w.r.t.* Θ *by* E_Θ. *Then*

$$\lim_{N \to \infty} E_\Theta f_N(a_1^p(\Theta), \cdots, a_n^p(\Theta)) = \lim_{N \to \infty} \widetilde{f}_N(p, p, \cdots, p). \qquad (1.14)$$

Proof. We have for all N,

$$E_\Theta f_N(a_1^p(\Theta), \cdots, a_N^p(\Theta)) = \widetilde{f}_N(p, \cdots, p). \qquad (1.15)$$

(This follows (1.8), from Lemma 4, and the fact that $\widetilde{f}_N$ is affine on each atom, and agrees with f_N for the extreme points of the atom.) Since $\widetilde{f}_N(p, p, \cdots, p)$ is increasing in N by Lemma 5, the limit in N exists (it is possibly infinite). □

Definition 3. *We call the sequence* $\{a^p(\Theta)\}$ *the randomized bracket policy with rate* p.

1.3.1 Upper Bounds

Lemma 7. *Under assumptions* $< 1 >$ *and* $< 2 >$, *for every* $\theta \in [0, 1)$,

$$\overline{\lim_{N \to \infty}} \frac{1}{N} \sum_{n=1}^{N} f_n(a_1^p(\theta), \cdots, a_n^p(\theta)) \leq \lim_{N \to \infty} \widetilde{f}_N(p, p, \cdots, p). \qquad (1.16)$$

Proof. Define

$$f_m(\theta, p) \stackrel{\triangle}{=} f_m(a_1^p(\theta), \cdots, a_m^p(\theta)).$$

f_m is periodic (in θ) with period 1. Define

$$f'_m(\theta, p) \stackrel{\triangle}{=} f_m(a_{-m+1}^p(\theta), \cdots, a_0^p(\theta)).$$

Then we have

$$f'_m(\theta', p) = f_m(\theta, p) \quad \text{where} \quad \theta' = \theta + mp, \qquad (1.17)$$

Indeed,

$$\begin{aligned} f'_m(\theta', p) &= f_m(a_{-m+1}^p(\theta'), \cdots, a_0^p(\theta')) \\ &= f_m(a_{-m+1}^p(\theta + mp), \cdots, a_0^p(\theta + mp)) = f_m(\theta, p), \end{aligned}$$

where the last equality follows from the fact that $a_{-m+k}^p(\theta + mp) = a_k^p(\theta)$, $k = 1, \cdots, m$. f'_m is again periodic w.r.t. θ, with period 1, and is increasing in m so that the following limit exists (possibly infinite):

$$f'_\infty(\theta, p) \triangleq \lim_{m \to \infty} f'_m(\theta, p).$$

Moreover, we have that $E_\Theta f'_m(\Theta, p) = \tilde{f}_m(p, \cdots, p)$, where Θ be a random variable, uniformly distributed in $[0, 1]$ (this follows from (1.8), from Lemma 4, and fact that $\tilde{f}_n$ is affine on each atom, and agrees with f_n for the extreme points of the atom). Hence,

$$E_\Theta f'_\infty(\Theta, p) = \lim_{N \to \infty} \tilde{f}_N(p, p, \cdots, p). \tag{1.18}$$

Consider now the bracket sequence for fixed θ. Then

$$\frac{1}{N} \sum_{m=1}^{N} f_m(a_1^p(\theta), \cdots, a_m^p(\theta))$$

$$\leq \frac{1}{N} \sum_{m=1}^{N} f_N(a_{-N+m+1}^p(\theta), \cdots, a_0^p(\theta), \cdots, a_m^p(\theta))$$

$$\leq \frac{1}{N} \sum_{m=1}^{N} f'_\infty(-mp + \theta, p).$$

The last inequality follows from assumption $<2>$ for the functions f_k, as well as an argument similar to the one used in (1.17).

If p is irrational, applying the ergodic theorem of Weyl and Von Neumann ([102]), we have

$$\lim_{N \to \infty} \frac{1}{N} \sum_{m=1}^{N} f'_\infty(-mp + \theta, p) = E_\Theta f'_\infty(\Theta, p).$$

From Equation (1.18), we have $E_\Theta f'_\infty(\Theta, p) = \lim_{N \to \infty} \tilde{f}_N(p, p, \cdots, p)$. This implies that if p is irrational,

$$\lim_{N \to \infty} \frac{1}{N} \sum_{m=1}^{N} f'_\infty(-mp + \theta, p) = \lim_{N \to \infty} \tilde{f}_N(p, p, \cdots, p). \tag{1.19}$$

If p is rational, then $p = q/d$ where q and d are relatively prime and $d \geq 1$. This implies that the sequence $(a_{-(N-m)}^p(\theta), \cdots, a_0^p(\theta), \cdots, a_m^p(\theta))$ is constant if $\theta(\bmod 1) \in [j/d, (j+1)/d)$, for all j. Therefore, $f'_m(\theta, p)$ is also constant on these intervals and by passage to the limit, $f'_\infty(\theta, p)$ is constant on these intervals. Now, note that $\mathrm{Frac}(\theta - mp) \in [j/d, (j+1)/d)$ for exactly one value of m out of d consecutive values of m because q are d are relatively prime. Now, we have

$$\lim_{N \to \infty} \frac{1}{N} \sum_{m=1}^{N} f'_\infty(-mp + \theta, p) = \frac{1}{d} \sum_{m=1}^{d-1} f'_\infty(m/d, p) = E_\Theta f'_\infty(\Theta, p).$$

Equation (1.18) concludes this case as well.   □

1.3.2 Lower Bounds

In this subsection, we establish lower bounds for the discounted cost for all control sequences $\{a_k\}$. This then serves for obtaining a lower bound on the average cost. Here, we use the following assumption for the functions f_k.

$-<3>$ For any sequence $\{a_k\}$ $\exists$ a sequence $\{b_k\}$ such that
$\forall k, m$ with $k > m$, $f_k(b_1, \cdots, b_{k-m}, a_1, \cdots, a_m) = f_m(a_1, \cdots, a_m)$.

We use the notions defined in the previous sections.

Let us fix the sequence $\{a_k\}$, as well as some arbitrary integer, N. We define $p_\alpha \triangleq (1 - \alpha) \sum_{k=1}^{\infty} \alpha^{k-1} a_k$.

Now, using assumptions $<1>$ through $<3>$, we have

$$\sum_{n=1}^{\infty} (1-\alpha)\alpha^{n-1} f_n(a_1, a_2 \cdots, a_n)$$

$$\geq \sum_{n=1}^{N} (1-\alpha)\alpha^{n-1} f_N(b_1, \cdots, b_{N-n}, a_1, a_2, \cdots, a_n)$$

$$+ \sum_{n=N+1}^{\infty} (1-\alpha)\alpha^{n-1} f_N(a_{n-N+1}, a_{n-N+2}, \cdots, a_n)$$

$$= \sum_{n=1}^{N} (1-\alpha)\alpha^{n-1} \widetilde{f}_N(b_1, \cdots, b_{N-n}, a_1, a_2, \cdots, a_n)$$

$$+ \sum_{n=N+1}^{\infty} (1-\alpha)\alpha^{n-1} \widetilde{f}_N(a_{n-N+1}, a_{n-N+2}, \cdots, a_n)$$

$$\geq \widetilde{f}_N \left(\sum_{n=1}^{N} (1-\alpha)\alpha^{n-1}(b_1, \cdots, b_{N-n}, a_1, a_2, \cdots, a_n) \right.$$

$$\left. + \sum_{n=N+1}^{\infty} (1-\alpha)\alpha^{n-1}(a_{n-N+1}, a_{n-N+2}, \cdots, a_n) \right) \qquad (1.20)$$

$$= \widetilde{f}_N \left(b_1 \sum_{n=0}^{N-2} (1-\alpha)\alpha^n + \alpha^N p_\alpha, b_2 \sum_{n=0}^{N-3} (1-\alpha)\alpha^n + \alpha^{N-1} p_\alpha, \cdots p_\alpha \right). (1.21)$$

Equation (1.20) follows from Jensen's inequality, since by Theorem 1, the function $\widetilde{f}_N$ is convex, and since the coefficients $(1-\alpha)\alpha^{n-1}$ are nonnegative and sum to 1. Define

$$B(N, \alpha, p) \triangleq \widetilde{f}_N(b_1 \sum_{n=0}^{N-2} (1-\alpha)\alpha^n + \alpha^N p, b_2 \sum_{n=0}^{N-3} (1-\alpha)\alpha^n + \alpha^{N-1} p, \cdots, p).$$

Note that B is defined for a fixed sequence $\{a_k\}$. Also note that $B(N, \alpha, p)$ is lower semi-continuous in α and in p.

Using Lemma 10 given in Appendix (section 1.9), we derive the following lower bounds

Lemma 8. *Under assumptions* $< 1 >$, $< 2 >$ *and* $< 3 >$,

$$\varlimsup_{m \to \infty} \frac{1}{m} \sum_{n=1}^{m} f_n(a_1, \cdots, a_n) \geq \inf_{q \in \mathcal{L}} \widetilde{f}_N(q, \cdots, q),$$

where $\mathcal{L}$ *is the set of all limit points of* p_α *as* $\alpha \uparrow 1$.

Proof.

$$\varlimsup_{m \to \infty} \frac{1}{m} \sum_{n=1}^{m} f_n(a_1, \cdots, a_n) \geq \varlimsup_{\alpha \uparrow 1} (1 - \alpha) \sum_{n-1}^{\infty} \alpha^{n-1} f_n(a_1, \cdots, a_n)$$

$$\geq \varlimsup_{\alpha \uparrow 1} B(N, \alpha, p_\alpha)$$

$$\geq \inf_{q \in \mathcal{L}} B(N, 1, q), \tag{1.22}$$

The Lemma follows since for any given p, by definition of B, $B(N, 1, p) = \widetilde{f}_N(p, p, \cdots, p)$. $\qquad\square$

1.3.3 Optimality of the Bracket Sequences

Theorem 6. *Under assumptions* $< 1 >, < 2 >$ *and* $< 3 >$, *and given some* $p \in [0, 1]$, *if the functions* $f_k(a_1, \cdots, a_k)$ *are increasing in all* a_i, *then the bracket sequence* $a^p(\theta)$ *for any* $\theta \in [0, 1)$, *minimizes the average cost* $g(a)$ *over all sequences that satisfy the constraint:*

$$\varliminf_{N \to \infty} \frac{1}{N} \sum_{n=1}^{N} a_n \geq p.$$

Proof. We denote by

$$\underline{p} \triangleq \varliminf_{N \to \infty} \frac{1}{N} \sum_{n=1}^{N} a_n$$

By using Lemma 10 in the Appendix (section 1.9),

$$p \leq \underline{p} \leq \varliminf_{\alpha \uparrow 1} p_\alpha = \inf\{q, q \in \mathcal{L}\}.$$

If the functions $\{f_k\}$ are increasing, then B is increasing in p, therefore,

$$g(a) \geq \inf_{q \in \mathcal{L}} B(N, 1, q) \geq B(N, 1, p) = \widetilde{f}_N(p, \cdots, p), \tag{1.23}$$

by Lemma 8. If we let N go to infinity, we get

$$g(a) = \varlimsup_{N \to \infty} \frac{1}{N} \sum_{n=1}^{N} f_n(a_1, \cdots, a_n) \geq \varlimsup_{N \to \infty} \widetilde{f}_N(p, \cdots, p). \qquad (1.24)$$

Lemma 7 shows that $\varlimsup_{N \to \infty} \widetilde{f}_N(p, p, ..., p) \geq g(a^p(\theta))$. Thus $g(a) \geq g(a^p(\theta))$.

$\square$

When the functions f_k are decreasing, we have the analogous result.

Theorem 7. *Under assumptions $< 1 >, < 2 >$ and $< 3 >$, and given some $p \in [0, 1]$, and any $\theta \in [0, 1]$, if the functions $f_k(a_1, \cdots, a_k)$ are decreasing in all a_i, then the bracket sequence $a^p(\theta)$ minimizes the average cost $g(a)$ over all sequences that satisfy the constraint:*

$$\varliminf_{N \to \infty} \frac{1}{N} \sum_{n=1}^{N} a_n \leq p.$$

Proof. The proof is similar to the previous one, using the fact that if

$$\bar{p} \overset{\triangle}{=} \varliminf_{N \to \infty} \frac{1}{N} \sum_{n=1}^{N} a_n,$$

then

$$p \geq \bar{p} \geq \varlimsup_{\alpha \uparrow 1} p_\alpha = \sup\{q, q \in \mathcal{L}\}.$$

$\square$

1.4 The optimality of bracket policies for multiple criteria

In this section, we establish general conditions under which the bracket policy is optimal when the cost function depends on multiple criteria. While the single criterion framework has application in admission control, this multiple criteria approach has applications for routing control to several queues. For instance, for the routing to several identical $\cdot/GI/1$ queues, it is known that the round robin routing is optimal in separable-convex increasing order [82]. In order to obtain this type of results in our framework, we shall show the existence of optimal asymptotic fractions in a very general case.

From now on, we study the following general optimization problem. Consider K sequences of functions $f_n^i, i = 1, \cdots, K$. Each sequence of functions f_n^i will only depend on the sequence of the ith coordinates a^i in a, and will satisfy assumptions $< 1 >, < 2 >$ and $< 3 >$, as in Section 1.3.

A policy is a sequence $a = (a_1, a_2, \cdots)$, where a_n is a vector taking values in $\{0, 1\}^K$. We consider the additional constraint that for every integer j, only one of the components of a_j may be different from 0. A policy satisfying this constraint is called *feasible* .

Let h be a convex increasing function from $\mathbb{R}^K$ to $\mathbb{R}$. Define

$$g(a) \triangleq \varlimsup_{N \to \infty} \frac{1}{N} \sum_{n=1}^{N} h(f_n^1(a^1), \cdots, f_n^K(a^K)). \tag{1.25}$$

Following notations introduced in Section 1.3, we get a bounding function called $B_i(N, \alpha, p)$ for coordinate i. Here, we denote by

$$B_i(\alpha, p) \triangleq \sup_N B_i(N, \alpha, p),$$

and

$$B_i(p) \triangleq \sup_{\alpha \le 1} B_i(\alpha, p).$$

Note that by convexity of $\widetilde{f}_n^{(i)}$ and Lemma 5, $B_i(\alpha, p)$ and $B_i(p)$ are continuous from below in (α, p) and p, respectively.

Our objective is to minimize $g(a)$, with no constraints on the asymptotic fractions.

Theorem 8. *Assume that for all i, the functions f_n^i satisfy assumptions $<1>,<2>$ and $<3>$. The following lower bound holds for all policies:*

$$g(a) \ge \inf_{p_1 + \cdots + p_K = 1} h(B_1(p_1), \cdots, B_K(p_K)).$$

Proof. Due to Lemma 10 in the Appendix C (section 1.9), Jensen's inequality and Equation (1.21), we have

$$\varlimsup_{N \to \infty} \frac{1}{N} \sum_{n=1}^{N} h(f_n^1, \cdots, f_n^K)$$

$$\ge \varlimsup_{\alpha \to 1} (1 - \alpha) \sum_{n=1}^{\infty} \alpha^{n-1} h(f_n^1, \cdots, f_n^K)$$

$$\ge \varlimsup_{\alpha \to 1} h \left((1 - \alpha) \sum_{n=1}^{\infty} \alpha^{n-1} f_n^1, \cdots, (1 - \alpha) \sum_{n=1}^{\infty} \alpha^{n-1} f_n^K \right)$$

$$\ge \varlimsup_{\alpha \to 1} h \left(B_1(\alpha, p_1^a(\alpha)), \cdots, B_K(\alpha, p_K^a(\alpha)) \right), \tag{1.26}$$

where

$$p_i^a(\alpha) \triangleq (1 - \alpha) \sum_{k=1}^{\infty} \alpha^{k-1} a_k^i. \tag{1.27}$$

We note that $\sum_{i=1}^{K} p_i^a(\alpha) = 1$. Hence, one may choose a sequence $\alpha_n \uparrow 1$ such that the following limits exist:

$$\lim_{n \to \infty} p_i^a(\alpha_n) \triangleq p_i, \qquad i = 1, \cdots, K \tag{1.28}$$

and $\sum_{i=1}^{K} p_i = 1$. From the lower semi-continuity of $B_i(\alpha, p_i)$ in p_i and α we get from (1.26)

$$g(a) \geq h(B_1(1, p_1), \cdots, , B_K(1, p_K)) \qquad (1.29)$$
$$\geq \inf_{p_1 + \cdots + p_K = 1} h(B_1(1, p_1), \cdots, B_K(1, p_K)).$$

$\square$

Note that there exists some p^* that achieves the infimum

$$\inf_{p_1 + \cdots + p_K = 1} h(B_1(1, p_1), \cdots, B_K(1, p_K)), \qquad (1.30)$$

for $h(B_1(1, p_1), \cdots, B_K(1, p_K))$ is continuous from below in $p = (p_1, \cdots, p_K)$.

Balanced policies

Consider the *bracket* policy $a^{p^*}(\theta)$ given by

$$a_{k,i}^{p^*}(\theta) = \lfloor kp_i^* + \theta_i \rfloor - \lfloor (k-1)p_i^* + \theta_i \rfloor. \qquad (1.31)$$

There are some p^* for which the condition of feasibility of the policy $a^{p^*}(\theta)$ is satisfied, that is, there exists some $\theta = (\theta_1, \cdots, \theta_K)$, such that the policy $a^{p^*}(\theta)$ given in (1.31) is feasible. These p^* are called *balanceable* and we shall come back to them in more details in Sections 4.6.1 and 2.2.

Theorem 9. *Assume that for all i, the functions f_n^i satisfy assumptions $< 1 >, < 2 >$ and $< 3 >$. Assume that h is linear increasing and that p^* is balanceable. Then $a^{p^*}(\theta)$ is optimal for the average cost, i.e. it minimizes $g(a)$ over all feasible policies.*

Proof. The proof follows directly from Lemma 7 together with Theorem 8. $\square$

The balance condition on p^* is still not completely characterized, however, we can mention two simple cases for which p^* is balanceable. i.e. for which there exist some $\theta = (\theta_1, \cdots, \theta_K)$, such that $a^{p^*}(\theta)$ is feasible.

– **C1:** $K = 2$.
– **C2:** K criteria with symmetric costs, i.e. $h(x) = \sum_i x_i$ and all f^i (as functions of a^i) are equal.

Corollary 3. *(i) Consider the case **C1**. For p^* that achieves the minimum in (1.30), the bracket policy for rate p^* and some initial phase θ is feasible and optimal.*
*(ii) Consider the case **C2**. By symmetry, the bracket policy with $p = 1/K$ for some initial phase θ is feasible and optimal.*

Remark 2. Hajek gives an argument ([59] Remark (5) p. 554) that shows for $K = 2$ the optimality of the bracket policy with rate vector $(p, 1 - p)$ in the restricted class of policies a such that $\lim_{N \to \infty} \frac{1}{N} \sum_{n=1}^{N} a_n^1 = p$. Note that Corollary 3 (ii) shows the optimality of a bracket policy for all control sequences a.

Next, we restrict again to the case of a single objective ($K = 1$), and show that the results of the previous section can be extended. More precisely, we show that a bracket policy is optimal in a stronger sense.

Corollary 4. *Under the conditions of Theorem 6, given some $p \in [0,1]$, and any $\theta \in [0,1]$, the bracket policy $a^p(\theta)$ minimizes the average cost $g(a)$ over all policies that satisfy the constraint:*

$$\overline{\lim_{\alpha \to 1}} \, p_1(\alpha) \geq p_1. \tag{1.32}$$

where $p_1^a(\alpha)$ is defined in (1.27).

Note that the constraint $\underline{\lim}_{N \to \infty} \frac{1}{N} \sum_{n=1}^{N} a_n \geq p_1$ (in Theorem 6) implies (1.32), due to Lemma 10 in the Appendix (section 1.9). Therefore the minimization in Theorem 6 is over a subclass of the set of policies on which minimization is performed in Corollary 4. Thus, Corollary 4 implies that a policy a that satisfies (1.32) does not perform better than the bracket policy (with $p = p_1$) even if $\underline{\lim}_{N \to \infty} \frac{1}{N} \sum_{n=1}^{N} a_n < p_1$.
Proof of Corollary 4:

Proof. Choose an arbitrary policy a that satisfies (1.32). Choose a subsequence $\alpha_n \uparrow 1$ such that $\overline{\lim}_{n \to \infty} p_1^a(\alpha_n) \geq p_1$. The proof now follows by combining Lemma 7 with (1.32). $\qquad\square$

1.5 Application of the optimization theorems

In this section, we will briefly present some optimization problems that fit the framework presented in the previous sections. Other models that require a more extensive analysis can be found in future chapters.

Remark 3. It seems difficult to give a consistent meaning to the workload associated with a negative number of customers. One may even doubt that there is a satisfying way to do so that will also preserve multimodularity. This is why we have not considered the quantity $\mathbb{E}_{\sigma,T} W_k(a_1, \cdots, a_k)$ on $\mathbb{Z}^k$, but only on $\mathbb{N}^k$, which is a convex union of atoms. Therefore, all the optimization framework constructed in section 1.3 can be used in this case.

1.5.1 Applications in high-speed telecommunication systems

We consider a simple model composed of a controlled D/D/1 queue with service times $\sigma_n = \sigma$ and inter-arrival times $\tau_n = \tau$ all deterministic. Assume that the available actions are 0 (corresponding to rejecting an arriving customer) and 1 (corresponding to acceptance of an arriving customer).
    The type of problem we consider is typical in high speed telecommunications networks, and in particular, to the ATM (Asynchronous Transfer

Mode). The latter has been chosen by the standardization committee ITU-T[1]
[77] as the main standard for integration of services in broadband networks.
In order to handle efficiently a large variety of applications, such as voice,
data, video and file transfer, cells of fixed size are used, giving rise to our
model that uses fixed service times. Fixed inter-arrival times are typical for
isochronous applications (voice, video) and also for large file transfer.

Two important measures of quality of services in ATM networks are loss
probabilities (CLR - Cell Loss Ratios) and delays. According to the ATM
standard [77], when a CBR (Constant Bit Rate) session is established, the
network should provide a guarantee that these two measures are bounded by
given constants. Since the available sources are limited and, moreover, might
be shared with other applications, a typical objective of the network is to
minimize the delay of the CBR session while meeting the constraint on the
loss probabilities. Losses might be due either to overflow, or to deliberate
packet discarding by the network (e.g. to allow the resources to be available
for other applications). The problem can be formulated in our framework
as one of discarding cells so as to minimize the average queue size (i.e. the
workload in the system) which is known to be proportional to the average
sojourn time (due to Little's law), subject to a lower bound p on the average
cell discarding rate.

We now describe the state evolution of the system. If x_n denotes the
amount of workload in the system immediately after the nth arrival that
occurs after time 0, and the system is initially empty (at time 0), then

$$x_n = \max(x_{n-1} - \tau, 0) + a_n \sigma.$$

The solution of this recursion is given by the expansion of the Lindley's
equation:

$$x_n = f_n(a_1, \cdots, a_n) = \max\left\{0, \sum_{k=j}^{n-1}(a_k \sigma - \tau), j = 1, \cdots, n-1\right\} + a_n \sigma.$$

$$(1.33)$$

We show by a simple inductive argument that for all n, x_n satisfies:

$$x_n(a + v_1) + x_n(a + v_2) \geq x_n(a) + x_n(a + v_2 + v_1), \qquad (1.34)$$

$$x_n(a + v_1) \vee x_n(a + v_2) \geq x_n(a) \vee x_n(a + v_2 + v_1). \qquad (1.35)$$

The function $x_1(a) = a_1 \sigma$ clearly satisfies (1.34) and (1.35).
    if v_1, v_2 are in $\mathcal{F} \backslash \{e_n, s_n\}$, then by induction

---

[1] The ITU-T is a group of the standardization organization ITU (International
Telecommunications Union), and is accessible through their WEB address:
http://www.itu.int/ITU-T/.

$$x_n(a + v_1) + x_n(a + v_2)$$
$$= (x_{n-1}(a + v_1) - \tau \vee 0) + a_n\sigma + (x_{n-1}(a + v_2) - \tau \vee 0) + a_n\sigma$$
$$= (x_{n-1}(a + v_1) + x_{n-1}(a + v_2) - 2\tau$$
$$\qquad \vee x_{n-1}(a + v_1) - \tau \vee x_{n-1}(a + v_2) - \tau \vee 0) + 2a_n\sigma$$
$$\geq (x_{n-1}(a) + x_{n-1}(a + v_2 + v_1) - 2\tau$$
$$\qquad \vee x_{n-1}(a) - \tau \vee x_{n-1}(a + v_2 + v_1) - \tau \vee 0) + 2a_n\sigma$$
$$= x_n(a) + x_n(a + v_2 + v_1).$$

$$x_n(a + v_1) \vee x_n(a + v_2) = (x_{n-1}(a + v_1) - \tau \vee x_{n-1}(a + v_2) - \tau \vee 0) + a_n\sigma$$
$$= ((x_{n-1}(a + v_1) \vee x_{n-1}(a + v_2)) - \tau \vee 0) + a_n\sigma$$
$$\geq x_n(a) \vee x_n(a + v_2 + v_1).$$

If $v_1 = s_n$ and v_2 is in $\mathcal{F}\backslash\{e_n, s_n\}$, then Equation (1.34) is obtained similarly, by induction. As for Equation (1.35), the proof is slightly different.

$$x_n(a + v_1) \vee x_n(a + v_2)$$
$$= (x_{n-1}(a + e_{n-1}) - \tau - \sigma \vee -\sigma \vee x_{n-1}(a + v_2) - \tau \vee 0) + a_n\sigma$$
$$= (x_{n-1}(a) - \tau \vee x_{n-1}(a + v_2) - \tau \vee 0) + a_n\sigma$$
$$\geq x_n(a) \vee x_n(a + v_2 + v_1).$$

If $v_1 = e_n$, $x_n(a + e_n) + x_n(a + v_2) = x_n(a) + x_n(a + v_2) + \sigma = x_n(a) + x_n(a + v_2 + e_n)$, and

$$x_n(a + e_n) \vee x_n(a + v_2) = x_n(a) + \sigma$$
$$\geq x_n(a) \vee x_n(a + v_2) + \sigma$$
$$= x_n(a) \vee x_n(a + v_2 + v_1).$$

Our goal is now to obtain a policy a^* that minimizes an expected average cost related to the amount of work in the system at arrival epochs. The cost to be minimized is thus

$$g(a) \triangleq \overline{\lim_{N \to \infty}} \frac{1}{N} \sum_{n=1}^{N} f_n(a_1, \cdots, a_n),$$

subject to the constraint:

$$\underline{\lim_{N \to \infty}} \frac{1}{N} \sum_{n=1}^{N} a_n \geq p^*.$$

Consider first the case of a queue with infinite capacity. Then, it follows from Theorem 6 that a bracket policy with rate p^* and arbitrary θ is optimal. The assumptions of the Theorem indeed hold:

- f_n (in (1.33)) is indeed monotone increasing in a_i;
- Property $< 3 >$ (in Subsection 1.3.2) holds by choosing $b_k = 0$, since

$$f_k(a_1, \cdots, a_k) = f_m(\underbrace{0, \cdots, 0}_{m-k}, a_1, \cdots, a_k), \quad k < m; \qquad (1.36)$$

- By combining (1.36) with the first monotonicity property, we get

$$f_{k-1}(a_2, \cdots, a_k) = f_k(0, a_2, \cdots, a_k) \leq f_k(a_1, \cdots, a_k),$$

which establishes Property $< 2 >$ (in the beginning of Section 1.3).

Consider now a queue with a finite storage capacity for the workload, i.e. the workload at the queue at each time instant is bounded by C. When the queue is full, the overflow workload is lost. The bracket policy need not be optimal anymore, as the following example shows.

Counter-example 10 *(Non optimality of a bracket policy)*
Let $\tau = 1, \sigma = 100, C = 100, p^* = 0.01$. Assume that the cost to be minimized is the average queue length. The bracket policy with rate 0.01 achieves an average queue length of 50.5 for any θ. Consider now the periodic policy of period 200 that accepts 2 consecutive customers and rejects all following ones. After the second acceptance, the amount of work in the system is 100 due to the limit on the queue capacity, and there is loss of workload (of 99 units). The average queue length is 25.75. Thus the new policy achieves half the queue length as the previous one.

Although the bracket policy in the above counter example results in a larger queue, it has the advantage over the other policy of not creating losses. As we now show, a bracket policy is optimal if we restrict to policies with the additional constraint that no losses are allowed. Thus, consider the class of policies that satisfy the constraint:

$$x_n(a_1, \cdots, a_n) \leq C$$

where x_n is given by (1.33).

Lemma 9. *The set $A \triangleq \{(a_1, \cdots, a_n) \in \mathbb{N}^n : x_n(a_1 \cdots, a_n) \leq C\}$ defines a convex union of atoms of $\mathbb{Z}^n$.*

Proof. By definition of x_n, the set A can also be defined as

$$A = \{(a_1, \cdots, a_n) : \forall j, \quad a_j \in \mathbb{N}, \quad a_j \geq 0, \quad \sum_{k=j}^{n} a_k \leq \frac{C + (n-j)\tau}{\sigma}\}.$$

$$(1.37)$$

Since A is only formed of integer points, we can also write

$$A = \{(a_1, \cdots, a_n) : \forall j, \quad a_j \in \mathbb{N}, \quad a_j \geq 0, \quad \sum_{k=j}^{n} a_k \leq \lfloor \frac{C + (n-j)\tau}{\sigma} \rfloor\}.$$

(1.38)

Now, let us consider the constraints one by one.

- The constraints $a_j \geq 0$ restrict A to $\mathbb{N}^n$ which is made of a convex union of atoms.
- The constraint $a_n \leq \lfloor \frac{C}{\tau} \rfloor$ also restricts A on a union of atoms.
- Now, let us look at a general constraint $\sum_{k=j}^{n} a_k \leq \lfloor \frac{C+(n-j)\tau}{\sigma} \rfloor$. On the projection over the last $n-j$ coordinates, this constraint is a convex union of faces of atoms. Therefore, on the whole set, this constraint is a union of atoms.

To finish the proof, remark that the intersection of convex union of atoms is a convex union of atoms.   □

Using now Corollary 1 and Theorem 6, we conclude that a bracket policy is again optimal.

In the above admission control we considered only the possibility of accepting or rejecting the whole arriving batch (of 100). In practice, arriving batches may correspond to cells originating from different sources, and it is often possible to reject only a part of the batch.

Assume, thus, that the available actions are $a \in \{0, 1, \cdots, \overline{N}\}$, where $a = i$ means accepting $i(100/\overline{N})$ units of workload. Assume that the batch size of 100 is an integer multiple of $\overline{N}$. We can thus split an arrival batch and accept only a fraction of it; more precisely, we can either reject it, or accept $1/\overline{N}$th of the batch, or $2/\overline{N}$th, etc... The smallest unit of batch which we can accept (i.e. $\overline{N}^{-1}$) is called a mini batch.

Consider now the bracket policy $a^*[\overline{N}]$ that is given in (1.13) corresponding to $p = p^*\overline{N}$. In other words, instead of considering a target fraction p of the whole batch to be accepted, which is smaller than (or equal to) 1, the new target corresponds to the average number of mini batches to be accepted, and can be any real number between 0 and $\overline{N}$ (in particular, $p = \overline{N}$ will correspond to accepting $\overline{N}$ mini-batches, i.e. the whole original batch).

We may repeat the above calculation and show that this policy is optimal for the cases of (i) the infinite queue and (ii) the bounded queue, restricting to policies that do not generate losses. Moreover, for both cases, this policy is better than the one that consists of accepting or rejecting the whole batch according to the policy a^* defined above, since a^* is a feasible policy in our new problem, for which $a^*[\overline{N}]$ is optimal (Theorem 6).

In order to illustrate the last point, consider $\overline{N} = 10$. A bracket policy corresponds to acceptance of a mini-batch of 10 units, once every 10 time slots. The average queue length obtained by that policy is 5.5, i.e. about ten times less than the one obtained when the whole batch was to be accepted or rejected.

1.6 Clustering versus Smoothing

We have already seen in the previous section an example where a bracket policy is not optimal, and only when adding a further constraint we obtain the optimality of a bracket policy. We present here yet another admission control example in which a policy that tends to cluster acceptances performs better than any bracket policy.

Consider discrete time; assume that every time unit there is a potential arrival (it becomes an actual arrival if it is accepted). Its service time takes 2.5 units. The fraction of accepted customers is not allowed to be below 1/2.

We consider two types of regimes of the system:

(i) The preemptive regime: the packet in service is lost if another one arrives while it is served.

(ii) The nonpreemptive policy: the arriving customer is lost if it finds a packet in service.

The best bracket policy accepts every second arrival. If the preemptive regime is used, then under this bracket policy, all accepted packets are lost. If the system operates under the nonpreemptive regime, then under this bracket policy, every second admitted packet is lost.

If we use the sequence $(110100)^\infty$ instead of $(101010)^\infty$ we get both in the preemptive as well as in the nonpreemptive case an improvement: now only 1/3 of the admitted packets are lost. Thus a non bracket policy performs better than the bracket ones.

1.7 Appendix A: proof of Lemma 2

Proof. (a) We shall show that for any $w, v \in \mathcal{F}$, there is a one to one map $z : \mathbb{Z}^m \to \mathbb{Z}^m$ such that

$$\Delta_v \Delta_w f(x) = f(z(x)) - f(z(x) + v + w) - f(z(x) + v) - f(z(x) + w). \quad (1.39)$$

Hence the multimodularity implies $\Delta_v \Delta_w f \leq 0$. Since the map is one to one, we get also the converse. In the next four equations we illustrate (1.39) which establishes the proof.

Consider first $w = s_i$, $v = s_j$ $(v \neq w)$. Then

$$\Delta_v \Delta_w f(x) = (\Delta_{i-1} - \Delta_i)(\Delta_{j-1} - \Delta_j) f(x)$$
$$= (\Delta_{i-1} - \Delta_i)(f(x + e_{j-1}) - f(x + e_j))$$
$$= f(x + e_{j-1} + e_{i-1}) - f(x + e_{j-1} + e_i)$$
$$- \quad f(x + e_j + e_{i-1}) + f(x + e_j + e_i)$$

$$(1.40)$$

$$= f(z + s_j + s_i) - f(z + s_j) - f(z + s_i) + f(z), \quad (1.41)$$

where $z \stackrel{\triangle}{=} x + e_j + e_i$.

Let $v = e_m, w = -e_1$. Then

$$\Delta_v \Delta_w f(x) = \Delta_m(f(x - e_1) - f(x))$$
$$= f(x + e_m - e_1) - f(x - e_1) - f(x + e_m) + f(x) \quad (1.42)$$

Let $v = e_m, w = s_j$.

$$\Delta_v \Delta_w f(x) = \Delta_m(\Delta_{j-1} - \Delta_j)f(x)$$
$$= \Delta_m(f(x + e_{j-1}) - f(x + e_j))$$
$$= f(x + e_{j-1} + e_m) - f(x + e_{j-1}) - f(x + e_j + e_m) + f(x + e_j)$$
$$= f(z + s_j + e_m) - f(z + s_j) - f(z + e_m) + f(z) \quad (1.43)$$

where $z = x + e_j$.

Let $v = -e_1, w = s_j$.

$$\Delta_v \Delta_w f(x) = \Delta_v(\Delta_{j-1} - \Delta_j)f(x)$$
$$= \Delta_{-e_1}(f(x + e_{j-1}) - f(x + e_j))$$
$$= f(x + e_{j-1} - e_1) - f(x + e_{j-1}) - f(x + e_j - e_1) - f(x + e_j)$$
$$= f(z + s_j - e_1) + f(z + s_j) + f(z - e_1) + f(z) \quad (1.44)$$

where $z = x + e_j$.

(b.i) For any $w \in \mathcal{F}$,

$$\Delta_{e_1} \Delta_w f(x) = \Delta_w \Delta_{e_1} f(x) = -\Delta_w \Delta_{-e_1} f(z)$$

where $z = x + e_1$. The result follows from Lemma 2 (a).

(b.ii) Without loss of generality, assume that $i \le j$. Then

$$\Delta_i \Delta_j f = \left(\Delta_{e_1} - \sum_{k=2}^{i} \Delta_{s_k}\right)\left(\sum_{l=j+1}^{m} \Delta_{s_l} + \Delta_{e_m}\right) f$$

The proof of (b.ii) is established by applying Lemma 2 (a) and Lemma 2 (b.i).

For $i < j$ we have

$$(\Delta_j - \Delta_i)\Delta_j f = \left(-\sum_{k=i+1}^{j} \Delta_{s_k}\right)\left(\sum_{l=j+1}^{m} \Delta_{s_l} + \Delta_{e_m}\right) f$$

and (b.iii) is established by applying Lemma 2 (a). For $i > j$ we have

$$\Delta_j = \Delta_i - \sum_{k=i+1}^{m-1} \Delta_{s_k} - \Delta_m + \Delta_{e_1} - \sum_{k=2}^{j} \Delta_k.$$

Hence

$$\Delta_j(\Delta_j - \Delta_i)f = \left(\Delta_{e_1} - \sum_{k=2}^{j} \Delta_{s_k}\right)\left(-\sum_{k=i+1}^{m} \Delta_{s_k} - \Delta_m + \Delta_{e_1} - \sum_{k=2}^{j} \Delta_k\right)f.$$

Again, (b.iii) is established by applying Lemma 2 (a) and 2 (b.i).
By taking $i = j - 1$ in Lemma 2 (b.iii), we obtain (b.iv).
(b.v) For $j < i$,

$$\Delta_{e_i}\Delta_{s_j}f = \left(\sum_{k=i}^{m} \Delta_{s_k} + \Delta_{e_m}\right)\Delta_{s_j}f.$$

For $j > i$,

$$\Delta_{e_i}\Delta_{s_j}f = \left(\Delta_{e_1} - \sum_{k=2}^{i} \Delta_{s_k}\right)\Delta_{s_j}f.$$

For both cases, the proof is established by applying Lemma 2 (a), and in the second case we use also Lemma 2 (b.i).
(b.vi)

$$\Delta_{s_1}\Delta_{s_i}f = (\Delta_{e_m} - \Delta_{e_1})\Delta_{s_i}f$$

The proof is established by applying Lemma 2 (a) and 2 (b.i).
(b.vii)

$$\Delta_{s_i}\Delta_{s_i}f = \Delta_{s_i}\left(-\sum_{j\neq i} \Delta_{s_j}\right)f \tag{1.45}$$

For $i \neq 1$, the proof is established by applying Lemma 2 (a) and 2 (b.i), where we replace in the summation Δ_{s_1} by $\Delta_{e_m} - \Delta_{e_1}$.
For $i = 1$ it follows from part (b.vi) of this Lemma and setting $i = 1$ in (1.45).
(c) $\Delta_{-e_1}\Delta_{e_2} \leq 0$ due to (1.4); $\Delta_{-e_1}\Delta_{s_2} = \Delta_{-e_1}(\Delta_{e_1} - \Delta_{e_2}) \leq 0$, and $\Delta_{e_2}\Delta_{s_2} = \Delta_{e_2}(\Delta_{e_1} - \Delta_{e_2}) \leq 0$ due to (1.5). Hence f is multimodular by Lemma 2 (a). □

1.8 Appendix B: Proof of Lemma 5

Proof. Let $z = (z_1, \cdots, z_k) \in \mathbb{R}_+^k$. This point belongs to an atom, say $S(z)$, made by the extreme points $x^0, x^1, \cdots, x^k$. The numbering of the extreme points of the atom is chosen such that according to the base $\mathcal{F}^k = (-e_1^k, s_2^k, \cdots, s_k^k, e_k^k)$, $x^1 = x^0 - e_1^k$. The other indices are arbitrary. This implies that $x_j^0 = x_j^1$ for all $j \geq 2$. If we call P the projection of $\mathbb{R}_+^k$ onto $\mathbb{R}_+^{k-1}$ along the first coordinate,

$$P(s_j^k) = P(s_j^{k-1}) \text{ if } 2 < j \leq k$$
$$P(s_2^k) = -e_1^{k-1}$$
$$P(e_1^k) = 0$$
$$P(e_k^k) = e_k^{k-1}$$
$$P(x^i) = (x_2^i, \cdots, x_k^i)$$

These equalities imply that $P(x^0) = P(x^1)$ and $P(x^1), \cdots, P(x^k)$ form an atom in $\mathbb{R}_+^{k-1}$, as follows from the definition of an atom, and $P(z)$ belongs to this atom. Let

$$(z_1, z_2, \cdots, z_k) = \left(1 - \sum_{i=1}^k \alpha_i\right) x^0 + \alpha_1 x^1 + \cdots + \alpha_k x^k,$$

then

$$(z_2, \cdots, z_k) = P(z_1, z_2, \cdots, z_k)$$
$$= \left(1 - \sum_{i=1}^k \alpha_i\right) P(x^0) + \alpha_1 P(x^1) + \cdots + \alpha_k P(x^k)$$
$$= \left(1 - \sum_{i=2}^k \alpha_i\right) P(x^1) + \cdots + \alpha_k P(x^k).$$

Now,

$$\widetilde{f}_k(z_1, z_2, \cdots, z_k) = \left(1 - \sum_{i=1}^k \alpha_i\right) f_k(x^0) + \alpha_1 f_k(x^1) + \cdots + \alpha_k f_k(x^k)$$
$$\geq \left(1 - \sum_{i=1}^k \alpha_i\right) f_{k-1}(P(x^0)) + \cdots + \alpha_k f_{k-1}(P(x^k))$$
$$= \left(1 - \sum_{i=2}^k \alpha_i\right) f_{k-1}(P(x^1)) + \cdots + \alpha_k f_{k-1}(P(x^k))$$
$$= \widetilde{f}_{k-1}(z_2, \cdots, z_k).$$

$\square$

1.9 Appendix C: average and weighted costs

The following Lemma is often used in applications of optimal control (or games) with an average cost criteria (see e.g. [105]), yet it is not easy to find its proof in the literature in the format in which we want to apply it.

Lemma 10. *Consider a sequence a_n of real numbers all having the same sign. Then,*

$$\varlimsup_{n \to \infty} \frac{1}{N} \sum_{n=1}^N a_n \geq \varlimsup_{\alpha \to 1} (1 - \alpha) \sum_{k=1}^{\infty} \alpha^{k-1} a_k \tag{1.46}$$

$$\geq \varliminf_{\alpha \to 1} (1 - \alpha) \sum_{k=1}^{\infty} \alpha^{k-1} a_k \geq \varliminf_{n \to \infty} \frac{1}{N} \sum_{n=1}^N a_n \overset{\triangle}{=} p \tag{1.47}$$

Proof. Note that

$$\frac{1}{1-\alpha}\sum_{k=1}^{\infty}\alpha^{k-1}a_k = \sum_{k=1}^{\infty}\left(\sum_{l=1}^{k}a_l\right)\alpha^{k-1} \tag{1.48}$$

$$\frac{1}{(1-\alpha)^2} = \sum_{k=1}^{\infty}k\alpha^{k-1}. \tag{1.49}$$

Hence

$$(1-\alpha)\sum_{k=1}^{\infty}\alpha^{k-1}a_k - p = (1-\alpha)^2\sum_{k=1}^{\infty}\left(\frac{1}{k}\sum_{l=1}^{k}a_l - p\right)k\alpha^{k-1}. \tag{1.50}$$

For any $\epsilon > 0$, choose N_ϵ such that

$$\frac{1}{N}\sum_{n=1}^{N}a_n \geq p - \epsilon$$

for all $N \geq N_\epsilon$. Then the right-hand side of (1.50) is bounded below by

$$(1-\alpha)^2\left(\sum_{k=1}^{N_\epsilon-1}\left(\frac{1}{k}\sum_{l=1}^{k}a_l - p\right)k\alpha^{k-1} - \epsilon\sum_{k=N_\epsilon}^{\infty}k\alpha^{k-1}\right)$$

$$\geq (1-\alpha)^2\left[\left(N_\epsilon\max_{1\leq k\leq N_\epsilon}\left|\sum_{l=1}^{k}a_l - kp\right|\right) - \epsilon(1-\alpha)^{-2}\right] \geq -2\epsilon$$

for α sufficiently close to 1. This establishes (1.47), and (1.46) is obtained similarly. $\qquad\square$

2 Balanced Sequences

2.1 Introduction

In this chapter, we will introduce the notion of balanced sequences, which is closely related to the notion of *Sturmian* sequences [85] as well as exactly covering sequences.

Those sequences are binary sequences where the ones are distributed "as evenly as possible" while satisfying a density constraint.

This presentation is not exhaustive and many other related articles and books can be consulted for further investigation on this topic [113, 75, 16, 28, 107, 46, 84]. Although the chapter is self-contained and presents several results which are of interest by their own, we mostly focus on the *rate problem* (see Problem 2), which will be used in the application section (§ 6.2).

2.1.1 Organization of the chapter

This chapter is structured as follows. In Section 2.2, we introduce a formal definition of *balanced sequence* and we present an overview on their properties. Section 2.3 defines the constant gap sequences and addresses the problem of finding the rates admissibles for constant gap sequences. Section 2.4 shows how balanced sequences can be characterized using constant gap sequences. Section 2.5 addresses the problem of finding balanceable rates for several cases.

2.2 Balanced sequences and bracket sequences

Let $\mathcal{A}$ be a finite alphabet, $\mathcal{A}^{\mathbb{Z}}$ is the set of bi-infinite sequences and $\mathcal{A}^{\mathbb{N}}$ is the set of infinite sequences defined on $\mathcal{A}$.

If $u \in \mathcal{A}^{\mathbb{Z}}$ or $u \in \mathcal{A}^{\mathbb{N}}$, then a *word* W of u is a finite subsequence of consecutive letters in u: $W = u_a u_{a+1} \cdots u_{a+k-1}$. The integer k is the *length* of W and will be denoted $|W|$.

If $a \in \mathcal{A}$, $|W|_a$ is the number of a's in the word W.

Definition 4. *The sequence $u \in \mathcal{A}^{\mathbb{Z}} \cup \mathcal{A}^{\mathbb{N}}$ is a-balanced if for any two words W and W' in u of same length,*

$$-1 \le |W|_a - |W'|_a \le 1.$$

The sequence u is balanced if it is a-balanced for each $a \in \mathcal{A}$.

If $a \in \mathcal{A}$, and $u \in \mathcal{A}^{\mathbb{N}}$, we also define the *indicator* in a u of the letter a as the function $\mathbf{1}_a(u) : \mathcal{A}^{\mathbb{N}} \to \{0,1\}^{\mathbb{N}}$ by, $\mathbf{1}_a(u)_i = 1$ if $u(i) = a$ and 0 otherwise. The definition is similar in the bi-infinite case. The *support* in u of the letter a is the set $\mathcal{S}_a = \{i \in \mathbb{N} : \mathbf{1}_a(u)_i = 1\}$.

As usual, for any real number x, $\lfloor x \rfloor$ will denote the largest integer smaller or equal to x and $\lceil x \rceil$ will denote that smallest integer larger or equal to x.

Lemma 11. *If a sequence $u \in \mathcal{A}^{\mathbb{N}}$ (resp. $u \in \mathcal{A}^{\mathbb{Z}}$) is balanced, then for any $a \in \mathcal{A}$, there exists a real number p_a, such that*

$$\lim_{n \to \infty} \frac{1}{n} \sum_{i=0}^{n} \mathbf{1}_a(u)_i \left(resp. = \lim_{n \to -\infty} \frac{1}{n} \sum_{i=n}^{0} \mathbf{1}_a(u)_i\right) = p_a.$$

p_a *is called the rate (or slope) of $\mathbf{1}_a(u)$.*

Proof. Let us define $c_n = \sum_{i=0}^{n} \mathbf{1}_a(u)_i$ and remark that $c_n + c_m - 1 \le c_{n+m} \le c_n + c_m + 1$. The rest of the proof is classical by sub-additivity arguments. The proof for $\{\mathbf{1}_a(u)_n\}_{n \le 0}$ is similar. The fact that both limits coincide is obvious.   □

Note that the sum of the rates for all letters in a sequence u is one.

$$\sum_{a \in \mathcal{A}} p_a = 1.$$

Now, we can present the main result which is the foundation of the theory of balanced sequences. We follow the presentation given in [109] and [85].

We now extend the definition of bracket sequences, as given in Definition 2.

Definition 5. *A sequence $u \in \{0,1\}^{\mathbb{Z}}$ (or $u \in \{0,1\}^{\mathbb{N}}$) is bracket (resp. ultimately bracket) if there exist two real numbers θ and p (and an integer k, respectively) such that for all $n \in \mathbb{Z}$ (or $n \in \mathbb{N}$) (resp. $n \ge k$), $u_n = \lfloor (n+1)p + \theta \rfloor - \lfloor np + \theta \rfloor$.*

Note that $u_n = \lfloor (n+1)p + \theta \rfloor - \lfloor np + \theta \rfloor$ is equivalent to $\mathcal{S}_1 = \{\lceil n\frac{1}{p} + \phi \rceil\}_{n \in \mathbb{Z}}$ with $\phi = 1 - \theta/p$, where $\lceil x \rceil$ denotes the smallest integer no smaller than x.

The following two theorems show the relation that exists between balanced sequences and *bracket* sequences.

Theorem 11 (Morse and Hedlund). *A sequence $a \in \{0,1\}^{\mathbb{Z}}$ is balanced with asymptotic rate $0 < p = 1/\alpha$ for letter 1 if and only if the support $\mathcal{S}_1$ of a satisfies one of the following cases.*

(a)(irrational case) p is irrational and there exists $\phi \in \mathbb{R}$ such that

$$S_1 = \{\lfloor i\alpha + \phi \rfloor\}_{i \in \mathbb{Z}} \quad or \quad S_1 = \{\lceil i\alpha + \phi \rceil\}_{i \in \mathbb{Z}}.$$

(b)(periodic case) $p \in \mathbb{Q}$ and there exists $\phi \in \mathbb{Q}$ such that

$$S_1 = \{\lfloor i\alpha + \phi \rfloor\}_{i \in \mathbb{Z}}.$$

(c)(skew case) $p \in \mathbb{Q}$ $(p = k/n,\ k, n \in \mathbb{N})$ and there exists $m \in \mathbb{Z}$ such that

$$S_1 = \{\lfloor in/k + m \rfloor\}_{i < k} \cup \{\lfloor in/k - 1/k + m \rfloor\}_{i > 0}$$

or

$$S_1 = \{\lfloor in/k + m \rfloor\}_{i > 0} \cup \{\lfloor in/k - 1/k + m \rfloor\}_{i < k}$$

Theorem 12 (Morse and Hedlund). *A sequence* $a \in \{0, 1\}^{\mathbb{N}}$ *is balanced with asymptotic rate* $0 < p = 1/\alpha$ *for letter* 1 *if and only if the support* S_1 *of a satisfies one of the following cases.*

(a)(irrational case) p is irrational and there exists $\theta \in \mathbb{R}$ such that

$$S_1 = \{\lfloor i\alpha + \theta \rfloor\}_{i \in \mathbb{N}} \quad or \quad S_1 = \{\lceil i\alpha + \theta \rceil\}_{i \in \mathbb{N}}.$$

(b)(periodic case) $p \in \mathbb{Q}$ and there exists $\theta \in \mathbb{Q}$ such that

$$S_1 = \{\lfloor i\alpha + \theta \rfloor\}_{i \in \mathbb{N}}.$$

As for the case $p = 0$, it includes two balanced sequences, namely the sequence where S_1 is the empty set and the sequence where S_1 is a singleton.

The case for bi-infinite and infinite binary balanced are subtly different. Indeed, the skew case does not appear in the infinite case while cannot be overlooked in the bi-infinite case.

The irrational case is the easiest case and can be characterized (see Theorem 13). The rational cases are more difficult to study.

Note that the skew sequences are not periodic but are ultimately periodic. In our applications (see sections 6.2 and 6.3), we will mainly be using infinite sequences, so that the skew case will not be considered. However, even in the bi-infinite case, we have the following property.

Proposition 1. *Let* $u \in \mathcal{A}^{\mathbb{Z}}$.
(i)- If $1_a(u)$ *is bracket for all* $a \in \mathcal{A}$, *then u is balanced.*
(ii)- If u is balanced, then $1_a(u)$ *is ultimately bracket for all* $a \in \mathcal{A}$.

Proof. (i) is straightforward.
(ii) is a direct consequence of Theorem 11, since in all three cases, the sequence $1_a(u)$ is ultimately bracket. An elementary proof of (ii) which does not use Theorem 11 can be found in [106]. □

Infinite balanced sequences with irrational rates are aperiodic and are sometimes called Sturmian words [94, 84].

2.3 Constant gap sequences

Constant gap sequences are strongly balanced sequences, in the following sense.

Definition 6. *A sequence G is constant gap if for any letter a, $\mathbf{1}_a(G)$ is periodic, with a period of the form $0 \cdots 010 \cdots 0$.*

Note that this explains the fact that G is said to have constant gaps for the letter a, since each a is separated from the next a in G by a constant number of letters.

Proposition 2. *Constant gap sequences are balanced.*

Proof. For each letter a, $\mathbf{1}_a(G)$ is of the form $(0^n 10^m)^\omega$. Therefore, $\mathbf{1}_a(G)$ is bracket with $p = 1/(m + n + 1)$ and $\theta = m/(m + n + 1)$. Using the characterization of balanced sequences given in Theorem 1, this shows that G is balanced. □

Proposition 3. *Constant gap sequences are periodic.*

Proof. For each letter a, $\mathbf{1}_a(G)$ is periodic with period p_a. The period of G is $lcm(p_a, a \in \mathcal{A})$. □

In the next lemma, we give a characterization of constant gap sequences that stresses the fact that constant gap is some kind of strong balance.

Proposition 4. *G is constant gap if and only if, for any two finite words, W and W' included in G with $||W| - |W'|| \leq 1$, then for each letter a, $||W|_a - |W'|_a| \leq 1$.*

Proof. Let a be a letter in the alphabet.
First, assume that G is constant gap. If $|W|_a - |W'|_a \geq 2$, then, necessarily, $|W| - |W'| \geq 2$. Conversely, let $W = aUa$ and $W' = aU'a$ be any two words in G with no a in the subwords U and U'. If $|U| \geq |U'| + 1$, then we have $||U| - |W'|| \leq 1$ and $||U|_a - |W'|_a| = 2$. This is a contradiction. Therefore $|U| = |U'|$ and G is constant gap. □

Since a constant gap sequence is balanced, each letter appears with a given rate in the sequence. note however that since a constant gap sequence is necessarily periodic, the rate of each letter is rational.

As we will do in Section 2.5 for the case of balanced sequences, we now address the following question:

Problem 1. Given a set $(p_1, \cdots, p_K)$, with $p_1 + \cdots + p_K = 1$ is it possible to construct a constant gap sequence on K letters with rates $(p_1, \cdots, p_K)$?.

We will not solve this problem for a general K, since it is NP-complete (see [26]).

Remark 4. One can use the following facts to show that the set of rates solving 1 is finite. Let $(p_1, \cdots, p_K)$ be a set of rates solving problem 1. We order the rates so that $1 > p_1 \geq \cdots \geq p_K$. Then necessarily,

– For all i, p_i^{-1} is an integer.
– For all i, $p_{i-1} \geq p_i \geq \frac{1-(p_1+\cdots+p_{i-1})}{K-i+1}$.

Note that combining both items leaves only a finite number of possibilities for $(p_1 \cdots, p_K)$.

We will now give some properties of the set $(p_1, \cdots, p_K)$ which will be useful in the following. A characterization of such $(p_1, \cdots, p_K)$ is given in [115], under the name *exact covering sequences*, but it does not provide an algorithm.

Definition 7. *The set of couples $\{(\theta_i, g_i), i = 1 \cdots K\}$ is called an exact covering sequence if for every nonnegative integer n, there exists one and only one $1 \leq i \leq K$ such that $n = \theta_i \mod g_i$.*

As a general remark, note that $(p_1, \cdots, p_K)$ are rational numbers of the form $p_i = k_i/d$, with d the smallest period of G. Therefore, we have, $\sum_i k_i/d = 1$ and for each i, k_i divides d. By definition of the rates, we also have $p_a = 1/q_a$ for all letters.

We have the following result.

Proposition 5. *The rates $(p_1, \cdots, p_K)$ are constant gap if there exists K numbers called phases, $\theta_1, \cdots \theta_K$ such that the couples $\{(\theta_i, q_i), i = 1 \cdots K\}$ form an exact covering sequence.*

Proof. This property is a simple rewriting of the fact that each letter a_i in a constant gap sequence appears every $\{\theta_i + kq_i, \quad k \in \mathbb{N}\}$. □

Now, suppose that $\{(\theta_i, q_i), i = 1 \cdots K\}$ is an exact covering sequence. Then in the series

$$S(x) \stackrel{\text{def}}{=} \sum_{i=1}^{K} \sum_{k \geq 0} x^{\theta_i + kq_i},$$

the coefficient of x^n in this series is equal to 1, $\forall n \geq 0$. Therefore, we have

$$S(x) = \sum_{i=1}^{K} \frac{x_i^\theta}{1 - x^{q_i}} = \frac{1}{1-x}.$$

Using this characterization we have the following interesting property which was proved in [93].

Lemma 12. *Assume* $\{(\theta_i, q_i), i = 1 \cdots K\}$ *is an exact covering sequence and that* $P = \max_i q_i$. *Then* P *appears at least twice in the set* $q_1, \cdots, q_K$.

Proof. The proof given here is similar to the discussion in [115] on exact covering sequences. Let $w \stackrel{\text{def}}{=} e^{2i\pi/r}$ for some integer $r > 1$. By definition, w is a primitive r-th root of one. We have:

$$(w - x)S(x) = \sum_{i=1}^{K} \frac{(w - x)x_i^{\theta}}{1 - x^{q_i}} = \frac{w - x}{1 - x}.$$

Let $x \to w$. This yields

$$\sum_{i:r|q_i} \frac{-w^{\theta_i}}{-q_i w^{q_i-1}} = 0. \tag{2.1}$$

Now, take $r = P$. The set $\{i : P|q_i\}$ is exactly the set $\{i : P = q_i\}$. Equation (2.1) specified for $r = P$ can be written

$$\sum_{i:q_i=P} w^{\theta_i} = 0.$$

This implies that the set $\{i : P = q_i\}$ cannot be reduced to a single point since w is not zero. □

To give some concrete examples, we consider the cases where K is small. First, note that in the case where the k_i are not all equal, (assume k_1 is the largest of all), we have

$$\sum_{i \neq 1} k_i/k_1 = d/k_1 - 1 = l_1,$$

where l_1 is the gap between two letters a_1. This implies,

$$l_1 \leq K - 2. \tag{2.2}$$

Proposition 6. *There exists a constant gap sequence* G *with rates* $(p, 1 - p)$ *if and only if* $p = 1/2$.

Proof. Let a be a letter in G with gap l. Since the alphabet contains only two letters, $l = 1$. This means $p = 1/2$. □

Proposition 7. *There exists a constant gap sequence* G *with rates* (p_1, p_2, p_3) *if and only if* $(p_1, p_2, p_3) \in \{(1/3, 1/3, 1/3), (1/2, 1/4, 1/4)\}$ *(up to a permutation).*

Proof. Assume that $(p_1, p_2, p_3) \neq (1/3, 1/3, 1/3)$ (otherwise, $G = (abc)^\omega$ is constant gap). Using Inequality (2.2), $l_1 = 1$ and $p_1 = 1/2$. Therefore, the sequence obtained from G when removing all the letters a_1 is constant gap. Applying Lemma 6 shows that $p_2 = p_3$. The only solution is $p_1 = 1/2, p_2 = 1/4$ and $p_3 = 1/4$. The associated constant gap sequence is $(a_1 a_2 a_1 a_3)^\omega$. □

Proposition 8. *A constant gap sequence G with rates (p_1, p_2, p_3, p_4) exists if and only if (p_1, p_2, p_3, p_4) belongs to the set (up to a permutation),*

$$\{(\frac{1}{4}, \frac{1}{4}, \frac{1}{4}, \frac{1}{4}), (\frac{1}{2}, \frac{1}{4}, \frac{1}{8}, \frac{1}{8}), (\frac{1}{2}, \frac{1}{6}, \frac{1}{6}, \frac{1}{6}), (\frac{1}{3}, \frac{1}{3}, \frac{1}{6}, \frac{1}{6})\}.$$

Proof. We give a sketch of an elementary proof of this fact. If the rates are all equal, then $(p_1, p_2, p_3, p_4) = (1/4, 1/4, 1/4, 1/4)$. Now, note that Equation 2.2 implies that if the rates are not all equal $l_1 \leq 2$. We consider any letter $a_i, i \neq 1$, assume that the number of a_1's in between two a_i's is not constant and takes values m and n, $m > n$. then we have $l_i \geq (n-1)l_1 + n$ on one hand and $l_i \leq n(l_1) - 3$ on the other hand. This is impossible since $l_1 \leq 2$. Therefore, the number of a_1's in between two a_i's is constant. This is true for all i. The sequence formed by removing all a_1's is still constant gap. It has rates of the form $(1/3, 1/3, 1/3)$ or $(1/2, 1/4, 1/4)$. From this point a case analysis shows that the original sequence has rates $(1/2, 1/4, 1/8, 1/8), (1/2, 1/6, 1/6, 1/6) or (1/3, 1/3, 1/6, 1/6)\}$ by inserting the letter a_1 in a constant gap sequence over the letters a_2, a_3, a_4. $\square$

These few examples of constant gap sequences illustrate the fact that there are very few rates that can be achieved by constant gap sequence.

2.4 Characterization of balanced sequences

Several studies have been recently done on balanced (or bracket) sequences [83, 107, 106, 109, 90, 89]. In [55, 75], a characterization involving constant gap sequences is given.

The proof of the following results, Proposition 9 and Theorem 13 was given by Graham [55] for bracket sequences. An independent later proof can be found in [75] for balanced sequences. The relation between balanced and bracket sequences given in Theorem 1 makes both proofs more or less equivalent.

Proposition 9. *Let U be a balanced sequence on the alphabet $\{0, 1\}$. Construct a new sequence S by replacing in U, the subsequence of zeros by a constant gap sequence G on an alphabet A_1, and the subsequence of ones by a constant gap sequence H on a disjoint alphabet A_2. Then S is balanced on the alphabet $A_1 \cup A_2$.*

Proof. We give a proof similar to Hubert's proof ([75]). Let a be a letter in A_1 (the proof is similar for a letter in A_2). Let W and W' be two words of S of the same length. Then, the corresponding words X and X' in U verify $||W|_0 - |W'|_0| \leq 1$ since U is balanced. If we keep only the 0's in X and X', then the corresponding Z and Z' words in G satisfy $||Z| - |Z'|| \leq 1$. Since G is constant gap, and using Lemma 4, $||Z|_a - |Z'|_a| \leq 1$. We end the proof noting that the construction of Z and Z' implies $|Z|_a = |W|_a$ and $|Z'|_a = |W'|_a$. $\square$

Conversely, we have the following theorem.

Theorem 13. *Let $u \in \mathcal{A}^{\mathbb{Z}}$ be balanced and non ultimately periodic. Then there exists a partition of $\mathcal{A}$ into two sets $\mathcal{A}_1$ and $\mathcal{A}_2$ such that the sequence v defined by:*

$$v_n = 1 \ if \ u_n \in \mathcal{A}_1, \tag{2.3}$$
$$v_n = 0 \ if \ u_n \in \mathcal{A}_2, \tag{2.4}$$

is balanced. Furthermore, the sequences z_1 and z_2 constructed from u by keeping only the letters from $\mathcal{A}_1$ and $\mathcal{A}_2$ respectively have constant gaps.

2.5 Balanceable rates

Let us formulate precisely the problem which we will study in this section.

Problem 2. Given a set $(p_1, \cdots, p_K)$, is it possible to construct a balanced sequence on K letters with rates $(p_1, \cdots, p_K)$?

We will see in the following that this construction is not possible for all values of the rates $(p_1, \cdots, p_K)$. If a K-tuple $(p_1, \cdots, p_K)$ makes the construction possible, such a tuple is said to be *balanceable*. A similar problem has been addressed in [55, 90, 107, 46], where relations between the rates of balanced sequences are studied.

2.5.1 The case $K = 2$

This case is well known and balanced sequences with two letters have been extensively studied (see for example [32, 85]). The following result is known under many different forms.

Theorem 14. *For all p, $0 \leq p \leq 1$, the set of rates $(p, 1 - p)$ is balanceable.*

Proof. The proof is similar to the proof of the first part of Theorem 1. We construct a sequence S as the support of the function $s(n) = \lfloor pn \rfloor - \lfloor p(n-1) \rfloor$. S is a balanced sequence because the interval $]k, k + m]$ contains exactly $e = \lfloor pk + pm \rfloor - \lfloor pk \rfloor$ elements of S. Now, $\lfloor pm \rfloor + \lfloor pk \rfloor - \lfloor pk \rfloor \leq e \leq \lceil pm \rceil + \lfloor pk \rfloor - \lfloor pk \rfloor$. This shows the value of e can differ by at most one when k varies so S is a balanced sequence. If S' is the complementary set of S, then it should be clear that S' has asymptotic rate $1 - p$ and S' is balanced because $S'_{]k,k+m]}$ contains $m - e$ elements. $\square$

Note that the proof of Theorem 14 also gives a construction of a balanced sequence with the given rates.

2.5.2 The case $K = 3$

The case $K = 3$ is essentially different from the case $K = 2$. In the case $K = 2$, all possible rates are balanceable while when $K = 3$, there is only one set of distinct rates which is balanceable. This result, when formulated under this form, was partly proved and conjectured in [83] and proved in [107]. In earlier papers by Morikawa, [90], a similar result is proved for bracket sequences. If Theorem 1 is used, then the result of Morikawa can be used directly to prove the following theorem.

Theorem 15. *A set of rates (p_1, p_2, p_3) is balanceable if and only if,*

$$(p_1, p_2, p_3) = (4/7, 2/7, 1/7)$$

or two rates are equal.

Proof. The proof of Morikawa is very technical since it does not use the balanced property for bracket sequences. If the balanced property is used, then the proof becomes easier. We give a proof slightly simpler than the proof in [107]. First, assume that $p_1 = p_2$. Then, let S be a balanced sequence with two letters $\{a, b\}$ constructed with the rates $(p_1 + p_1, p_3)$. In S, replace alternatively the "a"s by the letters a_1, a_2, we get a sequence S' on alphabet $\{a_1, a_2, b\}$ with rates (p_1, p_1, p_3). Let us show that S' is balanced. Since S is balanced, the number of "a"s in an interval of length m is k or $k + 1$, for some k. Now, for S', the number of "a_1"s (resp. "a_2"s) in such an interval is either $(k - 1)/2$ or $(k + 1)/2$ if k is odd and $k/2$ or $k/2 + 1$ if k is even. This proves that S' is balanced.

Now, assume that (p_1, p_2, p_3) are three different numbers. We assume that $p_1 > p_2 > p_3$. We will try to construct a sequence W with these respective rates on the alphabet $\{a, b, c\}$.

step 1: the sequence "aca" must appear in W.
There exists a pair of consecutive "a" with no "b" in between since $p_1 > p_2$. This means that a sequence "aa" or "aca" appears. If "aa" appears, then a "c" is necessarily surrounded by two "a"s.

step 2: the sequences "baab" and "abaaba" must appear in W.
There exist a pair of consecutive "b" with no "c" in between. This sequence is of the form "ba^nb". Now, $n \leq 1$ is not possible because of the presence of "aca" and b-balance. $n \geq 3$ implies the existence of "$a^{n-1}ca^{n-1}$" by a-balance which is incompatible with "ba^nb" because of b-balance. Therefore $n = 2$. Note that this also implies the existence of "aa" and of "abaaba".

step 3: the sequence "abacaba" appears in W.
the sequence W must contain a "c". This "c" is necessarily surrounded by two "a"s since "aa" exists by step 2. This group is necessarily surrounded by two "b"s since "baab" exists, and consequently, necessarily surrounded by two "a"s because "abaaba" exists. We get the sequence "abacaba".

Last step: $W = (abacaba)^\omega$.

No letter around this word can be a "c" because "$baab$" exists. None can be a "b" since "aca" exists. Therefore, they have to be two "a"s. Then note that the two surrounding letters cannot be "c" (because of the existence of "$abaaba$"), nor "a" (because of the existence of "bac") so they are "b", then followed necessarily by "a" (because "aa" exists). At this point, we have the sequence "$\star abaabacabaaba\star$". Both $\star$s are necessarily "c"s.

To end the proof, note that we have obtained the configuration around every "c" and this determines the whole sequence. The sequence W is periodic of the form $(abacaba)^\omega$.   □

2.5.3 The case $K = 4$

For distinct rates, the case $K = 4$ is very similar to the case $K = 3$. However when two rates are equal, this case is more complicated. Again, a similar result for bracket sequences is contained in [90] under a weaker form since, the proof is only done for rates of the form $(a_1/c, \cdots, a_4/c)$. The following theorem gives the result in its full generality. Again, using the balanced property helps to keep the proof rather elementary.

Theorem 16. *A rate tuple* (p_1, p_2, p_3, p_4) *with four distinct rates is balance-able if and only if* $(p_1, p_2, p_3, p_4) = (8/15, 4/15, 2/15, 1/15)$.

Proof. We suppose that $p_1 > p_2 > p_3 > p_4$ and we show that there is only one balanced sequence with frequencies $p_1 > p_2 > p_3 > p_4$ and those frequencies are $(8/15, 4/15, 2/15, 1/15)$.

As a preliminary remark, note that if $p_i > p_j$, then there exists at least one word "$a_i \cdots a_i$" that does not contain any a_j. This fact will be used several times in the following arguments.

The proof involves different steps.

Step 1: W contains the words "aca" *or* "ada" *or* "$acda$" *or* "$adca$".

There exist two consecutive "a"s with no "b" in between because $p_1 > p_2$. Therefore, either "aa" or "aca" or "ada" or "$acda$" or "$adca$" exist. If "aa" exists, then, a "c" is surrounded by two "a"s.

Step 2: W contains the word "$baab$"

First, we show that if a word "$ba^n b$" exists, then $n = 2$. Indeed, step 1 makes "bb" and "bab" impossible. On the other hand, if $n \geq 3$, the existence of "$a^{n-1}ca^{n-1}$" is necessary by a-balance and is incompatible with the existence of "$ba^n b$" because of b-balance.

Now, if no word of the form "$ba^n b$" exists, then there exist two consecutive "b"s with one "d" and no "c" in between. Such a word is of the form: "$ba^j da^k b$" and is surrounded by some "a"s, so that we get a word called s_1 and equal to "$a^i ba^j da^k ba^l$". Note that the numbers i, j, k, l may be equal to zero but $j + k \geq 1$ by b-balance.

There also exist two consecutive "c"s with no "d" in between. In between those two "c" we must have some "a"s and some "b"s. In fact we have exactly one "b" since no word "ba^nb" exists by assumption. We pick such a word called s_2 which is of the form: "ca^mba^nc". Note that i, j, k, l, m, n are integers that can differ by at most one. As for the length of s_1, we have $|s_1| \geq \max(4, 2(n+m)+1) \geq n+m+3 = |s_2|$. This is impossible by c-balance.

step 3: the word "abacabaabacaba" exists in W. There exist two consecutive "c"s with no "d" in between. From step 3 in the proof of Lemma 15, we know that a "c" is necessarily surrounded by the word "aba". Moreover, from step 4 in the proof of Lemma 15, we have: "$abacabaabaUcaba$", where U is a word that contains no "d" and no "c". U cannot start with an "a" (because of "$bacab$") and cannot start with a "b" (because of "aca"). Therefore U has to be empty.

Step 4: W is uniquely defined and is periodic of period "abacabadabacaba". Somewhere, W contains a "d". From this point on, we can extend the word uniquely around this "d" as the word: "$abacabadabacaba$", and, on the other hand, the word "$abacabaabacaba$" has to be surrounded by two "d"s. This ends the proof. □

To complete the picture, it is not difficult to see that,

Proposition 10. *if the tuple (p_1, p_2, p_3, p_4) is made of no more than two distinct numbers, then it is balanceable.*

Proof. First, if the rates are all equal, they are obviously balanceable. If three of them are equal, say $p_1 = p_2 = p_3$, then, we can construct a balanced sequence with rates $(3p_1, p_4)$ and we construct a balanced sequence with rates (p_1, p_1, p_1, p_4) by using Proposition 9 (where we take G the constant gap sequence $(a_1a_2a_3)^\omega$ and $H \stackrel{\text{def}}{=} (a_4)^\omega$). If two pairs of rates are equal, say $p_1 = p_2$ and $p_3 = p_4$, then we construct a balanced sequence with rates $(2p_1, 2p_3)$ and we apply Proposition 9.

□

If the tuple (p_1, p_2, p_3, p_4) is made of exactly three distinct numbers, then this is a more complex case which is not studied here.

2.5.4 The general case

In this section, we are interested in the case of arbitrary K. First, note that Proposition 10 easily generalizes to any dimension.

Proposition 11. *If the tuple $(p_1, p_2, p_3, \cdots, p_K)$ is made of less than two distinct numbers, then it is balanceable.*

Proof. The proof is similar to the proof of Proposition 10. □

Proposition 12. *If the tuple $(p_1, p_2, \cdots, p_K)$ is balanceable, then the tuple $\underbrace{(p_1/k, \cdots, p_1/k}_{k}, p_2, \cdots, p_K)$ is balanceable.*

Proof. The proof is very similar to that of Proposition 9. If W is a balanced sequence with letters $\{a_1, \cdots, a_K\}$, consider the sequence W' constructed starting from W and replacing each "a_1" by an element of (b_1", "b_2", $\cdots$, "b_k") in a cyclic way. Note that W' has the following set of rates, $(p_1/k, \cdots, p_1/k, p_2, \cdots, p_K)$.

Next, we show that W' is balanced. Since W is balanced, for an arbitrary integer m, the number of "a_1"s in an interval of length m is n or $n+1$, for some n. Now, for W', the number of "b_i"s in such an interval is either $\lfloor (n-1)/k \rfloor$ or $\lfloor (n+1)/k \rfloor$. This proves that W' is balanced. □

For the general case and distinct rates, it is natural to give the following conjecture (due to Fraenkel for bracket sequences):

Conjecture 1. A set of distinct rates $\{p_1, \cdots, p_K\}$ with $K > 2$ is balanceable if and only if

$$\{p_1, \cdots, p_K\} = \{2^{K-1}/(2^K - 1), \cdots, 2^{K-i}/(2^K - 1), \cdots, 1/(2^K - 1)\}.$$

We have not been able to prove this fact. Morikawa has also given some insight in this conjecture. Very recently, in [108] the cases $K = 5$ and $K = 6$ are proven using techniques inspired by those introduced here. However, it seems clear that a different approach is needed in order to complete the proof in the general case. Here, we only have partial results given in the following propositions.

Proposition 13. *The rates* $(2^{K-1}/(2^K-1), \cdots, 2^{K-i}/(2^K-1), \cdots, 1/(2^K-1))$ *are balanceable, for all* $K \in \mathbb{N}$.

This proposition is the "if" direction of the conjecture.

Proof. We construct a balanced sequence w_K in the following inductive way. $w_1 = a_1$, $w_K = w_{K-1}a_K v_{K-1}$ and $w_K = (v_K)^\omega$. First note that w_K has rates $(2^{K-1}/(2^K - 1), \cdots, 2^{K-i}/(2^K - 1), \cdots, 1/(2^K - 1))$. Then, we show that w_K is balanced by induction. In the sequence w_K, any letter (say letter j) appears 2^{K-j} times in one period and is of the form of $2^{K-j} - 1$ intervals of the same length (2^j) and one of length $2^j - 1$.

By construction of w_{K+1}, these properties still hold and therefore, w_{K+1} is balanced. □

Proposition 14. *Let* $K > 2$ *and* w *be balanced with rates* $p_1 > \cdots > p_K$, *then,* w *is ultimately periodic. In particular, this means that* $p_i \in \mathbb{Q}$, $\forall 1 \leq i \leq K$.

Proof. If w is not ultimately periodic, Theorem 13 says that w is composed of two constant gap sequences. At least one of these sequences has at least two letters, and therefore two letters have rates which are equal by Lemma 12. Therefore, the rates in w of these two letters are also equal. □

Proposition 15. *Let w be balanced with rates $p_1 > \cdots > p_K$, with the following property: for any $1 \leq i \leq K$, there exists two consecutive letters "a_i" with no a_j in between, with $j > i$. Then, $(p_1, \cdots, p_K) = (2^{K-1}/(2^K - 1), \cdots, 2^{K-i}/(2^K - 1), \cdots, 1/(2^K - 1))$ and w is uniquely defined, up to a shift.*

This proposition is a partial "only if" result for the conjecture.

Proof. The proof holds by induction. Let v_k denote the period of the balanced sequence with rates $(2^{k-1}/(2^k - 1), \cdots, 2^{k-i}/(2^k - 1), \cdots, 1/(2^k - 1))$ given in Proposition 13. We recall that according to the construction in the proof of proposition 13, $v_k = v_{k-1}a_kv_{k-1}$. We will prove by induction that w is periodic with period v_K.

We prove by induction on k that w contains "$v_{k-1}a_kv_{k-1}v_{k-1}a_kv''_{k-1}$" and that for $j \geq k$, each letter in w, "a_j", is surrounded by v_{k-1}'s, for all possible $1 < k \leq K$. For the first step of the induction ($k = 1$), note that according to the property on w, w contains the factor "a_1a_1" which is the same as "v_1v_1". Therefore any other letter is surrounded by two "a_1"s. This also implies the existence of "$a_1a_2a_1a_1a_2a_1$" by using a similar argument as step 2 in the proof of Theorem 16. This ends the case $k = 1$.

For the general case, by the induction assumption, a_k is surrounded by v_{k-2}, and w contains the word "$v_{k-2}a_kv_{k-2}$". The existence of the factor "$v_{k-2}a_{k-1}v_{k-2}v_{k-2}a_{k-1}v_{k-2}$" proves that this word is surrounded by two "a_{k-1}"s. Therefore, two consecutive a_k form the word

$$v_{k-2}a_{k-1}v_{k-2}a_kv_{k-2}a_{k-1}v_{k-2}Uv_{k-2}a_{k-1}v_{k-2}a_kv_{k-2}a_{k-1}v_{k-2},$$

where U does not contain any letter "a_j", $j \geq k$.

- If U does not contain "a_{k-1}" and contains a letter "a_i" with $i < k-1$ then U is reduced to this letter, because of a_{k-1}-balance and the existence of the word $a_{k-1}v_{k-2}v_{k-2}a_{k-1}$ (induction assumption). But now the construction of v_{k-2} implies the presence of "$a_iv_{i-1}a_i$" and contradicts the existence of "$a_iv_{i-1}v_{i-1}a_i$".
- If U contains "a_{k-1}", then by a_{k-1}-balance, U must contain $v_{k-2}a_{k-1}v_{k-2}$. therefore U is of the form $Xv_{k-2}a_{k-1}v_{k-2}Y$. The arguments used for U can be applied to X and Y. If they are not empty, they must both contain $v_{k-2}a_{k-1}v_{k-2}$. Eventually, we have the existence of the word $v_{k-2}v_{k-2}a_{k-1}v_{k-2}v_{k-2}$ which contradicts obviously the existence of the word $v_{k-2}a_{k-1}v_{k-2}a_kv_{k-2}a_{k-1}v_{k-2}$ by a_1-balance.

Therefore, U is empty. This implies the existence of $v_{k-1}a_kv_{k-1}v_{k-1}a_kv_{k-1}$.

Now, we finish the proof by noticing that the letter a_K is surrounded by v_{K-1} and by noting that $v_{K-1}v_{K-1}$ is necessarily surrounded by "a_K". $\qquad \square$

Proposition 16. *The projection w' of a sequence w over the alphabet $\mathcal{A}-\{a\}$ is w where all a's have been removed. If $p_a \geq 0.5$, then w is balanced implies that w' is balanced.*

Proof. Choose two words v_1', v_2' of length n in w'. Let v_1 and v_2 be any two words in w whose projections over the alphabet $\mathcal{A} - \{a\}$ are v_1' and v_2', respectively. Assume, furthermore, that the first and last letters in v_1 and v_2 are not a. Let $k = |v_1| - n$ and $l = |v_2| - n$ denote the number of appearances of the letter a in v_1 and v_2, resp.

Step 1: If $l = k$ then the difference in the number of occurrences of any letter $b \neq a$ in v_1' and in v_2' is at most 1, since w is balanced, and since the number of b's in v_1 (resp. v_2) is the same as its number in v_1' (resp. v_2').

Step 2: Assume that $l > k + 1$.

- Let $\hat{v}_2$ be the word obtained from v_2 by truncating the first and last letter. Then $|\hat{v}_2| = n + l - 2$, and the number of a's in $\hat{v}_2$ is l.

- Let $\hat{v}_1$ be the word obtained from v_1 by concatenating to it the next $m = l - k - 2$ letters that appear after v_1 in the sequence w. Then $|\hat{v}_1| = n + l - 2 = |\hat{v}_2|$, and the number of a's in $\hat{v}_1$ is not larger than $k + m = l - 2$. This is a contradiction with the fact that w is balanced.

Step 3: It remains to check the case $l = k + 1$. Add to v_1 the next letter that occurs in w to its right, to form the new word $\overline{v}_1$. If it is not a then we have two successive letters that are not a, which contradicts the fact that a has an asymptotic frequency of at least $1/2$. If it is a, then $\overline{v}_1$ and v_2 have the same number of a's. We can now apply the same argument as in step 1 and conclude that the number of occurrences of any letter b in v_1' and in V_2' differs by at most 1.

Combining the above steps, we conclude that w' is balanced. $\qquad\square$

2.5.5 Extensions of the original problem

So far we have only analyzed the case where all the rates add up to one. The different results tend to prove that very few sets of rates are balanced.

Now let us look at a generalization when all the rates do not add up to one. Assume that S is a sequence on the alphabet $\{a_1, a_2, \cdots, a_k, *\}$. We only require that S is balanced for the letters $a_1, \cdots, a_k$, but not for the special letter $*$.

On a more practical point of view, the question can be viewed as whether this allows more possibilities for rates to be balanced when "losses" are allowed (represented by the letter $*$). Then again, in general, the rates are not balanced, even if the total sum is very small as illustrated by the following proposition.

Proposition 17. *For an arbitrary $\varepsilon > 0$, there exists two real numbers p_1 and p_2 such that $p_1 + p_2 < \varepsilon$ and such that there is no sequence S on the alphabet $\{a, b, *\}$ with asymptotic rate p_1 for letter a and p_2 for letter b which is balanced for a and b.*

Proof. Choose two irrational numbers p_1 and p_2 with $p_1 + p_2 < \varepsilon$ such that p_1, p_2 and 1 are not linearly dependent on $\mathbb{Z}$. Now assume that there exists a sequence S on $\{a, b, *\}$ with asymptotic rate p_1 for letter a and p_2 for letter b which is balanced for a and b. By Theorem 1, then there exists two real numbers x, y such that $\mathbf{1}_a(S)(n) = 1$ if $x + p_1 n \mod 1 \in [1 - p_1, 1]$ and 0 otherwise. $\mathbf{1}_b(S)(n) = 1$ if $y + p_2 n \mod 1 \in [1 - p_2, 1]$ and 0 otherwise. In the cube $[0, 1]^2$, the set of points $(x, y) + n(p_1, p_2) \mod (1, 1)$ is dense (see for example, Weyl's ergodic theorem [102]) and therefore hits the rectangle $[(1 - p_1, 1 - p_2), (1, 1)]$. This is not possible. $\qquad\square$

More on this kind of problems can be found in [106].

To end this short overview on balanced sequences, we must mention on the positive side that "usual" rates, such as $(1/k, 1/k, \cdots, 1/k)$ are often balanceable. In Appendix 2.6, some examples of balanced sequences and their rates are given.

2.6 Appendix

In this appendix, we shall give a collection of balanceable set of rates which can be put into two classes. Some of them are *composite*: they can be constructed using Proposition 12 (once or several times) starting from a smaller balanceable set. The ones which cannot be constructed that way are called *primitive*.

We shall first give a list that contains some primitive balanceable rates with $K = 4$ (as well as other cases with different values of K).

- $(1/11, 2/11, 4/11, 4/11)$ is balanceable and $S = (abcababcabd)^\omega$.
- $(1/11, 2/11, 2/11, 6/11)$ is balanceable and $S = (abacaabacad)^\omega$.
- $(1/11, 1/11, 3/11, 6/11)$ is balanceable and $S = (acabaabadab)^\omega$.
- for all K, $((2^{K-1}/(2^K - 1), \cdots, 2^{K-i}/(2^K - 1), \cdots, 1/(2^K - 1)))$ is balanceable. The associated balanced sequence is constructed recursively as in Proposition 13.

Here are other balanceable sets of rates when $K = 4$ which are composite.

- $(1/14, 1/14, 4/14, 8/14)$ is balanceable and $S = (abacabaabadaba)^\omega$. It is composite since it comes from $(1/7, 2/7, 4/7)$ where the smallest rate is split into two.
- For each real number $0 < p \leq 1$, the rates $(1 - p, p/4, p/4, p/2)$ are balanceable, with a corresponding balanced sequence constructed from a bracket sequence with rate p where all 1 are replaced in turn by the sequence $(abac)^\omega$ and each 0 by the letter d. It is composite, originating from $(1-p, p)$ and split twice.
- $(1/k, \cdots, 1/k)$ is balanceable. A balanced sequence is: $S = (a_1 a_2 a_3 \cdots a_k)^\omega$. This is composite, coming from (1) split once into k rates.

$-$ $(p, \cdots, p, \beta, \cdots, \beta)$ is balanceable. A balanced sequence with those rates is constructed in the following way: Choose a balanced sequence S on letters, (A, B) with rate $(p_1 = \sum_{i=1}^{k} p, p_2 = \sum_{i=1}^{h} \beta)$. In S replace all the A (resp. B) by $a_1, a_2, \cdots, a_k$ (resp. $b_1, \cdots, b_h$) in a round robin fashion to get a balanced sequence with the required rates. This is also composite, coming from (p_1, p_2), where p_1 is split into k rates and p_2 is split into h rates.

3 Stochastic Event Graphs

3.1 Introduction

This chapter introduces the timed Petri net models with a special focus on timed event graphs with stochastic timings.

The first goal on the chapter is to give a precise definition of the semantics of timed Petri nets with timings associated to the transitions. This is used to show that the evolution of the firing epochs in an event graph can be written under the form of a max,plus linear system, as proved in [23].

We give several examples from queuing theory, manufacturing systems and communication networks of event graph models in order to illllustrate their modelling power as well as their limitations.

The (max,plus) evolution equations are used to derive a vectorial form of Lindley's equation for waiting times and sojourn times in networks of queues [25]. This equation will be used in Chapter 4 to show that the sojourn time of a customer in a timed event graph is a multimodular function of the admission actions. Thus, timed event graphs will constitute the most general class of networks treated in this book for which the general theorems proved in Chapter 1 will apply.

3.1.1 Organization of the chapter

This chapter is organized as follows. Section 3.2 introduces the definitions and notations foir stochastic Petri nets. Section 3.3 deals with the dynamic behavior of event graphs and shows that the evolution equations of event graphs can be put under the form of a linear equation in the (max,+) formalism. Section 3.4 shows several examples of discrete event systems taken from queuing theory, manufacturing and telecommunication with their even graph modeling and the associated (max,+) evolution equation. Finally, Section 3.5 derives a vectorial Lindley's equation for (max,+) linear systems.

3.2 Stochastic Petri Nets

Petri nets constitute a model of discrete systems that combines concurrency and competition at the node level. They were primarily used for analyzing logical properties of systems involving parallelism and synchronizations, see for example [44]. More recently, the notion of time has been introduced in Petri nets to make performance analysis possible as in [31, 23].

Definition 8. *A Petri net is a bi-partite graph given by the tuple* $\mathcal{G} = (\mathcal{P}, \mathcal{Q}, \mathcal{E}, \mathcal{M}_0)$, *where* $\mathcal{P} = \{p_1, \cdots, p_P\}$ *is the set of places,*
$\mathcal{Q} = \{q_1, \cdots, q_Q\}$ *is the set of transitions,*
$\mathcal{E}$ *is a subset of* $(\mathcal{Q} \times \mathcal{P}) \cup (\mathcal{P} \times \mathcal{Q})$ *and is the set of edges. We will denote by* $p^\bullet$ *(*$^\bullet p$, $q^\bullet$, $^\bullet q$*), the set of downstream (upstream) transitions (places) of place (transition)* p *(*q*).*
$M^0 : \mathcal{P} \to \{0, 1, 2, \cdots M\}$ *is the initial number of tokens in each place.*

Since $\mathcal{G}$ is an oriented graph, then the following definitions are classical.

- *Paths:* a path is a set of transitions $\{q_1, \cdots, q_n\}$ and a set of places $\{p_1, \cdots, p_{n-1}\}$ such that for all $i = 1 \cdots n - 1$, (q_i, p_i) and (p_i, q_{i+1}) are arcs in $\mathcal{A}$.
- *Circuits:* a circuit is a path such that $q_1 = q_n$.
- *Strongly connected component:* a strongly connected component is a maximal sub-net $\mathcal{C}$ of $\mathcal{G}$ such that for each pair of transitions q_1, q_2 in $\mathcal{C}$, there exists a path in $\mathcal{C}$ from q_1 to q_2.
- *inputs and outputs:* an input transition q verifies $^\bullet q$ is empty. An output transition q is such that $q^\bullet$ is empty. Note that an input transition (resp. an output transition) is a strongly connected component by itself.

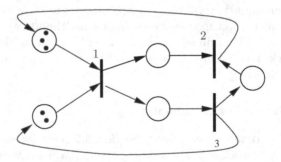

Fig. 3.1. A petri net

A Petri net is displayed in Figure 3.1. Places are drawn as circles and transitions as thin rectangles. The initial marking is displayed with tokens in the places.

Definition 9. *A timed Petri net is a Petri net with in addition, firing times associated with the transitions* [1]

---

[1] Time is sometimes also attached to places. It is possible convert a Petri net with time associated with places into a Petri net with time on transitions and vice-versa.

$-\ \sigma = \{\sigma_q(n), q \in Q, n \in \mathbb{N}\}$ *is the set of the firing times of the transitions,* *where* $\sigma_q(n)$ *is the nth firing of transition* q.

In the following, we will assume that the stochastic processes $\{\sigma_i(n)\}_{n \in \mathbb{N}}$ are stationary with finite expectations. In addition, we assume that the sequences $\{\sigma_i(n)\}$ and $\{\sigma_j(n)\}$ are mutually independent for all i and j in Q.

The marking in place p is a right-continuous function $M_p(t) : \mathbb{R}_+ \to \mathbb{N}$ and evolves according to the semantics of a timed Petri net which follows.

Enabling - A transition q is *enabled* at time t if each input place p of q $(p \in {}^\bullet q)$ contains a token at time t.

Firing - If a transition q starts its n-th firing at time t, then one token is "frozen" in each input places at time t. The transition ends the firing at time $t + \sigma_q(n)$ while the frozen tokens are removed and one token is added in all output places of q at time $t + \sigma_q(n)$.

The marking $M(t) \stackrel{\text{def}}{=} (M_1(t), \cdots M_P(t))$ of a Petri net changes according to this *firing rule:* [2] if transition q fires for the n-th time at time t,

$$M_p(t + \sigma_q(n)) = \begin{cases} M_p(t + \sigma_q(n)_-)) - 1 & \text{if } p \in {}^\bullet q, \\ M_p(t + \sigma_q(n)_-) + 1 & \text{if } p \in q^\bullet, \\ M_p(t + \sigma_q(n)_-) & \text{otherwise.} \end{cases} \qquad (3.1)$$

Initial conditions - The initial marking is the marking at time 0, and no tokens are frozen yet.

$$\forall p \in \mathcal{P}, \quad M_p(0_-) = M_p^0. \qquad (3.2)$$

Input transitions - Note that the firings of input transitions cannot obey the firing rule defined above. Input transition would fire at any time since they are always enabled. To overcome this difficulty, firing times of the input transitions are given as extra data.

The sequence of firing epochs of input transition k is given beforehand and denoted $\{U_k(n)\}_{n \in \mathbb{N}}$. The sequence $\{U_k(n)\}_{n \in \mathbb{N}}$ is non-negative and increasing, $0 \leq U_k(1) \leq U_k(1) \leq \cdots \leq U_k(n) \leq \cdots$, and $U_k(n)$ is the epoch of the n-th firing of input transition k. At the same time, one token is released in each output place of k.

These input transitions are very useful for modeling purposes. They model exogenous arrivals of tokens into an open system (see section 3.4 for some examples).

Several pictures of the firing of a transition are displayed in Figure 3.2. At time 0_-, the initial marking is displayed in Figure 3.2(a). At time 0,

---

[2] the notation $f(x_-)$, for a right continuous functions f is the limit of $f(y)$ when y goes to x, $y < x$

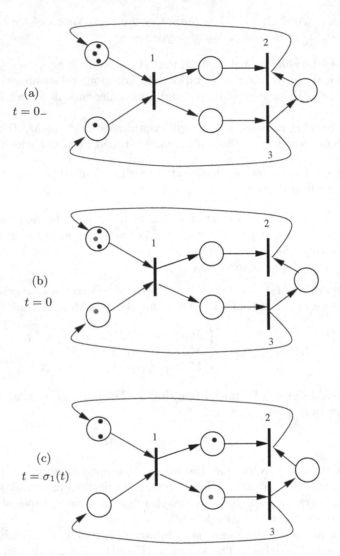

Fig. 3.2. Three snapshots of a Petri net taken at time (a) 0_-, (b) 0 and $\sigma_1(1)$. Frozen tokens are grey, available tokens are black.

transition 1 becomes enabled and freezes one token in each input place as it starts its first firing. The situation remains unchanged up to time $\sigma_1(1)_-$ (Figure 3.2(b)). At time $\sigma_1(1)$, the frozen tokens are removed. Tokens are put in the output places of transition 1. At the same time, transitions 3 becomes enabled and starts firing by freezing the new token (Figure 3.2(c)).

3.2.1 Properties

The behavior of a (timed) Petri net may be classified according to the fact that all transitions fire infinitely often, or some starve after some time or even everything stops in a situation where no transition can fire.

More precisely, a Petri net is *live* if any transition q may eventually be fired starting from any marking M reachable from M^0: there exists a sequence of transitions $q_1, \cdots q_n$ such that firing this sequence, starting in M yields a marking M' enabling q.

A petri net is in a *deadlock* M if after reaching the marking M from M^0, no transition is enabled anymore.

A petri net with input transitions is in an *input deadlock* M if after reaching the marking M, no transitions is enabled except the input transitions (which are always enabled by definition).

In [22], it has been shown that liveness (or presence of deadlocks) does not depend on the timings of the Petri net. This is a structural property which only depends on the topology, the routing of tokens in places with several output (see [48]) and the initial marking of the net.

Checking if a Petri net is live is difficult in general. It has been proven co-NP-Complete for the class of Free choice nets (see [44]).

However, testing if a Petri net is live is polynomial for some other sub-classes, such as state machines or event graphs (see § 3.2.2).

3.2.2 Event graphs

Definition 10. *An event graph is a Petri net where each place has one in-coming transition and one out-coming transition:*

$$\forall p \in \mathcal{P}, |{}^\bullet p| = |p^\bullet| = 1. \tag{3.3}$$

The net displayed in Figure 3.1 is an event graph: all places have one incoming and one outgoing transition.

The modeling power of an event graph is limited but its dynamic is simple and can be put in a linear from in the (max,plus) semi-ring (see § 3.3.3).

This subclass of Petri net can be used to model several systems on communication, manufacturing or queuing which are highly synchronized and where the routes are all pre-defined. Some examples will be given in Section 3.4.

Liveness - As mentioned above, testing if an event graph is live is easy.

Lemma 13. *An event graph is live if and only if all circuits are marked under the initial marking.*

For a proof of this lemma, see for example [44]. This provides a polynomial test for liveness of event graphs.

Input connectedness - We consider an event graph with input transitions. We say that the graph is *input-connected* (*i.e.* for each transition q in the net, there exists a path from one input transition to q).

For each transition (or node) q in $\mathcal{G}$, we consider all paths π from input transitions to q. This set is denoted by $\mathcal{P}(q)$. We also denote by $M^0(\pi) \stackrel{\text{def}}{=} \sum_{s \in \pi} \mathcal{M}^0(s)$, the sum of the initial tokens on the path π. Now, we define

$$L(q) = \min_{\pi \in \mathcal{P}(q)} M(\pi). \tag{3.4}$$

Lemma 14. *For an event graph which is input connected, the $n + L(q)$-th firing of transition q of $\mathcal{G}$ involves a token produced by the n-th firing of an input transition to which q is connected.*

Proof. Let h_q be a shortest path from one input transition q_0 to q with $L(q)$ tokens. The length of h_q is called the "distance" from q_0 to q. The proof holds by induction on the length of h_q. If $h_q = 0$, then $q = q_0$ and the result is true. Suppose that the result is true for all transitions at "distance" $k - 1$ from q_0. Choose q at distance k, then the transition q' preceding q on the path h_q is at distance $k - 1$ from q_0 and induction applies to q'. Now the place (q', q) contains m tokens. By definition of q', $L(q') = L(q) - m$, and by induction, the $n + L(q')$-th firing of transition q' uses the token number n and since the buffer place between q' and q is only fed by q' and emptied by q by the event graph assumption, the $n + L(q)$-th firing of q will use the same token (n-th token produced by q_0). $\qquad\square$

3.3 Dynamics of Event Graphs

In the following, we will consider a timed event graph $\mathcal{G} = (\mathcal{P}, \mathcal{Q}, \mathcal{E}, \mathcal{M}^0, \sigma)$. We further assume that there is at most one place between two transitions in an event graph. This assumption does not restrict the modeling power of event graphs. Now, if there is a place p between transitions i and j, this place is unique and can be denoted (ij), with initial marking $M_{ij}^0 \stackrel{\text{def}}{=} M_p^0$. The maximal initial marking of all the places is denoted by $m \stackrel{\text{def}}{=} \max_{p \in \mathcal{P}} M_p^0$.

3.3.1 State variables: the firing epochs

In the Markovian case (all $\sigma_q(n)$ have an exponential distribution), the marking $M(t)$ is a Markov process. The main problems with this approach are that the exponential distribution assumption may not be appropriate in many cases[3] also the state space is potentially infinite (or very large) and the trajectory of the system is not easy to retrieve.

In the following, it is merely assumed that the firing times are stationary processes. Here, the marking $M(t)$ is no longer the appropriate state variable. Instead, let the functions $\{X_i(n), i \in \mathcal{Q}, n = 1, \cdots\}$ be the epoch of the beginning of the n-th firing of transition i.

Note that the marking can be retrieved from the $X_i(n)$ by the following formula:

$$M_p(t) = M_p(0) + \sum_{n=1}^{\infty} \mathbf{1}\{X_{\bullet p}(n) + \sigma_{\bullet p}(n) \leq t\} - \mathbf{1}\{X_{p\bullet}(n) + \sigma_{p\bullet}(n) < t\} \quad (3.5)$$

Transformation of the graph Any event graph can be transformed into another event graph with at most one token per place in its initial marking and with the same behavior. This is done by expanding places with more than one initial token into several places with one initial token as illustrated by Figure 3.3.

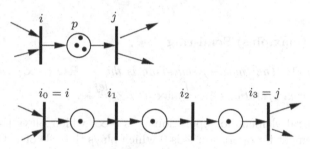

Fig. 3.3. Expansion of a place with 3 initial tokens into three places with one initial token

If $\mathcal{G} = (\mathcal{P}, \mathcal{Q}, \mathcal{E}, M^0, \sigma)$ is an event graph with more than one token in place π, then $\mathcal{G}' = (\mathcal{P}', \mathcal{Q}', \mathcal{E}', M^{0'}, \sigma')$ is a event graph where place π is replaced by $M^0(\pi) - 1$ dummy transitions and $M^0(\pi)$ dummy places with one token initially. More precisely, let $k \stackrel{\text{def}}{=} M^0(\pi)$ be the marking in place π and i and j be the input and output transition of π respectively.

---

[3] The Markovian approach can be extended to the case where the firing times are either exponential or deterministic [41]. This makes the marking $M(t)$ semi-Markovian but the state space problem and the difficulty to study the trajectories are even more acute.

$$P' = P \setminus \{\pi\} \cup \{p_1, \cdots p_k\},$$
$$Q' = Q \cup \{q_1, \cdots q_{k-1}\},$$
$$\mathcal{E}' = \mathcal{E} \setminus \{(\pi, j), (i, \pi)\} \bigcup_{n=0}^{k-1} (q_n, p_{n+1}), (p_{n+1}, q_{n+1}),$$
$$M_p^{0'} = M_p^0, \quad \text{if } p \in P \setminus \{\pi\}$$
$$M_p^{0'} = 1, \quad \text{if } p \in P' \setminus P$$
$$\sigma_q'(n) = \sigma_q'(n), \quad \text{if } q \in Q,$$
$$\sigma_q'(n) = 0, \quad \text{if } q \in Q' \setminus Q,$$

with the convention that $q_0 = i$ and $q_k = j$.

By repeating the expansion for all places with more than one token initially, one gets an event graph $\tilde{\mathcal{G}} = (\tilde{\mathcal{P}}, \tilde{\mathcal{Q}}, \tilde{\mathcal{E}}, \tilde{M}^0, \tilde{\sigma})$ with a maximal marking $m = 1$ and such that the behavior of the initial graph is preserved. For each original transitions $i \in Q$ and for all n, the firing epochs are the same under $\mathcal{G}$ and under $\tilde{\mathcal{G}}$: $X_i(n) = \tilde{X}_i(n)$.

In the following, we will assume that the initial marking in each place is at most one.

3.3.2 The (max,plus) Semi-ring

Definition 11. *The (max, +) semi-ring is the set $\mathbb{R} \cup \{-\infty\}$ equipped with the two internal operations $\oplus \overset{def}{=} \max$ and $\otimes \overset{def}{=} +$.*

The neutral element for operation $\oplus$ is $-\infty$, which plays the role of 0. The unit element for operation $\otimes$ is 0 which plays the role of 1. Operation $\otimes$ distributes with respect to $\oplus$.

The main difference with the conventional ring $(\mathbb{R}, +, \times)$ is the fact that the first operation $\oplus$ is idempotent: $a \oplus a = a$ and does not have an inverse (hence the denomination semi-ring).

The operations can be extended to matrices with the classical construction.

If A and B are both matrices of size $n \times m$, then $A \oplus B$ is a matrix of size $n \times m$ with

$$(A \oplus B)_{ij} \overset{def}{=} A_{ij} \oplus B_{ij}, \quad i = 1 \cdots n, j = 1 \cdots m. \tag{3.6}$$

As for the product, if A and B are matrices of size $n \times \ell$ and $\ell \times m$ respectively, then $A \otimes B$ is a matrix of size $n \times m$ where

$$(A \otimes B)_{i,j} \overset{def}{=} \bigoplus_{k=1}^{\ell} A_{ik} \otimes B_{kj}, \quad i = 1 \cdots n, j = 1 \cdots m. \tag{3.7}$$

Linear equations - Solving linear systems in (max,+) is rather different from the classical linear case. However, there exists one important class of linear systems where the solution exists and is unique.

Lemma 15. *A vectorial linear equation of the form* $X = A \otimes X \oplus b$ *has a minimal solution of the form* $X = A^* \otimes b$ *where* $A^* \overset{def}{=} \oplus_{i=0}^{\infty} A^i$. *This solution is unique if all the entries of A are finite.*

For a proof of this result, see for example [23]. An important special case is presented in the following corollary.

Corollary 5. *If matrix A is acyclic, then* A^* *has all its entries finite and the vectorial linear equation* $X = A \otimes X \oplus b$ *has a unique solution* $X = A^* \otimes b$.

3.3.3 Evolution Equation in the (max,+) semi-ring

The autonomous case - Let us consider the case of a timed event graph $\mathcal{G} = (\mathcal{P}, \mathcal{Q}, \mathcal{E}, \mathcal{M}^0, \sigma)$ with no input transitions.

We assume that this net is *live* and that the initial marking has at most one token per place.

We also assume that the net satisfies a local *FIFO assumption* in each transition. Namely, the n-th firing to start is the n-th firing to finish.

This local FIFO assumption can be enforced by constraining the firing times. For example if the firing time $\sigma_q(n)$ of a transition q are non-decreasing in n, then q satisfies the FIFO assumption. However, this kind of assumptions on the firing time process are often too restrictive, since we want to deal with firing times forming stationary sequences. The local FIFO assumption can also be enforced by topological constraints. The easiest one is by *recycling* the transitions, as illustrated in Figure 3.4.

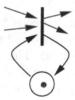

Fig. 3.4. A recycled transition

In a recycled transition only one firing can take place at any given time, since the token in the recycled place is frozen and will only be available once the current firing is over. This enforces the FIFO assumption.

Now, under these assumptions, and according to the firing rule, a transition i in $\mathcal{Q}$, starts its n-th firing at time

$$X_i(n) \stackrel{\text{def}}{=} -\sigma_i(1), \quad \forall n \le 0, \tag{3.8}$$

$$X_i(n) = \max_{j \in \bullet\bullet i} X_j(n - M_{ji}^0) + \sigma_j(n - M_{ji}^0), \quad \forall n > 0. \tag{3.9}$$

Using the (max,+) notation, one gets

$$X_i(n) = \bigoplus_{j \in \bullet\bullet i} X_j(n - M_{ji}^0) \otimes \sigma_j(n - M_{ji}^0). \tag{3.10}$$

This equation can be seen as a linear equation between the variables $X_i(n)$, with coefficients $\sigma_j(n - M_{ji}^0)$.

When written in vectorial form, it becomes

$$X(n) = A(0, n) \otimes X(n) \oplus A(1, n) \otimes X(n - 1), \quad \forall n > 0. \tag{3.11}$$

where $X(n)$ is the vector $(X_1(n), \cdots X_Q(n))^t$ and for $k \in \{0, 1\}$, $A(k, n)$ is a $Q \times Q$ matrix defined by

$$A(k, n)_{ij} \stackrel{\text{def}}{=} \begin{cases} \sigma_j(n - k) & \text{if } M_{ji}^0 = k, \\ -\infty & \text{otherwise.} \end{cases} \tag{3.12}$$

Equation 3.11 is implicit but can be made explicit. Since $\mathcal{G}$ is live, then all circuits are marked initially. This implies that matrix $A(0, n)$ does not contain any circuit. Therefore, $A(0, n)^*$ is finite and the solution of Equation 3.11 is unique, by applying corollary 5.

$$X(n) = A(0, n)^* \otimes A(1, n) \otimes X(n - 1), \forall n > 0. \tag{3.13}$$

If one define $A(n) \stackrel{\text{def}}{=} A(0, n)^* \otimes A(1, n)$, then the standard form of the evolution equation of an autonomous event graph becomes:

$$X(n) = \begin{cases} -\sigma(1) & \forall n \le 0, \\ A(n) \otimes X(n - 1) & \forall n > 0. \end{cases} \tag{3.14}$$

The non-autonomous case - Now, we consider an event graph which is live with at most one initial token per place and which contains transitions with no incoming places (*input transition*). This case will also be referred to as the "open case". Then the set of all the transitions $\mathcal{Q}$ is split into two parts: input transitions (set $\mathcal{Q}_I$ of size Q_I) and all other transitions, (set $\mathcal{Q}_N$ of size Q_N).

Recall that firing of input transitions does not obey the standard firing rule since they have no input places. The sequence of firing epochs of input transition k is given beforehand and denoted $U_k(n)$.

For the dynamics of the other transitions (transitions in $\mathcal{Q}_N$), we define the matrix B of size $Q_N \times Q_K$ which gives the connections of the inputs with the other transitions. More precisely, the entry i, j in B is 0 if there is a place (initially empty) between the input transition j and transition i, and is $-\infty$ otherwise.

The evolution equation of a non-autonomous event graph is then of the form

$$X(n) = A(0,n) \otimes X(n) \oplus A(1,n) \otimes X(n-1) \oplus B \otimes U(n), \quad \forall n > 0. \quad (3.15)$$

Where $X(n)$ is a vector of size Q_N and $A(0,n)$ as well as $A(1,n)$ are matrices of size $Q_N \times Q_N$ and are defined as in the autonomous case (see Equation 3.12).

Since the graph is live, it does not contain any empty circuit and matrix $A(0,n)^*$ exists and is finite. By using Lemma 15 once again, the unique solution of Equation 3.15 is for all $n > 0$,

$$X(n) = A(0,n)^* \otimes A(1,n) \otimes X(n-1) \oplus A(0,n)^*B \otimes U(n). \quad (3.16)$$

Let $A(n) \overset{\text{def}}{=} A(0,n)^* \otimes A(1,n)$ and $B(n) = A(0,n)^*B$.
Then, the standard evolution equation in the open case becomes

$$X(n) = A(n) \otimes X(n-1) \oplus B(n) \otimes U(n). \quad (3.17)$$

In this framework, whenever we refer to an open (max,plus) linear system, we refer to a discrete event system for which the evolution equation can be written under the form of Equation 3.17. Most of the time, these systems can be modeled by (stochastic) event graphs as we just did. The following section presents several examples of such systems.

3.4 Queuing networks

The aim of this section is to give some practical examples of systems from queuing, manufacturing and communication that fall in the class of (max,plus) linear systems. For each case we will exhibit the stochastic event graphs model as well as the corresponding (max,plus) standard evolution equation.

3.4.1 The G/G/1 queue

This first example that is going to be detailled is the G/G/1 queue.

Figure 3.5 shows the event graph model of a G/G/1 queue. This event graph has two transitions, (one input transition modeling the arrivals, and one transition modeling the service) and two places (one for the infinite buffer and one to impose the mono-server semantics which also imposes the FIFO assumption). Note that tokens have different meanings according to the place where they are in. A token in place p_1 represents a customer (a frozen token in place p_1 is a customer being served, an available token in place p_1 is a

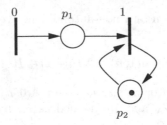

Fig. 3.5. Event graph model of a G/G/1 queue

customer waiting for service). However the token in place p_2 represents the server (if frozen, the server is busy, if not, the server is idle).

Should the initial marking in the recycling place p_2 be k, this becomes a G/G/k queue. If the recycling place is removed, this becomes a G/G/∞ queue. Note that only the G/G/1 queue is (max,+) linear for general firing times, since it is the only one complying with the FIFO assumption.

For the G/G/1 queue, the evolution equation is given in Equation 3.18:

$$X_1(n) = A(n) \otimes X_1(n-1) \oplus B(n) \otimes U(n), \qquad (3.18)$$

where

- $X_1(n)$ is the epoch of the n-firing time of transition 1 (*beginning of the n-th service*),
- $U(n)$ is the the epoch of the n-firing time of transition 0 (*arrival time of the n-th customer*),
- $A(n) = A(1,n)$ is a 1×1 matrix equal to the n-th firing time of transition 1 (*duration of the $n-1$-th service*), $A(n) = \sigma_1(n-1)$,
- $B(n)$ is a 1×1 matrix which gives the traveling time of a token from its entry in the system till reaching transition 1: $B(n) = 0$.

Since the equation is scalar in this case, it can be written using classical notation:

$$X_1(n) = \max(\sigma_1(n-1) + X_1(n-1), U(n)),$$

which is classical in queuing theory.

The advantage of the (max,+) notation will better appear in the following example where a network of queues is considered.

3.4.2 Queues in tandem

Consider a network of Q G/G/1 queues in tandem, with all queues initially empty.

The event graph model of the network of queues has one input transition and Q ordinary transitions modeling respectively the arrival of the customers

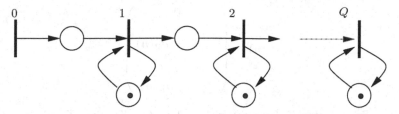

Fig. 3.6. Event graph model of queues in tandem

and the Q servers. An example of the event graph model of queues in tandem is given in Figure 3.6.

In this case, we get

$$
A(0,n) = \begin{pmatrix}
-\infty & -\infty & \cdots & -\infty & -\infty & -\infty \\
\sigma_1(n) & -\infty & \cdots & -\infty & -\infty & -\infty \\
-\infty & \sigma_2(n) & \cdots & -\infty & -\infty & -\infty \\
\vdots & & & & & \\
-\infty & -\infty & \cdots & \sigma_{Q-1}(n) & -\infty & -\infty \\
-\infty & -\infty & \cdots & -\infty & \sigma_Q(n) & -\infty
\end{pmatrix},
$$

$$
A(1,n) = \begin{pmatrix}
\sigma_1(n-1) & -\infty & \cdots & -\infty & -\infty \\
-\infty & \sigma_2(n-1) & \cdots & -\infty & -\infty \\
\vdots & & & & \\
-\infty & -\infty & \cdots & \sigma_{Q-1}(n-1) & -\infty \\
-\infty & -\infty & \cdots & -\infty & \sigma_Q(n-1)
\end{pmatrix},
$$

$$
B = \begin{pmatrix}
0 \\
-\infty \\
-\infty \\
\vdots \\
-\infty \\
-\infty
\end{pmatrix}.
$$

The matrix $A(0,n)$ is clearly acyclic and applying Corollary 5, the matrix $A(0,n)^*$ exists. Therefore, the matrix $A(n) = A(0,n)^* \otimes A(1,n)$ is given by :

$$
(A(n))_{ij} = \begin{cases}
-\infty & \text{if } i < j, \\
\sum_{k=j}^{i-1} \sigma_k(n) + \sigma_j(n-1) & \text{if } i \geq j.
\end{cases} \tag{3.19}
$$

and the vector $B(n) = A(0,n)^* \otimes B$ is given by

$$B(n) = \begin{pmatrix} 0 \\ \sigma_1(n) \\ \sigma_1(n) + \sigma_2(n) \\ \vdots \\ \sum_{k=1}^{Q-1} \sigma_k(n) \end{pmatrix}.$$

The firing instants of the transitions 1 to Q , $X(n) = (X_1(n), \cdots X_Q(n))$ corresponds to the start of the service times in the queues. The exogenous arrival instants are denoted $U(n)$ and the evolution equation is $X(n) = A(n) \otimes X(n-1) \oplus B(n) \otimes U(n)$ for all $n > 1$.

3.4.3 Kanban systems

Consider a system of single server queues in tandem as depicted in Figure 3.7.

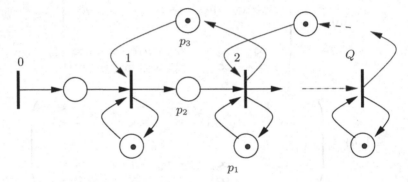

Fig. 3.7. Kanban system with blocking before service

The first queue is fed by the external arrivals and has an infinite capacity buffer. All the other queues have a finite capacity buffer. The mechanism for service is blocking before service: A customer can start its service at queue i if the buffer as queue $i + 1$ is not full. Otherwise, it stays in buffer i until some customer leaves queue $i + 1$ so that it can start its service.

In Figure 3.7, the places of type p_1 impose the mono-server semantics as previously, places of type p_2 represent the buffers in front of each server and places of type p_3 enforce the finite capacity of the buffers. The initial number of tokens in a place p_3 is the buffer size for that queue (which is one in the figure). The fact that the total number of tokens remains constant in the circuit made of places p_2 and p_3 imposes the finite buffer and the blocking. When the buffer is full, place p_3 is empty and no service can start in the previous queue. As soon as a customer leaves the queue, one token is released in place p_3 and the previous queue can start one service, should a customer be waiting for service there.

The (max,+) equation of that system is again of dimension Q, of the usual form

$$X(n) = A(n) \otimes X(n-1) \oplus B(n) \otimes U(n).$$

In that case, we get the same matrix $A(0,n)$ as for tandem queues with infinite buffers (see Section 3.4.2).

$$A(0,n) = \begin{pmatrix} -\infty & -\infty & \cdots & -\infty & -\infty & -\infty \\ \sigma_1(n) & -\infty & \cdots & -\infty & -\infty & -\infty \\ -\infty & \sigma_2(n) & \cdots & -\infty & -\infty & -\infty \\ & \vdots & & & & \\ -\infty & -\infty & \cdots & \sigma_{Q-1}(n) & -\infty & -\infty \\ -\infty & -\infty & \cdots & -\infty & \sigma_Q(n) & -\infty \end{pmatrix},$$

However, the matrix $A(1,n)$ is different:

$$A(1,n) = \begin{pmatrix} \sigma_1(n-1) & \sigma_2(n-1) & -\infty & \cdots & -\infty & -\infty \\ -\infty & \sigma_2(n-1) & \sigma_3(n-1) & \cdots & -\infty & -\infty \\ & \vdots & & & & \\ -\infty & -\infty & \cdots & \sigma_{Q-1}(n-1) & \sigma_Q(n-1) \\ -\infty & -\infty & \cdots & -\infty & \sigma_Q(n-1) \end{pmatrix},$$

$$B = \begin{pmatrix} 0 \\ -\infty \\ -\infty \\ \vdots \\ -\infty \\ -\infty \end{pmatrix}.$$

Therefore, the matrix $A(n) = A(0,n)^* \otimes A(1,n)$ is given by :

$$(A(n))_{ij} = \begin{cases} -\infty & \text{if } i < j - 1, \\ \sigma_j(n-1) & \text{if } i = j - 1, \\ \sum_{k=j}^{i-1} \sigma_k(n) + \sigma_j(n-1) & \text{if } i \geq j. \end{cases} \quad (3.20)$$

and the matrix $B(n) = A(0,n)^* \otimes B$ is given by

$$B(n) = \begin{pmatrix} 0 \\ \sigma_1(n) \\ \sigma_1(n) + \sigma_2(n) \\ \vdots \\ \sum_{k=1}^{Q-1} \sigma_k(n) \end{pmatrix}.$$

One can also analyze tandem queue with manufacturing blocking (blocking after service), which can be modeled by event graphs as well. The main characteristic needed to keep the (max,+) linearity is that no customer is ever lost and the preservation of the FIFO property.

3.4.4 Window flow control

This example is taken from communication networks. If a node *Sender* sends information to *Receiver* through a packet switched network made of Q nodes (counting *Receiver* but not *Sender*), a flow control is a mechanism which slows *Sender* in order to avoid overloads. A simple mechanism which is used by many protocols (such as TCP) is the window flow control. This is used to limit the total number of packets in the network. At reception of the n-th packet, *Receiver* sends an acknowledgment to *Sender*. As for *Sender*, it sends its n-th packet only if the acknowledgment for packet number $n - W$ has already been received. The parameter W is called the window size.

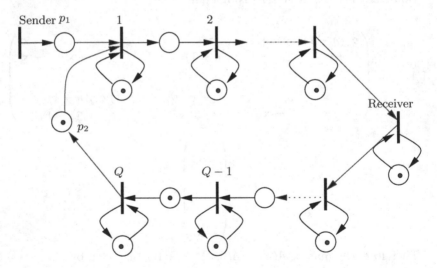

Fig. 3.8. Window flow control model

The event graph model of a simple window flow control mechanism is displayed in Figure 3.8. Transition *Sender* is an input transition and models the sender. Transition *Receiver* models the receiver. The transitions from *Sender* to *Receiver* are the nodes (with a mono-server semantic) forming the route in the network for the packets from *Sender* to *Receiver*. The transitions from *Receiver* to *Sender* model the nodes crossed by the acknowledgments. Tokens in place p_1 represent the packets that a has produced which are waiting for a new acknowledgment to come to be sent. Tokens in place p_2

represent the number of acknowledgments that have been received and which allow sending new packets in the network. When the place p_2 is empty, W packets have been sent without any reception of acknowledgments and the next packet to be sent has to wait before heading to the network. The initial number of tokens in the whole circuit is W and remains constant. To ease the matrix representation, the initial marking is such that each place (starting with place p_2 and going backward) contains at most one token. If the total number of nodes in the circuit is smaller than W, this is done by adding "dummy" nodes, with no service time, just as described in 3.3.

Since $W > 0$, the matrix $A(0, n)$ is acyclic (lower diagonal).

$$
A(0,n) = \begin{pmatrix}
-\infty & -\infty & \cdots & -\infty & -\infty & \cdots & -\infty \\
\sigma_1(n) & -\infty & \cdots & -\infty & -\infty & \cdots & -\infty \\
-\infty & \sigma_2(n) & \cdots & -\infty & -\infty & \cdots & -\infty \\
\vdots & & \ddots & & & & \\
-\infty & -\infty & \cdots & \sigma_{Q-W}(n) & -\infty & \cdots & -\infty \\
-\infty & -\infty & \cdots & -\infty & -\infty & \cdots & -\infty \\
\vdots & & & & & & \\
-\infty & -\infty & \cdots & -\infty & -\infty & \cdots & -\infty
\end{pmatrix},
$$

The matrix $A(1, n)$ captures the recycling places with one initial token and all those W places containing one initial token. We have

$$
A(1, n+1) = \begin{pmatrix}
\sigma_1(n) & -\infty & \cdots & -\infty & -\infty & \cdots & \sigma_Q(n) \\
-\infty & \sigma_2(n) & \cdots & -\infty & -\infty & \cdots & -\infty \\
\vdots & & \ddots & -\infty & -\infty & & \vdots \\
-\infty & -\infty & \cdots & \sigma_{Q-W+1}(n) & -\infty & \cdots & -\infty \\
-\infty & -\infty & \ddots & \sigma_{Q-W+1}(n) & \sigma_{Q-W+2}(n) & \ddots & -\infty \\
\vdots & & & & & \ddots & \ddots \\
-\infty & -\infty & \cdots & & \cdots & \sigma_{Q-1}(n) & \sigma_Q(n)
\end{pmatrix},
$$

and the vector B is

$$
B = \begin{pmatrix}
0 \\
-\infty \\
-\infty \\
\vdots \\
-\infty \\
-\infty
\end{pmatrix}.
$$

The matrix $A(n) = A(0,n)^* \otimes A(1,n)$ is given by :

$$(A(n))_{ij} = \begin{cases} \sum_{k=j}^{i-1} \sigma_k(n) + \sigma_j(n-1) & \text{if } 1 \le j \le i \le Q - W + 1, \\ \sum_{k=1}^{i-1} \sigma_k(n) + \sigma_Q(n-1) & \text{if } 1 \le i \le Q - W + 1 \text{ and } j = Q, \\ \sigma_j(n-1) & \text{if } i > Q - W + 1 \text{ and } j \le i \le j+1, \\ -\infty & \text{otherwise.} \end{cases}$$

$$(3.21)$$

and the vector $B(n) = A(0, n)^* \otimes B$ is given by

$$B(n)_i = \begin{cases} \sum_{k=1}^{i-1} \sigma_k(n) & \text{if } 1 \le i \le Q - W + 1, \\ -\infty & \text{otherwise.} \end{cases}$$

$$(3.22)$$

Finally, the standard evolution equation of the system is again $X(n) = A(n) \otimes X(n-1) \oplus B(n) \otimes U(n)$, with $A(n)$ and $B(n)$ as above and $U(n)$ being the time when the n-th packet is emitted by *Sender*.

3.4.5 Leaky buckets

This example is also taken from communication networks. In large networks, leaky buckets are a commun device often used to cut off bursts in the entry traffic. They have been extensively studied, on a stochastic [111] as well as deterministic point of view [39, 110].

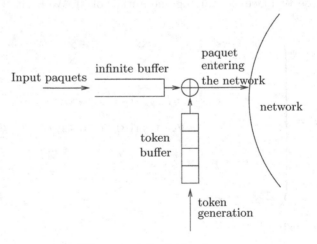

Fig. 3.9. A Leaky bucket

A leaky bucket is represented in Figure 3.9. The input stream is buffered in an infinite buffer and each paquet has to get a token in order to enter the network. Tokens are generated according to a process with intensity ρ and are kept in a buffer with capacity σ. Therefore arrival bursts of size larger than σ cannot enter the network and are filtered by the leaky bucket.

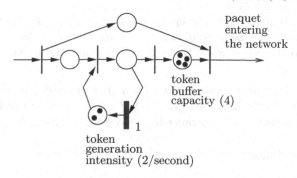

Fig. 3.10. Event graph model of a leaky bucket

If we assume that the token generation is done according to a deterministic process which lets ρ tokens enter the token buffer every unit of time, then a leaky bucket can be modeled by an event graph, represented in Figure 3.10.

Again, one can construct matrices $A(n)$ and $B(n)$ such that the standard evolution equation of the system is $X(n) = A(n) \otimes X(n-1) \oplus B(n) \otimes U(n)$, with $U(n)$ being the time when the n-th packet arrives in the leaky bucket and $X_a(n)$ being the time when it enters the network.

3.5 Lindley's equation for (max,+) systems

In this section, we will show how the evolution equation on the firing epochs can help to establish an equation for the response time of a (max,plus) system. By analogy with the G/G/1 case, this equation will be called the Lindley's equation.

First, let us state the assumptions that are made on the system under study. We consider an open stochastic event graph $\mathcal{G}$, which satisfies all the assumption which as necessary to have a linear evolution equation of the form 3.17, *i.e.* it is live and locally FIFO. Here, we further assume that $\mathcal{G}$ have only one input transition (denoted q_0) and is input-connected. We also assume that the event graph is initially in an input deadlock: no transition can fire (except the input transition). These last assumptions imply that for every transition q in $\mathcal{G}$, there exists a path from q_0 to q that contains no tokens initially. More formally, this can be written using 3.4 as,

$$L(q) = 0, \forall q \in S.$$

Note however, that the system being in an initial input deadlock does not mean that the initial marking is zero in all the places. For instance, the network depicted in Figure 3.7 is in an initial deadlock, since $L(1) = L(2) = \cdots = L(Q) = 0$. However, some places as for example the recycling places or the places of type p_3 have a positive initial marking.

We recall that $X_q(n)$ is the epoch when transition q starts to fire for the n-th time. Note that because of Lemma 14 and using the fact that $L(q) = 0$, $X_q(n)$ is also the epoch when the n-th token generated by the input is fired by transition q.

Now, let W_n be a vector , with its q-th component equal to: $W_n^q \overset{\text{def}}{=} X_q(n) - U(n)$. Using Lemma 14 again, W_n^q can be seen as the *traveling time* for customer n between its entrance in the system and its passage in transition q.

Using the standard evolution equation for $X(n)$, namely

$$X(n+1) = A(n) \otimes X(n) \oplus B(n) \otimes U(n+1),$$

and if we consider only the component $X_i(n)$ and the row i in matrix $A(n)$ (denoted $A_i(n)$), we can subtract $U(n+1)$ (a scalar since we have a single input) on each side of this equation and we get:

$$W_{n+1}^i = A_i(n) \otimes X(n) \oplus B(n)_i \otimes U(n+1) - U(n+1).$$

We can rewrite this as

$$\begin{aligned} W_{n+1}^i &= A_i(n) \otimes (X(n) - U(n+1)) \oplus B(n)_i \\ &= (A_i(n) - \tau_n) \otimes (X(n) - U(n)) \oplus B(n)_i, \end{aligned}$$

with $\tau_n \overset{\text{def}}{=} U(n+1) - U(n)$

If we write this last equality for W_{n+1} in vectorial form, we get

$$W_{n+1} = A(n) \otimes D(-\tau_n) \otimes W_n \oplus B(n),$$

where $D(h)$ is the diagonal matrix with h on the diagonal and $-\infty$ everywhere else.

This recursion is a generalization of the Lindley's equation in the case of a network. This equation was also derived in [25].

By using elementary matrix operations in the (max,+) algebra, the equation can also be developed into:

$$W_{n+1} = B(n) \oplus \bigoplus_{i=1}^{n} C_i, \tag{3.23}$$

with

$$C_i = \bigotimes_{j=i}^{n} (A(j) \otimes D(-\tau_j)) \otimes B(n-i-1),$$

where we define for convenience $B(0) \overset{\text{def}}{=} (-\infty, \cdots, -\infty)^t$.

Part II

Admission and routing control

This part shows how the general theorems presented in Part I can be applied in networks of queues or Petri nets which form timed event graphs.

The first two chapters focus on the problem of admission control where the controller is positioned at the entrance of the network and decides
-either to accept (packets are allowed in the network and continue their routes to the destination)
- or to reject the packet (in which case, the packet is lost forever).

Chapter 4 gives the form of the optimal admission control in open loop for a stochastic event graph.

Chapter 5 discusses the applicability of this technique in realistic telecommunication networks.

The next chapter (6) considers the routing control problem where the controller decides the route that the packet must follow rather than it should accept or reject the incoming packet. The results in the general case are much weaker than for the admission problem. The optimal policy is shown to have rates for each system when a balanced policy is admissible. However the computation of the optimal rates is not done in the general case. Even when the optimal rates are known, the optimal policy remains difficult to find.

Only special cases are solved in the following. The case where the parallel systems have the same service distributions is solved completely in Chapter 6. The case of two deterministic systems in parallel is solved in Chapter 7.

These examples illustrate the intrinsic difficulty of the computation of the optimal rates as well as the determination of the optimal policy in the general case. It seems doubtful that a general construction, merely based on the parameters of the system can always be given in closed form.

4 Admission control in stochastic event graphs

4.1 Introduction

The work on admission control in queuing systems can be split into two main domains, depending on the the information available for control decisions. When some knowledge on the state is available (closed loop systems), then the optimal control policy is usually based on dynamic programming techniques [52, 95] and more recently in [100]. When no state information is available (open loop control), then the control is often transformed into a problem of assigning an optimal constant input rate, see Section II in [104]. Instead of a dynamic control problem we are then faced with an optimization of a single parameter. This presentation falls in between since it belongs to the open loop framework but remains dynamic in some sense since control decisions will be taken for each arriving customer.

The general result which is proved in this chapter is that when the admission rate is fixed, then the individual arrivals have to be distributed evenly over time in order to minimize the average waiting time (or the workload). The property of the cost function used here in order to prove the optimization result is multimodularity.

We focus on systems with the following properties which can be consider as rather realistic for classiacl networks models (more on this in Chapter 5).

1. We consider a queuing network with one input node. This network is assumed to be a stochastic event graph (queues in tandem for example, fit in this framework, see Section 3.3).
2. The service times of the customers admitted in the network may be any stationary process.
3. We look at performance measures such as the average workload or the average traveling time. More generally, any performance measure which is a weighted average of expectations of convex functions of the workload (or traveling time) to any of the queues can be considered. (see Lemmas 22 and 5 for a precise formulation).
   In this framework,

we show that the following result holds: under all admission policies, with an asymptotic fraction p of acceptance, the balanced policy with rate p is optimal. This policy can be given by:

$$a_n = \lfloor np \rfloor - \lfloor (n-1)p \rfloor,$$

where $\lfloor X \rfloor$ denotes the largest integer smaller or equal to X, and $a_n = 1$ (0) if the n-th arrival is accepted (rejected). The intuitive idea behind it is that for stationary arrivals, balanced admissions spread out the entrance times of the admitted jobs "most equally".

The main objective of the chapter is to verify the conditions of Lemma 22 (or Lemma 5). The main point is the verification of the multimodularity property, its proof for the single server case relies on Lindley's equation. For the stochastic event graph, we use the vectorial Lindley's equation established in Chapter 3. Although technically and notationally more involved, it turns out that the verification of multimodularity in this general setting has the same structure as for the one-server queue with a general service distribution and "first in first out" service discipline. Via a counter example, we point out in Section 4.2.1 that the multimodularity property does not hold sample path wise (when we consider a deterministic sequence of service times which are not equal). Therefore, we consider the expected workloads (expected traveling times) and we apply a coupling of the service times (inter-arrival times). See Section 4.3 for details.

4.1.1 Organization of the chapter

The chapter is structured as follows. Section 4.2 introduces the definitions and Section 4.3 studies the particular case of a FIFO (First In First Out) queue, to give an idea of the proof in the general case. The main goal of Section 4.4 is to show that the workload in a (max,+) linear system is multimodular with respect to the arrival sequence. In Section 4.5, we give the proof of multimodularity for the traveling time of a customer in the system. Finally, in Section 4.6, we show that a balanced admission policy is optimal among all open-loop policies with a given admission rate by applying our results from Chapter 1.

4.2 Multimodularity and admission control

4.2.1 Admission policy: the time slot approach

In the rest of the chapter, we will use the following notations:

let $\{T_i\}_{i \in \mathbb{N}}$ be the instants of arrival opportunities, with the convention that $T_1 = 0$. Since all the rest of the notations are based on the original sequence, this can be considered as a time driven approach.

We denote by δ_i the ith *interval* length, that is: $\delta_i = T_{i+1} - T_i$. We assume that $\delta_0 = 0$. From now, the sequence T is fixed (and hence the sequence δ).

As for the customers, they arrive by batches. Their arrival is defined through an *arrival sequence* which is a sequence of integer numbers, $a = (a_1, a_2, \cdots, a_N, \cdots)$, where a_i gives the number of customers entering to the queue at time T_i. For convenience, we introduce the value $a_0 = 0$. In Section 4.6 the sequence a will be seen as a control sequence over the arrivals but will be used in the same way.

Furthermore, each individual customer carries a load. The load of the $j-$th customer is denoted σ_j .

The *counting function* $\kappa(i)$ is the number of individual arrivals by time T_i:

$$\kappa(i) \stackrel{\text{def}}{=} \sum_{j=1}^{i} a_j. \tag{4.1}$$

The function $\nu(i)$ is the *number of time-intervals* elapsed when the ith individual customer enters the system. We assume that $\nu(0) = 0$ and

$$\nu(i) \stackrel{\text{def}}{=} \min\{m : \sum_{j=1}^{m} a_j \geq i\}. \tag{4.2}$$

The function $\tau_i(a)$ is the time elapsed between the i-th and $(i+1)$-th individual arrivals, defined by:

$$\tau_i(a) \stackrel{\text{def}}{=} \sum_{j=\nu(i)}^{\nu(i+1)-1} \delta_j, = T_{\nu(i+1)} - T_{\nu(i)} \quad i < \kappa(N), \quad \text{and} \tag{4.3}$$

$$\tau_{\kappa(N)}(a) \stackrel{\text{def}}{=} T_N - T_{\nu(\kappa(N))}.$$

Figure 4.1 illustrates all these preliminary definitions.

4.3 The FIFO queue

In this section, we assume that the arrival stream enters a single FIFO queue. This is a simple case and the proofs presented here are typical of what happens in the general framework. The system considered is a $G/G/1$ queue with batch arrivals. Note that according to Section 3.4.1, this system is (max,+) linear.

Here, if $a_i = k$, then this means that k customers enter the queue at time T_k. We denote by $W_j(a)$, the workload in the system at the arrival time of the j-th individual arrival, $T_{\nu(\kappa(j))}$, and caused by arrivals up to and including the $(j-1)$-th individual arrival, for $j \leq \kappa(N)$. Also, $W_{\kappa(N)+1}(a)$ is the workload just after time T_N under the arrival sequence $(a_1, \cdots, a_N)$. The function $W_j(a)$ satisfies the following recurrence equation.

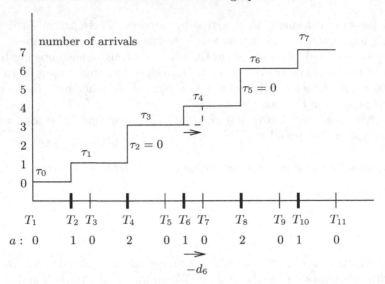

Fig. 4.1. an example

$$W_{j+1}(a) = (W_j(a) + \sigma_j - \tau_j) \vee 0 \quad j \leq \kappa(N). \tag{4.4}$$

The solution of this equation is given by the expanded Lindley's equation. For $W_{\kappa(N)+1}(a)$, we get

$$W_{\kappa(N)+1}(a) = \max\left(0, \max_{j=1}^{\kappa(N)} \sum_{i=j}^{\kappa(N)} (\sigma_i - \tau_i)\right). \tag{4.5}$$

For convenience, we denote $W_N(a) \stackrel{\text{def}}{=} W_{\kappa(N)+1}(a)$ and

$$w_j \stackrel{\text{def}}{=} \sum_{h=j}^{\kappa(N)} (\sigma_h - \tau_h).$$

Using this definition, we have:

$$W_N(a) = \max\left(0, w_{\kappa(N)}, \cdots, w_1\right).$$

4.3.1 Coupling of the service times with the customers

The rest of the section is devoted to proving multimodularity of the workload. Unfortunately, this does not hold on sample paths (as illustrated by the following example). We have to use a coupling of the service times with the customers entering the queue, and then, by assuming that the service times are stationary, we will prove multimodularity of the expected workload.

First, we will illustrate the difficulty of attaching the service times to customers while insuring that multimodularity holds, through an example.

An example We consider a single queue with a sequence of service times $\sigma(n) = (4, 1, 1, 1, 1, 1, \cdots)$ and with the integer points as arrival epochs.

We focus on the workload immediately after time $T_8 = 7$ under the arrival streams:

$$a = (0, 0, 1, 0, 0, 1, 1, 0) \tag{4.6}$$
$$a + e_1 = (1, 0, 1, 0, 0, 1, 1, 0) \tag{4.7}$$
$$a - s_7 = (0, 0, 1, 0, 0, 0, 2, 0) \tag{4.8}$$
$$a - s_7 + e_1 = (1, 0, 1, 0, 0, 0, 2, 0) \tag{4.9}$$

The workload satisfies: $W_8(a) = 1, W_8(a + e_1) = 0, W_8(a + e_1 - s_7) = 1, W_8(a - s_7) = 1$, as shown in Figure 4.2.

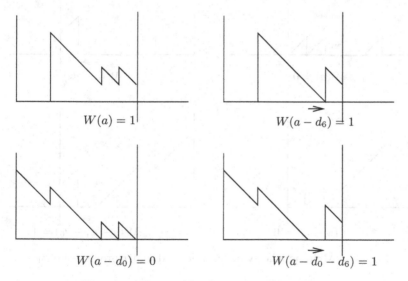

$$W(a) = 1 \qquad\qquad W(a - d_6) = 1$$

$$W(a - d_0) = 0 \qquad\qquad W(a - d_0 - d_6) = 1$$

Fig. 4.2. The workload is not multimodular.

This shows that the function W_8 is not multimodular, since $W_8(a) + W_8(a + e_1 - s_7) > W_8(a + e_1) + W_8(a - s_7)$.

However, under a proper coupling of the service times with the customers, then W can be made multimodular.

We couple the service times with the customers entering the queue in the following way.

$$
\begin{array}{rl}
a & = 0\ 0\ 1\ 0\ 0\ 1\quad 1\quad 0 \\
\sigma & = \quad\ \sigma_1\quad\ \ \sigma_2\ \ \sigma_3 \\[4pt]
a + e_1 & = 1\ 0\ 1\ 0\ 0\ 1\quad 1\quad 0 \\
\sigma & = \sigma_0\quad \sigma_1\quad\ \ \sigma_2\ \ \sigma_3 \\[4pt]
a - s_7 & = 0\ 0\ 1\ 0\ 0\ 0\quad 2\quad 0 \\
\sigma & = \quad\ \sigma_1\quad\ \ \ \sigma_2, \sigma_3 \\[4pt]
a + e_1 - s_7 & = 1\ 0\ 1\ 0\ 0\ 0\quad 2\quad 0 \\
\sigma & = \sigma_0\quad \sigma_1\quad\ \ \ \sigma_2, \sigma_3
\end{array}
$$

Under this coupling c, the workloads become as in Figure 4.3. We see that $W_8^c(a) = 0, W_8^c(a + e_1) = 0, W_8^c(a + e_1 - s_7) = 1, W_8^c(a - s_7) = 1$, and W_8^c satisfies the multimodular inequality (which is an equality here),

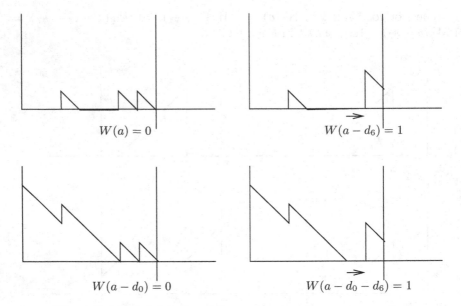

Fig. 4.3. Multimodularity of the workload under proper coupling.

General coupling In general, the coupling of the service times with the arrival stream is done in the following way.

– Let a be an arbitrary arrival sequence in $\mathbb{N}^N$: $a = (a_1, a_2, \cdots, a_N)$. The service times are coupled with the customers entering the queue in the following fashion: With arriving batch a_i, we attach the service times:

$$
\sigma_{\kappa(i-1)+1}, \sigma_{\kappa(i-1)+2}, \cdots, \sigma_{\kappa(i)}.
$$

– With arrival sequence, $a + e_1$, the service time sequence for batch $a_i, i \neq 1$, is not modified and the batch attached with a_1 becomes:

$$\sigma_0, \sigma_1, \sigma_2, \cdots, \sigma_{\kappa(1)},$$

where σ_0 is a new service time.

– If $a - s_{j+1} \in \mathbb{N}^N$, $1 \leq j < N$, the service time sequence for batch $a_i, i \neq j, i \neq j+1$, is not modified and the batch attached with a_j (which is not empty) becomes:

$$\sigma_{\kappa(j-1)+1}, \sigma_{\kappa(j-1)+2}, \cdots, \sigma_{\kappa(j)-1}$$

and the batch attached with a_{j+1} becomes:

$$\sigma_{\kappa(j)}, \sigma_{\kappa(j)+1}, \cdots, \sigma_{\kappa(j+1)}.$$

In other words, the service time $\sigma_{\kappa(j)}$ is moved from the j-th batch to the $j+1$-th batch.

– If $a - e_N \in \mathbb{N}^N$, the service time sequence for batch $a_i, i \neq N$, is not modified and the batch attached with a_N (which was not empty) becomes:

$$\sigma_{\kappa(N-1)+1}, \sigma_{\kappa(N-1)+2}, \cdots, \sigma_{\kappa(N)-1}.$$

The last service time $\sigma_{\kappa(N)}$ has been removed.

– If $a + u + v \in \mathbb{N}^N$, $u, v \in \overline{\mathcal{F}}$, $u \neq v$, then the coupling of the service times is obtained by composing the modification of the coupling associated with a induced by u and v. This construction is commutative.

Note that this coupling strongly depends on the initial choice of a. If one changes the starting point a, then other service times will be chosen. Also, the coupling is not defined on a's which do not lay in the positive orthant, $\mathbb{N}^N$.

4.3.2 Multimodularity

We choose a point a in $\mathbb{N}^N$ and we construct the associated coupling c. We will denote by $W_N^c(.)$ the workload immediately after time T_N under this coupling.

Now, using the notations given in the preliminaries, we have the following results.

Property 1. Let $0 < i < N$. If $a_i > 0$, $\tau_{\kappa(i)-1}(a - s_{i+1}) = \tau_{\kappa(i)-1}(a) + \delta_{i+1}$, $\tau_{\kappa(i)}(a - s_{i+1}) = \tau_{\kappa(i)}(a) - \delta_i$. All other τ_j are unchanged.

Proof. The addition of $-s_{i+1}$ corresponds to delaying the last acceptance at time T_i to time T_{i+1}. $\tau_{\kappa(i)}$ is the time interval between T_i and the arrival instant of the next customer that arrives after time T_i. Hence, delaying the arrival from T_i to time T_{i+1} results in increasing $\tau_{\kappa(i)-1}$ by $T_{i+1} - T_i = \delta_i$. $\tau_{\kappa(i)}$ decreases by that value. See Figure 4.1 for an illustration of this proof. $\square$

Property 2. Let $0 < i < N$. If $a_i > 0$, $w_{\kappa(i)}(a - s_{i+1}) = w_{\kappa(i)} + \delta_i$. All others w_j are left unchanged.

Proof. Follows from Property 1 and the definition of w_j. $\qquad\square$

Under the previous coupling, we have the following results.

Lemma 16. *Let $h : \mathbb{R} \to \mathbb{R}$ be a nondecreasing convex function. Then*

$$h \circ W_N^c(a) + h \circ W_N^c(a + u + v) \leq h \circ W_N^c(a + u) + h \circ W_N^c(a + v),$$

for all $u, v \in \overline{\mathcal{F}}$ and all a such that $a + u + v$, $a + u$ and $a + v$ are in $\mathbb{N}^N$.

Proof.

– Let us first consider the case $0 < i, j < N$ and $a_i > 0, a_j > 0$. In this case, using the previous Property 2, we have:

$$W_N^c(a - s_{i+1}) = \max(W_N^c(a), w_{\kappa(i)} + \delta_i) \tag{4.10}$$
$$W_N^c(a - s_{j+1}) = \max(W_N^c(a), w_{\kappa(j)} + \delta_j) \tag{4.11}$$
$$W_N^c(a - s_{i+1} - s_{j+1}) = \max(W_N^c(a), w_{\kappa(i)} + \delta_i, w_{\kappa(j)} + \delta_j). \tag{4.12}$$

Therefore, we have

$$\max(W_N^c(a), W_N^c(a - s_{i+1} - s_{j+1})) = \max(W_N^c(a - s_{i+1}), W_N^c(a - s_{j+1})),$$
$$W_N^c(a) + W_N^c(a - s_{i+1} - s_{j+1}j) \leq W_N^c(a - s_{i+1}) + W_N^c(a - s_{j+1}) \tag{4.13}$$

This means that the two dimensional vector $(W_N^c(a), W_N^c(a - s_{i+1} - s_{j+1}))$ is weakly majorized by $(W_N^c(a - s_{i+1}), W_N^c(a - s_{j+1}))$. By using Theorem B.2, p.109 in [88], for any non-decreasing convex function h, $h(W_N^c(a)) + h(W_N^c(a - s_{i+1} - s_{j+1})) \leq h(W_N^c(a - s_{i+1})) + h(W_N^c(a - s_{j+1}))$.

– Let us now assume that $i = 0$. This case corresponds to the arrival of an extra customer at time $T_1 = 0$. This customer brings a load that we denote σ_0. In this case, we have:

$$W_N^c(a + e_1) = \max(W_N^c(a), w_1 + (\sigma_0 - \tau_0)). \tag{4.14}$$

This case is treated similarly to the case above since $W_N^c(a + e_1)$ is of the form $\max(W_N^c(a), X)$, where $X = w_1 + (\sigma_0 - \tau_0))$.

– If $i = N$ and $a_N > 0$, then we have:

$$W_N^c(a - e_N) = (W_N^c(a) - \sigma_{\kappa(N)})^+. \tag{4.15}$$

Here, for some X (see the first two cases)

$$W_N^c(a - v) = \max(W_N^c(a), X), \tag{4.16}$$

and

$$W_N^c(a - v - e_N) = (\max(W_N^c(a), X) - \sigma_{\kappa(N)})^+. \tag{4.17}$$

By a case analysis, we see that if $W_N^c(a-v-e_N) = 0$ then, $W_N^c(a-e_N) = 0$ and by monotonicity of h,

$$h \circ W_N^c(a - e_N) + h \circ W_N^c(a - v) \geq h \circ W_N^c(a) + h \circ W_N^c(a - v - e_N).$$

If $W_N^c(a - v - e_N) > 0$, then,

$$W_N^c(a - v) - W_N^c(a - v - e_N) = \sigma_{\kappa(N)}.$$

This yields,

$$W_N^c(a) + W_N^c(a - v - e_N) \leq W_N^c(a - e_N) + W_N^c(a - v).$$

On the other hand, it is direct to see that

$$\max(W_N^c(a), W_N^c(a - v - e_N)) \leq \max(W_N^c(a - e_N) + W_N^c(a - v)).$$

Again, by using the same Schur convexity property, for h convex and increasing,

$$h \circ W_N^c(a) + h \circ W_N^c(a - v - e_N) \leq h \circ W_N^c(a - e_N) + h \circ W_N^c(a - v).$$

This concludes the proof. □

Theorem 17. *Suppose $\{\sigma_n\}_{n=1}^\infty$ is a stationary sequence, then the function $\mathbb{E}_\sigma(h \circ W_N)$ is multimodular, where $\mathbb{E}_\sigma$ denotes the expectation w.r.t. the sequence $\{\sigma_n\}$.*

Proof. Let a be an arbitrary point in $\mathbb{N}^N$. Construct the associated coupling c of the services times. Under this coupling and for all i, j, Lemma 16 shows that

$$h \circ W_N^c(a + u) + h \circ W_N^c(a + v) - h \circ W_N^c(a) - h \circ W_N^c(a + u + v) \geq 0,$$

$u \neq v$, $u, v \in \overline{\mathcal{F}}$. Therefore,

$$\mathbb{E}_\sigma \Big(h \circ W_N^c(a+u) + h \circ W_N^c(a+v) - h \circ W_N^c(a) - h \circ W_N^c(a+u+v) \Big) \geq 0.$$

By the stationary assumption on the service times and using the fact that under coupling c, the service times involved in $W_N^c(.)$ are always consecutive,

$$\mathbb{E}_\sigma h \circ W_N^c(.) = \mathbb{E}_\sigma h \circ W_N(.),$$

for the points $a + u$, $a + v$, $a + u + v$ and a since the expectation is invariant with respect to the shift operator. Finally we get

$$\mathbb{E}_\sigma \Big(h(W_N(a + u)) + h(W_N(a + v)) - h(W_N(a)) - h(W_N(a + u + v)) \Big) \geq 0.$$

□

4.4 (max,+) systems with one input: multimodularity

This section will generalize the multimodularity properties to the case of an arbitrary network which is (max,+) linear, has one input, is connected to its input and initially input-deadlocked.

The main result established in this section is that the expectation of W_n^q is multimodular for all transition q. The proof is very similar to the case of the single queue and is made, surprisingly, even easier by using the vectorial form of the Lindley equation in the (max,+) algebra.

In the following, we will also often use the following transformation for notational convenience. If X is a vector of size Q, then $[X]$ is a diagonal matrix of size $Q \times Q$, with the vector X on the diagonal and $-\infty$ elsewhere.

The multidimensional coupling of the service times in each transition with the arriving customers is done similarly as in the one queue case.

We construct a coupling for transition q which is independent of the coupling for any other transition.

Let a be an arbitrary arrival sequence in $\mathbb{N}^N$: $a = (a_1, a_2, \cdots, a_N)$. The service times are coupled with the customers entering the system in the following fashion:

- With arriving batch a_i (if size $\kappa(i)$), we attach the respective service times: $\sigma_{\kappa(i-1)+1}^q, \sigma_{\kappa(i-1)+2}^q, \cdots, \sigma_{\kappa(i)}^q$.
- With arrival sequence, $a + e_1$, the service time sequence for batch $a_i, i \neq 0$ is not modified and the service times attached to a_1 now becomes under this coupling: $\sigma_0^q, \sigma_1^q, \sigma_2^q, \cdots, \sigma_{\kappa(1)}^q$ where σ_0^q is a new service time.
- If $a - s_{j+1} \in \mathbb{N}^N$, then with arrival sequence, $a - s_{j+1}, 1 < j < N$ the service time sequence for batch $a_i, i \neq j, j-1$ is not modified, the service times attached to a_{j-1} become: $\sigma_{\kappa(j-2)+1}^q, \sigma_{\kappa(j-2)+2}^q, \cdots, \sigma_{\kappa(j-1)-1}^q$ and the service times attached to batch a_j also become: $\sigma_{\kappa(j-1)}^q, \sigma_{\kappa(j-1)+1}^q, \cdots, \sigma_{\kappa(j)}^q$. In other words, the service time $\sigma_{\kappa(j-1)}^q$ is moved from the $(j-1)$-th batch to the j-th batch.
- If $a - e_N \in \mathbb{N}^N$, then with arrival sequence, $a - e_N$, the service time sequence for batch $a_i, i \neq N$ is not modified and the service times attached to a_N (which is not empty) become: $\sigma_{\kappa(N-1)+1}^q, \sigma_{\kappa(N-1)+2}^q, \cdots, \sigma_{\kappa(N)-1}^q$. The last service time $\sigma_{\kappa(N)}^q$ has been removed.
- If $a + u + v \in \mathbb{N}^N$, $u, v \in \overline{calF}$, then with arrival sequence, $a + u + v$, the coupling of the service times is obtained by composing the modification of the coupling associated with a induced by u and v. We note again that the construction is commutative.

Now, using this coupling, we consider the traveling time vector of a potential customer entering the system just after time T_N under arriving stream a and its associated coupling. We denote this vector by $\mathbf{W}_N(a)$. This vector is defined by the following equation:

$$\mathbf{W}_N(a) = B(\kappa(N)) \oplus \bigoplus_{i=1}^{\kappa(N)} C_i(a), \tag{4.18}$$

with

$$C_i(a) = \bigotimes_{j=i}^{\kappa(N)} (A(j) \otimes D(-\tau_j(a))) \otimes B(\kappa(N) - i - 1), \tag{4.19}$$

where $\tau_j(a)$ is defined in Equation 4.3.

We have similar lemmas as in the one queue case, (see Properties 1 - 2).

Lemma 17. *Let $0 < i < N$. If $a_{i-1} > 0$, then $C_{\kappa(i)}(a - s_{i+1}) = D(\delta_i) \otimes C_{\kappa(i)}(a)$. All other C_j are left unchanged.*

Proof. Using property 1, we have:

$$D(-\tau_{\kappa(i)-1}(a - s_{i+1})) = D(-\delta_i) \otimes D(-\tau_{\kappa(i)-1}(a))$$

and

$$D(-\tau_{\kappa(i)}(a - s_{i+1})) = D(\delta_i) \otimes D(-\tau_{\kappa(i)}(a))$$

with all others τ_j left unchanged.

- Now, for every $j > \kappa(i)$, $C_j(a - s_{i+1})$ does not involve $D(-\tau_{\kappa(i)}(a - s_{i+1}))$ or $D(-\tau_{\kappa(i)-1}(a - s_{i+1}))$, and therefore is left unchanged.
- If $j < \kappa(i)$ then $C_j(a - s_{i+1})$ involves $D(-\tau_{\kappa(i)}(a - s_{i+1}))$ and $D(-\tau_{\kappa(i)-1}(a - s_{i+1}))$. Since the matrices $D(x)$ commute with everything, and since $D(\delta_i) \otimes D(-\delta_i) = E$, the identity matrix, then, $C_j(a - s_{i+1})$ is left unchanged.
- Finally, if $j = \kappa(i)$, then $C_j(a - s_{i+1})$ involves $D(-\tau_{\kappa(i)}(a - s_{i+1}))$ but not $D(-\tau_{\kappa(i)-1}(a - s_{i+1}))$. Using the fact that $D(\delta_i)$ commutes with all the other matrices, we have $C_{\kappa(i)}(a - s_{i+1}) = D(\delta_i) \otimes C_{\kappa(i)}(a)$.

$\square$

In the following we will use, to simplify the equations, the matrices: $Z_i \overset{\text{def}}{=} D(\delta_i) \otimes C_{\kappa(i)}(a)$.

Lemma 18. *Let $h = (h_1, \cdots, h_Q)$ be such that for all q, $h_q : \mathbb{R} \to \mathbb{R}$ is an increasing convex function. Component-wise, we have for $u, v \in \overline{calF}, u \neq v$,*

$$h_q \circ \mathbf{W}_N^q(a+u) \otimes h_q \circ \mathbf{W}_N^q(a+v) \geq h_q \circ \mathbf{W}_N^q(a) \otimes h_q \circ \mathbf{W}_N^q(a+v+u). \tag{4.20}$$

Proof. First note as a general remark that for any matrix M and any positive number x, $D(x) \otimes M \oplus M = D(x) \otimes M$. Now, as in the case of a single queue, we have to distinguish three cases.

– The case where $0 < i, j < N$ and $a_i, a_j > 0$. By commutativity of the $\oplus$ operator, we have

$$\mathbf{W}_N(a - s_{i+1}) = \mathbf{W}_N(a) \oplus Z_{i+1} \tag{4.21}$$
$$\mathbf{W}_N(a - s_{j+1}) = \mathbf{W}_N(a) \oplus Z_{j+1} \tag{4.22}$$
$$\mathbf{W}_N(a - s_{j+1} - s_{i+1}) = \mathbf{W}_N(a) \oplus Z_{j+1} \oplus Z_{i+1}. \tag{4.23}$$

Now using the distributivity of $\otimes$ w.r.t. $\oplus$, we have $[W_N(a - s_{i+1})] \otimes [W_N(a - s_{j+1})] = [W_N(a)] \otimes [W_N(a - s_{j+1} - s_{i+1})] \oplus [Z]_j \otimes [Z]_i$. This last equation interpreted in the classical algebra says that for each server q, the traveling time at time T_N satisfies:

$$\mathbf{W}_N^q(a - s_{i+1}) + \mathbf{W}_N^q(a - s_{j+1}) \geq \mathbf{W}_N^q(a) + \mathbf{W}_N^q(a - s_{j+1} - s_{i+1}). \tag{4.24}$$

The fact that h is increasing and convex component-wise, and using a case analysis with Equation (4.24) shows that

$$h(\mathbf{W}_N^q(a - s_{i+1})) + h(\mathbf{W}_N^q(a - s_{j+1}))$$
$$\geq h(\mathbf{W}_N^q(a)) + h(\mathbf{W}_N^q(a - s_{j+1} - s_{i+1}))$$

– Now we examine the case where $u = e_1$. As in the single queue case, this corresponds under our coupling to the arrival of an extra customer at time $T_1 = 0$ that has a service time that we denote σ_0^q in queue q. In this case, we have with $n = \kappa(N)$:

$$\mathbf{W}_N(a + e_1) = \mathbf{W}_N(a) \oplus \bigotimes_{j=0}^{n} \left(A_j \otimes D(-\tau_j) \right) \otimes B_i(n). \tag{4.25}$$

This case is treated as the general case, since $\mathbf{W}_N(a + u)$ is of the form $\mathbf{W}_N(a) \oplus Z$, for some vector

$$Z \stackrel{\text{def}}{=} \bigotimes_{j=0}^{n} \left(A_j \otimes D(-\tau_j) \right) \otimes B_i(n).$$

– If $u = -e_N$ and $a_N > 0$. This case corresponds to the removal of the last customer in the arrival batch, which happens to have arrived at time T_N. In this case, we have:

$$\mathbf{W}_N(a - e_N) = S \otimes \mathbf{W}_N(a) \oplus O,$$

where O is a vector composed of zeros and S is a diagonal matrix with $S_{[q,q]} = -\sigma_{\kappa(N)}^q$. Since we also have

$$\mathbf{W}_N(a + v) = \mathbf{W}_N(a) \oplus Z$$

and

$$\mathbf{W}_N(a + v - e_N) = S \otimes (\mathbf{W}_N(a) \oplus Z) \oplus O = S \otimes \mathbf{W}_N(a) \oplus S \otimes Z \oplus O,$$

with $Z \overset{\text{def}}{=} Z_j$, we get using distributivity of $\otimes$ with respect to $\oplus$,

$$
\begin{aligned}
[W_N(a - e_N)] &\otimes [W_N(a + v)] \\
&= (S \otimes [W_N(a)] \oplus [O]) \otimes ([W_N(a)] \oplus [Z] \\
&= S \otimes [W_N(a)] \otimes [W_N(a)] \\
&\quad \oplus [O] \otimes [W_N(a)] \\
&\quad \oplus S \otimes [W_N(a)] \otimes [Z] \\
&\quad \oplus [O] \otimes [Z],
\end{aligned}
$$

and on the other hand,

$$
\begin{aligned}
[W_N(a)] &\otimes [W_N(a + v - e_N)] \\
&= [W_N(a)] \otimes S \otimes [W_N(a)] \oplus [W_N(a)] \otimes S \otimes [Z] \oplus [W_N(a)] \otimes [O].
\end{aligned}
$$

Since all the matrices involved in these equations are diagonal, they commute and we have:

$$[W_N(a - e_N)] \otimes [W_N(a + v)] = [W_N(a)] \otimes [W_N(a + v - e_N)] \oplus [O] \otimes [Z].$$

Since $[O] \otimes [Z]$ is a non-negative diagonal matrix, the result is established by rewriting this equation in the conventional algebra.

As with the function h, the proof is similar to the case of a single queue. $\square$

Let $\sigma_n, n \geq 1$ be the stochastic vectors with components $\sigma_n^q, q \in \mathcal{Q}$.

Theorem 18. *Suppose $\{\sigma_n\}$ is a stationary sequence of stochastic vectors. Then, the function $\mathbb{E}_\sigma h(\mathbf{W}_N^q(a))$ is multimodular, where $\mathbb{E}_\sigma$ denote the expectation w.r.t. the service times in all the nodes of the system.*

Proof. The proof is similar to the one queue case. The coupling c is compatible with the shift of stochastic vectors of the service times in all nodes. Therefore, the inequality given in Equation (4.20) implies the multimodularity of the expected traveling time w.r.t. all service times. $\square$

Note that this theorem proves that the traveling time for a customer arriving at time T_N, that is, the time between its entrance in the system and its service in queue q, is multimodular, for all q and N.

Also note that the case of event graphs with multiple or no entries as well as the case where the marked graph is not empty initially are intractable by this means. Indeed, the coupling of the service times to the customers is not feasible *a priori* in those cases. It will depend on the sequence of arrivals in the case of multiple entries, and on all the service times in the case of a closed system.

4.5 A dual policy: counting variable and waiting time

In the previous section, we were interested in the study of the workload which is a criterion related to the server. Here we will focus on performances related to the customers, namely, the waiting time of the customers entering the network. This approach will be dual (in some particular sense) to the previous one and may be a more important issue for practical applications, where the customer satisfaction is more important than resource optimization.

Previously, all quantities were indexed by n, the number of time slots. In this section, all quantities will rather be indexed by $\kappa(n)$, the number of arrivals.

The counting sequence b, will be given in the following way: $b_n = \nu(n) - \nu(n-1)$, $n > 1$. For example if $a = (0, 0, 1, 0, 0, 2, 0, 1, 1, 0, 0, 1)$, then $b = (3, 0, 2, 1, 3)$. Note that a and b represent the same information (up to the initial arrival); a_n gives the number of arrivals at time slot n (this is a time driven concept), and the dual variable b_k gives the number of time slots elapsed between the $(k-1)$-th arrival and the k-th arrival (this is a event driven concept).

4.5.1 Waiting time

Let $\mathcal{G}$ be a (max,+) linear system with a single input satisfying the assumptions given in Chapter 3.

Then, the traveling time of the k-th admitted customer to node i is denoted by $\mathcal{W}_k^i(b)$. The vector $\mathcal{W}_k(b)$ satisfies the vectorial Lindley equation, using the function τ defined in equation (4.3):

$$\mathcal{W}_{k+1}(b) = A(k) \otimes D(-\tau_k) \otimes \mathcal{W}_k(b) \oplus B(k),$$

This can be written:

$$\mathcal{W}_{k+1} = B(k) \oplus \bigoplus_{i=1}^{n} \mathcal{C}_i, \quad \text{with} \quad \mathcal{C}_i = \bigotimes_{j=i}^{k} \left(A(j) \otimes D(-\tau_j)\right) \otimes B(k-i-1).$$

This equation has essentially the same form as the equation (4.18).

4.5.2 Coupling

The coupling adapted in this case is essentially the dual of the coupling on the service times used previously. Here we rather couple the inter-arrival times, δ_i. This coupling uses the function ν instead of the function κ.

We build the coupling d in the following way.

— Let b be an arbitrary arrival sequence in $\mathbb{N}^N$: $b = (b_1, b_2, \cdots, b_N)$. The intervals are coupled with the customers entering the queue in the following fashion: With interval length b_i, we attach the intervals:

$$\delta_{\nu(i-1)+1}, \delta_{\nu(i-1)+2}, \cdots, \delta_{\nu(i)}.$$

— With $b-e_1$, the interval sequence for length $b_i, i \neq 1$ is not modified and the length attached with b_1 (which was not empty) becomes: $\delta_1, \delta_2, \cdots, \delta_{\nu(1)}$, where δ_0 has been removed.
— If $b + s_{j+1} \in \mathbb{N}^N$, then with arrival sequence, $b + s_j, 1 \leq j < N$ the interval sequence for length $b_i, i \neq j, j-1$ is not modified and the length attached with b_{j-1} becomes:

$$\delta_{\nu(j-2)+1}, \delta_{\nu(j-1)+2}, \cdots, \delta_{\nu(j-1)}, \delta_{\nu(j-1)+1}$$

and the length attached with b_j (which is not empty) becomes:

$$\delta_{\nu(j-1)+2}, \cdots, \delta_{\nu(j)}.$$

In other words, the interval $\delta_{\nu(j-1)+1}$ is moved from the j-th length to the $(j-1)$-th length.
— If $b + s_N \in \mathbb{N}^N$, then with arrival sequence, $b + s_N$, the interval sequence for length $b_i, i \neq N$ is not modified and the length attached with b_N becomes

$$\delta_{\nu(N-1)+1}, \delta_{\nu(N-1)+2}, \cdots, \delta_{\nu(N)-1}, \delta_{\nu(N)+1}.$$

The last interval $\delta_{\nu(N)+1}$, is a new interval.
— If $b + u + v \in \mathbb{N}^N$, then with arrival sequence, $b + u + v$, the coupling of the δ's is obtained by composing the modification of the coupling associated with b induced by u and v, $u \neq i$, $u, v \in \mathcal{F}$. This construction is commutative.

4.5.3 Multimodularity

In this case, we will use the direct base $\mathcal{F} = \{-e_1, s_2, \cdots, e_N\}$ rather than $\overline{\mathcal{F}} = \{e_1, -s_2, \cdots, -e_N\}$ used in the previous sections.

Property 3. Let $0 < i < N$. If $b_i > 0$, then $\tau_i(b + s_{i+1}) = \tau_i(b) + \delta_{\nu(i)+1}$ and $\tau_{i+1}(b + s_{i+1}) = \tau_{i+1}(b) - \delta_{\nu(i)+1}$. All others τ_j are unchanged.

Proof. By definition. See Figure 4.1   □

Lemma 19. *Let $0 < i < N$. If $b_i > 0$, then*

$$\mathcal{C}_{i+1}(b + s_{i+1}) = D(\delta_{\nu(i)+1}) \otimes \mathcal{C}_{i+1}(b).$$

All other $\mathcal{C}_j$ are left unchanged.

Proof. The proof is similar to the proof of Lemma 17 and follows from the Property 3 and the definition of $\mathcal{C}_j$ □

In the following we will simplify the equations using the variables:

$$Z_i \stackrel{\text{def}}{=} D(\delta_{\nu(i)+1}) \otimes C_{i+1}(b).$$

Under the previous coupling, we have the following results:

Lemma 20. *Let $h : \mathbb{R}^Q \to \mathbb{R}^Q$ be a increasing convex function. Then,*

$$h \circ W(b) + h \circ W(b + u + v) \leq h \circ W(b + u) + h \circ W(bv),$$

for all $u, v \in \mathcal{F}$ and all b such that $b + u + v$, $b + u$ and $b + v$ are in $\mathbb{N}^N$.

Proof. The proof is essentially similar to the proof of Lemma 18. The proof follows from the following equalities by using the same technique as in the previous case.

– If $1 < i, j \leq N$,

$$W(b + s_i) = W(b) \oplus Z_{i-1} \tag{4.26}$$
$$W(b + s_j) = W(b) \oplus Z_{j-1} \tag{4.27}$$
$$W(b + s_j + s_i) = W(b) \oplus Z_{j-1} \oplus Z_{i-1}. \tag{4.28}$$

If $u = -e_1$,

$$W(b - e_1) = W(b) \oplus Z_0. \tag{4.29}$$

If $u = e_n$,

$$W(b + e_N) = S \otimes W(b) \oplus \mathcal{O}, \tag{4.30}$$

where $\mathcal{O}$ is a vector composed of zeros and S is a diagonal matrix with $S_{[q,q]} = -\delta^q_{\nu(N)+1}$.

□

Theorem 19. *If the intervals $\{\delta_i\}_{i=0}^{\infty}$ form a stationary sequence, then the function $\mathbb{E}_\delta(h \circ W)$ is multimodular.*

The proof is essentially similar to the proof of Theorem 18.

4.6 Optimal admission sequence

In this section, we study the admission control in a (max,+) linear system, as presented in Chapter 3. We use the multimodularity property to derive the optimality of the bracket admission sequence.

We want to admit customers in a (max,+) linear system $\mathcal{G}$, under the constraint that in the long run, the fraction of customers lost be at most $1 - p$.

Now the problem is to find which admission policy minimizes the traveling time to node i of a customer admitted in the system, as illustrated

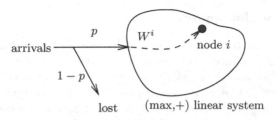

Fig. 4.4. Illustration of the control problem

in the figure 4.4. The admission is governed by a binary control sequence $a = (a_1, a_2, \cdots)$. If $a_n = 1$ then the n-th customer is admitted to the queue. If $a_n = 0$ then the n-th customer is lost.

Note that this *control sequence* can be seen as an *arrival sequence* as used in the previous sections.

In this model, the vectors $\sigma_n, n = 1, 2, \cdots$ are service times of the customers which are admitted to the system, they correspond precisely to the definition of the firing times introduced in Section 3.3. There are no service times attached to the customers who are not admitted to the queue. This framework is natural in the stochastic Petri net context (see for example [20]). Note also that if the service times (or firing times) are independent, then this assumption is not needed and service times can as well be attached to all customers.

4.6.1 Dual bracket sequences

In this part, we claim that the arrival sequence a is bracket if and only the counting sequence b is also bracket up to an adequate choice of the phase. The following lemma uses results about bracket sequences which are detailed in Chapter 2.

Lemma 21. *If sequence a is a bracket sequence with rate p and phase θ, with θ and p linearly independent over the rationals:*

$$a_n = \lfloor p(n+1) + \theta \rfloor - \lfloor pn + \theta \rfloor,$$

then b is also bracket with rate $1/p$ and phase $\phi = -\theta/p$:

$$b_k = \lfloor (k+1)/p + \phi \rfloor - \lfloor k/p + \phi \rfloor.$$

This lemma is a direct consequence of the form of the support of the bracket sequence a, given in Chapter 2, § 2.2 Note that θ and p are linearly independent over the rationals for almost all θ in $[0, 1)$.

4.6.2 The time slot approach

Now, we are ready to consider the problem to find the optimal admission policy with a given rate. We denote by $\mathbb{E}_{\sigma,\delta}(.)$ the expectation w.r.t. the inter-arrival times and the service times in all nodes in the system. Following the technical conditions given in Chapter 1, we have to make sure that the expected traveling times to node i under the admission sequence a, satisfy the properties given in Lemma 22. Let $h_i : \mathbb{R} \to \mathbb{R}, \quad 1 \le i \le Q$ be arbitrary non-decreasing convex functions. For simplicity we denote in the following $\mathbf{W}_n^i(a_1, \cdots, a_n)$ instead of $h_i \circ \mathbf{W}_n^i(a_1, \cdots, a_n)$.

In the next two lemmas, we use increasing and decreasing in the non-strict sense.

Lemma 22. *Assume that the inter-arrival times and the service times are stationary sequences, independent of each other. The following properties are true:*

i- $\mathbb{E}_{\sigma,\delta}\mathbf{W}_n^i(a_1, \cdots, a_n)$ *is increasing.*

ii- $\mathbb{E}_{\sigma,\delta}\mathbf{W}_n^i(a_{m-n+1}, \cdots, a_m) \le \mathbb{E}_{\sigma,\delta}\mathbf{W}_m^i(a_1, \cdots, a_m), \, n < m.$

iii- $\mathbb{E}_{\sigma,\delta}\mathbf{W}_n^i(a_1, \cdots, a_n) = \mathbb{E}_{\sigma,\delta}\mathbf{W}_m^i(0, \cdots, 0, a_1, \cdots, a_n), \, n < m.$

iv- $\mathbb{E}_{\sigma,\delta}\mathbf{W}_n^i(a_1, \cdots, a_n)$ *is multimodular.*

Proof. We prove the four properties in a convenient order.

i- Using the extended Lindley formula, the expected traveling time $\mathbb{E}_{\sigma,\delta}\mathbf{W}_n^i(a_1, \cdots, a_n)$ is increasing in a_k, $k = 1, \cdots n$.

iii- Let us fix n and m with $n < m$. Since the inter-arrival times are stationary, we can couple the δ's such that $\forall 1 \le j \le n$, $\delta_j^{(n)} = \delta_{j+m-n}^{(m)} = \delta_{j+m-n}$. Under this coupling and because the system is initially empty, we have:

$$\mathbf{W}_n^i(a_1, \cdots, a_n) = W_m^i(0, \cdots, 0, a_1, \cdots, a_n),$$

and,

$$\mathbb{E}_\sigma \mathbf{W}_n^i(a_1, \cdots, a_n) = \mathbb{E}_\sigma \mathbf{W}_m^i(0, \cdots, 0, a_1, \cdots, a_n).$$

Therefore, if $n < m$,

$$\mathbb{E}_{\sigma,\delta}\mathbf{W}_n^i(a_1, \cdots, a_n) = \mathbb{E}_{\sigma,\delta}\mathbf{W}_m^i(0, \cdots, 0, a_1, \cdots, a_n).$$

ii- This is a direct consequence of *i* and *iii*.

iv- Theorem 18 shows that $\mathbb{E}_\sigma \mathbf{W}_n^i(a_1, \cdots, a_n)$ is multimodular, thus integrating over δ, $\mathbb{E}_{\sigma,\delta}\mathbf{W}_n^i(a_1, \cdots, a_n)$ is multimodular as well. $\qquad\square$

Now, we choose the cost function $g^i(a)$ of the control sequence $a = (a_1, a_2, \cdots)$ to be the Cezaro sum of the expectation of an increasing convex function of traveling time to a given node i, of a potential customer admitted to the queue at time T_n. We have:

$$g^i(a) \stackrel{\text{def}}{=} \overline{\lim_{N \to \infty}} \frac{1}{N} \sum_{n=1}^{N} \mathbb{E}_{\sigma,\delta}\mathbf{W}_n^i(a_1, ..., a_n).$$

Note that $g^i(a)$ could be called the average traveling time at epochs $\{T_n\}$. In the case the sequence $\{T_n\}$ is a Poisson process, $g^i(a)$ is the time-average of the traveling times.

To find which admission policy minimizes the cost $g^i(.)$, we recall the notation for the bracket sequence with phase θ and rate p,

$$a_k^p(\theta) = \lfloor kp + \theta \rfloor - \lfloor (k-1)p + \theta \rfloor. \tag{4.31}$$

We can restate the general theorems given in Chapter 1, in the particular case studied here.

Theorem 20. *For each sequence a such that*

$$\lim_{N \to \infty} \frac{1}{N} \sum_{n=1}^{N} a_n \geq p,$$

it holds that,

$$g^i(a) \geq g^i(a^p(\theta)),$$

for any $\theta \in [0, 1]$.

Proof. First note that the bracket sequence $a_k^p(\theta)$ has an asymptotic rate equal to p. Then, the results from 1.3 can be applied, since all necessary assumptions on W are satisfied. $\qquad\square$

4.6.3 The counting approach, the bounded case

In this section, we use the counting sequence b rather than the admission sequence a. Therefore, we rather focus on the function $h_i \circ \mathcal{W}(i)_n(b_1, \cdots, b_n)$, which is denoted $\mathcal{W}_n^i(b_1, \cdots, b_n)$ for the sake of notation simplicity.

Lemma 23. *Assume that the service time vectors and the inter-arrival times are stationary sequences, independent of each other. Assume that the σ_k^q (resp. δ_k), for each $k \in \mathbb{N}$ and $q \in \mathcal{Q}$ is bounded from above (resp. from below) by R (resp. by D).*
i- $\mathbb{E}_{\delta,\sigma} \mathcal{W}_n^i(b_1, \cdots, b_n)$ is decreasing.
ii- $\mathbb{E}_{\delta,\sigma} \mathcal{W}_n^i(b_1, \cdots, b_n) \geq \mathbb{E}_{\delta,\sigma} \mathcal{W}_m^i(b_{n-m+1}, \cdots, b_n)$, $n > m$.
iii- $\mathbb{E}_{\delta,\sigma} \mathcal{W}_n^i(b_1, \cdots, b_n) = \mathbb{E}_{\delta,\sigma} \mathcal{W}_m^i(\beta, \cdots, \beta, b_{n-m+1}, \cdots, b_n)$, $n < m$, where $\beta = \lceil QR/D \rceil$.
iv- $\mathbb{E}_{\delta,\sigma} \mathcal{W}_n^i(b_1, \cdots, b_n)$ is multimodular.

Proof. The proof holds by using stationarity of the inter-arrival times and of the service time vectors.

 i- Using the extended Lindley's Formula, it is clear that $\mathbb{E}_{\delta,\sigma} \mathcal{W}_n^i(b_1, \cdots, b_n)$ is decreasing.

ii- Let us fix n and m with $n > m$. Since the service time vectors are stationary, we can couple the service times such that $\forall 1 \leq j \leq m$, $\sigma_{j+n-m}^{(n)} = \sigma_j^{(m)} = \sigma_j$. We also couple the δ's in the following way:

$$\forall 1 \leq j \leq \nu(m), \; \delta_{j+\nu(n-m)}^{(n)} = \delta_j^{(m)} = \delta_j.$$

Under this coupling and because the system is initially empty, we have:

$$\mathcal{W}_n^i(b_1, \cdots, b_n) \geq \mathcal{W}_m^i(b_{n-m+1}, \cdots, b_n).$$

Therefore, the inequality holds for the expected values as well,

$$\mathbb{E}_{\delta,\sigma} \mathcal{W}_n^i(b_1, \cdots, b_n) \geq \mathbb{E}_{\delta,\sigma} \mathcal{W}_m^i(b_{n-m+1}, \cdots, b_n).$$

Note that properties *i* and *ii* do not use the fact that the service time vectors and the inter-arrival times are bounded. This assumption is used for property *iii*.

iii- First, we show that under the sequence $(\beta, b_2, \cdots, b_n)$, the net is empty at the time of the second arrival.

Let us assume for a moment that the second arrival is infinitely delayed. Since the system is originally empty, then after the first arrival, each transition in the system will fire exactly once (this is a well known result for Event Graphs, see for example [38]). At this time, the system is again empty and no transition can fire. An upper bound of this time is given by QR. If the next customer arrives in the system more than QR units of time later, then it will find the system in the original empty state.

Now, note that $\tau_1 \geq \beta D$ by definition of D and taking $\beta \geq QR/D$ makes τ_1 larger than QR.

Finally, using the same coupling as for *ii*, we get

$$\mathbb{E}_{\delta,\sigma} \mathcal{W}_n^i(b_1, \cdots, b_n) = \mathbb{E}_{\delta,\sigma} \mathcal{W}_{n+1}^i(\beta, b_2, \cdots, b_n).$$

An easy induction on n gives the result.

iv- Theorem 19 shows that $\mathbb{E}_\delta \mathcal{W}_n^i(b_1, \cdots, b_n)$ is multimodular. Therefore, $\mathbb{E}_{\delta,\sigma} \mathcal{W}_n^i(b_1, \cdots, b_n)$ is multimodular as well.   □

Similarly to the previous case, we choose the cost function of any control sequence $b = (b_1, b_2, \cdots)$ to be the Cezaro sum of the expected traveling times of all customers admitted to the queue, to a given node i:

$$\gamma^i(b) = \overline{\lim_{N \to \infty}} \frac{1}{N} \sum_{n=1}^{N} \mathbb{E}_{\delta,\sigma} \mathcal{W}_n^i(b_1, ..., b_n).$$

We recall that the bracket policy with phase θ and rate p is

$$b_k^r(\theta) \stackrel{\text{def}}{=} \lfloor kr + \theta \rfloor - \lfloor (k-1)r + \theta \rfloor. \tag{4.32}$$

We can restate the general theorems given in Chapter 1 in the particular case studied here.

Theorem 21. *Under the foregoing assumptions, for all sequence b such that*

$$\overline{\lim_{N \to \infty}} \frac{1}{N} \sum_{n=1}^{N} b_n \leq r,$$

then $\gamma^i(b) \geq \gamma^i(b^r(\theta))$, for any $\theta \in [0,1]$.

Proof. Lemma 21 shows that the counting sequence associated with $b_k^r(\theta)$ is a bracket sequence with rate r.

Then, the optimization theory developed in Section 1.3 can be applied since all necessary assumptions on $\mathbb{E}_{\sigma,\delta} W$ are satisfied. □

Note that Lemma 21 says that the average workload and the average waiting time are both optimized by the same admission control sequence $a^p(\theta)$.

In the next section, we will prove that the same optimization result holds in the unbounded case. The proof uses strongly the result stated in Theorem 21.

4.6.4 The unbounded case

If the service times and the inter-arrival times are not bounded in the original stochastic event graph $\mathcal{G}$, then, we use a fixed quantity Z and we introduce a new system $\mathcal{G}^Z$

where all services times, σ_n^q are replaced by $\sigma_n^{Z,q} \overset{\text{def}}{=} \min(Z, \sigma_n^q)$. The inter-arrival times δ_k are also replaced by $\delta_k^Z \overset{\text{def}}{=} \max(1/Z, \delta_k)$. In the new system $\mathcal{G}^Z$, the service times are bounded from above by Z and the inter-arrival times are bounded from below by $1/Z$.

In $\mathcal{G}^Z$, $W_n^{Z,i}$ is the compose of the traveling time of the n-th customer to the i-th node by a convex increasing function.

If Θ is a random variable with uniform distribution on $[0,1)$,

$$b_k^r(\Theta) \overset{\text{def}}{=} \lfloor kr + \Theta \rfloor - \lfloor (k-1)r + \Theta \rfloor$$

is called the randomized bracket policy with rate r.

Following the results in Chapter 1, the time average of $\mathbb{E}_{\delta,\sigma} W^Z$, is minimized by the randomized bracket policy, (as well by the one with θ fixed at an arbitrary value). We have for all policy b with rate less than p,

$$\gamma^{Z,i}(b) = \overline{\lim_{N \to \infty}} \frac{1}{N} \sum_{n=1}^{N} \mathbb{E}_{\delta,\sigma} W_n^{Z,i}(b_1, ..., b_n)$$

$$\geq \overline{\lim_{N \to \infty}} \frac{1}{N} \sum_{n=1}^{N} \mathbb{E}_{\delta,\sigma,\Theta} W_n^{Z,i}(b_1^p(\Theta), ..., b_n^p(\Theta))$$

To prove the optimality of the randomized bracket sequence for the original system $\mathcal{G}$ where the service times and the inter-arrival times are unbounded, we need to let Z go to infinity in the previous inequality. For that, we need several technical lemmas.

Lemma 24. *The random sequence $b^r(\Theta)$ with Θ uniform on $[0,1)$ is stationary and $\mathbb{P}(b_k^r(\Theta) = 1) = r$, $\forall k$.*

Proof. By definition, $b_k^r(\Theta) = \lfloor kr + \Theta \rfloor - \lfloor (k-1)r + \Theta \rfloor$. The fact that $\int_0^1 \lfloor x + \theta \rfloor d\theta = x$ for any x implies that

$$\forall k, \quad \int_0^1 b_k^r(\theta) d\theta = r.$$

$\square$

Lemma 25. *If $\{\sigma_n\}$ is a stationary sequence of vectors, then $\{\min(\sigma_n, Z)\}$ and $\{\max(\sigma_n, Z)\}$ are also stationary.*

Proof. The proof follows by definition of stationary sequences $\qquad\square$

We define the variable $\beta_k \overset{\text{def}}{=} \sum_{i=1}^k b_i$. The inter-admission times satisfy

$$\tau_k = \sum_{i=\beta_k}^{\beta_{k+1}-1} \delta_i. \tag{4.33}$$

One sees that τ_k is a sum of a random number of random variables.

Lemma 26. *Assume that the process $\{b_k\}$ is stationary, and the process $\{\delta_i\}$ is stationary and independent of $\{b_k\}$. Then, the process $\{\tau_k\}$ is stationary.*

Proof. We compute the distribution of the joint process $\tau_2, \cdots, \tau_k$. It is determined by the probabilities:

$$P(\delta, b) \overset{\text{def}}{=} \mathbb{P}(b_2 = n_2, \cdots, b_k = n_k, \delta_{b_1+1} \le \hat{\delta}_1, \cdots, \delta_{b_1+n_1+\cdots+n_k} \le \hat{\delta}_{n_1+\cdots+n_k})$$
$$= \sum_i \mathbb{P}(b_1 = i, \cdots, b_k = n_k, \delta_{i+1} \le \hat{\delta}_1, \cdots, \delta_{i+n_1+\cdots+n_k} \le \hat{\delta}_{n_1+\cdots+n_k}).$$

Now,

$$P(\delta, b)$$
$$= \sum_i \mathbb{P}(b_i = i, \cdot, b_k = n_k) \mathbb{P}(\delta_{i+1} \le \hat{\delta}_1, \cdot, \delta_{i+n_1+\cdot+n_k} \le \hat{\delta}_{n_1+\cdot+n_k})$$
$$= \sum_i \mathbb{P}(b_1 = i, \cdot, b_k = n_k) \mathbb{P}(\delta_1 \le \hat{\delta}_1, \cdot, \delta_{n_1+\cdot+n_k} \le \hat{\delta}_{n_1+\cdot+n_k})$$
$$= \sum_i \mathbb{P}(b_0 = i, b_1 = n_2, \cdot, b_{k-1} = n_k) \mathbb{P}(\delta_1 \le \hat{\delta}_1, \cdot, \delta_{n_1+\cdot+n_k} \le \hat{\delta}_{n_1+\cdot+n_k})$$
$$= \mathbb{P}(b_1 = n_2, \cdot, b_{k-1} = n_k, \delta_1 \le \hat{\delta}_1, \cdot, \delta_{n_1+\cdot+n_k} \le \hat{\delta}_{n_1+\cdot+n_k}),$$

where the first equality follows from the independence of b and δ, the second follows from the stationarity of δ and the third from the stationarity of b.

This last expression gives the distribution of the joint process $\tau_1, \cdots \tau_{k-1}$. Since all of this holds for all k, the process $\{\tau_k\}$ is stationary. $\square$

Lemma 27. *Under the randomized bracket policy, if $\{\sigma_k\}$ and $\{\delta_k\}$ are stationary sequences, then $\{\sigma_k^Z\}$ and $\{\tau_k^Z\}$ are also stationary.*

Proof. The proof comes from a direct combination of the three previous lemmas. $\square$

Lemma 28. *If $\{\sigma_k^i\}$ and $\{\delta_k\}$ are stationary sequences and if the system $\mathcal{G}^Z$ is empty originally, then for all n,*

$$\mathcal{W}_n^{Z,i}(b^p(\Theta)) \leq_s \mathcal{W}_{n+1}^{Z,i}(b^p(\Theta)),$$

where $\leq_s$ denotes the stochastic order.

Proof. If the service time and the inter-arrival sequences are stationary (as it is the case here), this result is well known (see for example [24]). We can set $\mathcal{W}_0^{Z,i}(b^p(\Theta)) = 0$ and we have

$$\mathcal{W}_0^{Z,q}(b^p(\Theta)) \leq_s \mathcal{W}_1^{Z,i}(b^p(\Theta)).$$

Now, the Lindley formula for (max,+) systems can be used in this case.

$$\mathcal{W}_k^Z(b^p(\Theta)) = A(k-1) \otimes D(-\tau_{k-1}) \otimes \mathcal{W}_{k-1}^Z(b^p(\Theta)) \oplus B,$$

and

$$\mathcal{W'}_{k+1}^Z(b^p(\Theta)) = A'(k) \otimes D(-\tau_k') \otimes \mathcal{W'}_k^Z(b^p(\Theta)) \oplus B,$$

where the primes denote another sample path of the service and inter-admission times. By induction, we can assume that component-wise,

$$\mathcal{W}_k^Z(b^p(\Theta)) \geq_s \mathcal{W}_{k-1}^Z(b^p(\Theta)).$$

Now, stationarity of the service times in all queues and of inter-arrival times comes from Lemma 27. We can couple the service times and the inter-arrival times such that for all i, $\sigma_k^{i'} = \sigma_{k-1}^i$ and $\tau_k' = \tau_{k-1}$. Under this coupling, $A(k-1) = A'(k)$ and $D(-\tau_k') = D(-\tau_{k-1})$. Hence

$$\mathcal{W'}_{k+1}^Z(b^p(\Theta)) \geq \mathcal{W'}_k^Z(b^p(\Theta))$$

component-wise, under this coupling. $\square$

The optimality of the randomized bracket policy is established by the following theorem.

Theorem 22. *Consider the system $\mathcal{G}$ where the service time vectors and the inter-arrival times are stationary sequences, independent of each other. For each control sequence b such that*

$$\overline{\lim_{N \to \infty}} \frac{1}{N} \sum_{n=1}^{N} b_n \leq p,$$

it holds that

$$\gamma^i(b) \geq \overline{\lim_{N \to \infty}} \frac{1}{N} \sum_{n=1}^{N} \mathbb{E}_{\delta,\sigma,\Theta} \mathcal{W}_n^{Z,i}(b_1^p(\Theta), ..., b_n^p(\Theta)).$$

Proof. Remark that the quantity $\mathcal{W}_k^{Z,i}(b)$ is increasing in Z and

$$\lim_{Z \to \infty} \mathcal{W}_k^{Z,i}(b) = \sup_Z \mathcal{W}_k^{Z,i}(b) = \mathcal{W}_k^i(b).$$

The proof follows from the series of inequalities,

$$\gamma^i(b) = \overline{\lim_{N \to \infty}} \sup_Z \left(\frac{1}{N} \sum_{n=1}^{N} \mathbb{E}_{\delta,\sigma} \mathcal{W}_n^{Z,i}(b_1, ..., b_n) \right)$$

$$\geq \sup_Z \overline{\lim_{N \to \infty}} \frac{1}{N} \sum_{n=1}^{N} \mathbb{E}_{\delta,\sigma} \mathcal{W}_n^{Z,i}(b_1, ..., b_n)$$

$$\geq \sup_Z \overline{\lim_{N \to \infty}} \frac{1}{N} \sum_{n=1}^{N} \mathbb{E}_{\delta,\sigma,\Theta} \mathcal{W}_n^{Z,i}(b_1^p(\Theta), ..., b_n^p(\Theta)).$$

By Lemma 28, we know that $\mathbb{E}_{\delta,\sigma,\Theta} \mathcal{W}_n^{Z,i}(b_1^p(\Theta), ..., b_n^p(\Theta))$ is increasing in n. Therefore, the Cezaro limit equals the supremum on all n. We continue the previous inequalities:

$$\gamma^i(b) \geq \sup_Z \sup_n \mathbb{E}_{\delta,\sigma,\Theta} \mathcal{W}_n^{Z,i}(b_1^p(\Theta), ..., b_n^p(\Theta))$$

$$= \sup_n \sup_Z \mathbb{E}_{\delta,\sigma,\Theta} \mathcal{W}_n^{Z,i}(b_1^p(\Theta), ..., b_n^p(\Theta))$$

$$= \overline{\lim_{N \to \infty}} \frac{1}{N} \sum_{n=1}^{N} \mathbb{E}_{\delta,\sigma,\Theta} \mathcal{W}_n^i(b_1^p(\Theta), ..., b_n^p(\Theta)).$$

$\square$

Our last theorem shows that the (non-randomized) bracket policy is optimal for any initial phase θ.

Theorem 23. *Under the assumptions of Theorem 22, for each control sequence b such that*

$$\varlimsup_{N \to \infty} \frac{1}{N} \sum_{n=1}^{N} b_n \leq p,$$

it holds that

$$\gamma^i(b) \geq \gamma^i(b_p(\theta)),$$

for all $0 \leq \theta \leq 1$.

Proof. Define

$$f_m(\theta, p) \stackrel{\text{def}}{=} \mathbb{E}_{\delta, \sigma} \mathcal{W}_m^i(b_1^p(\theta), ..., b_m^p(\theta)).$$

Note that f_m is periodic (in θ) with period 1. Define

$$f_m'(\theta, p) \stackrel{\text{def}}{=} f_m(a_{-m+1}^p(\theta), ..., a_0^p(\theta)).$$

Then we have

$$f_m'(\theta', p) = f_m(\theta, p) \quad \text{where} \quad \theta' = \theta - mp,$$

Again f_m' is periodic w.r.t. θ, with period 1, and is increasing in m because of property *ii* in Lemma 22, so that the following limit exists,

$$f_\infty'(\theta, p) \stackrel{\text{def}}{=} \lim_{m \to \infty} f_m'(\theta, p).$$

Hence,

$$\mathbb{E}_\Theta f_\infty'(\Theta, p) = \lim_{m \to \infty} \mathbb{E}_\Theta f_m'(\Theta, p) = \lim_{m \to \infty} \mathbb{E}_\Theta f_m(\Theta, p). \tag{4.34}$$

Now, it follows from our results in Chapter 1 that under properties *ii* and *iv* of Lemma 22, the following holds:

$$\lim_{N \to \infty} \frac{1}{N} \sum_{n=1}^{N} \mathbb{E}_{\delta, \sigma} \mathcal{W}_n^i(b_1^p(\theta), ..., b_n^p(\theta)) = \mathbb{E}_\Theta f_\infty'(\Theta, p).$$

Combining this with Theorem 22 and Equation 4.34 finishes the proof. $\square$

5 Applications in queuing networks

5.1 Introduction

This chapter discusses the relevance of the assumptions made in the systems studied so far for the application of the theory for queuing networks.

Let us first consider a *circuit switched* network. We focus on a single connection in the network with nodes dedicated to a single class of packets, as illustrated in Figure 5.1.

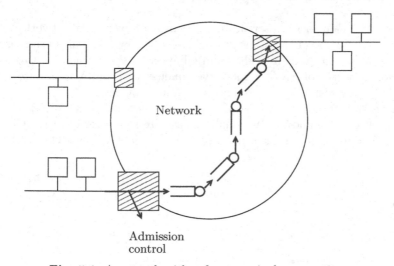

Admission
control

Fig. 5.1. A network with a focus on single connection

If we consider networks where nodes are not dedicated to a single connection (this is the case in most networks), then we have to take into account *cross traffic*, that is other connections using the same nodes inside the network.

In Figure 5.2, cross traffic has been added to the connection under study.

The goal of the rest of the chapter is to check when the general theory is applicable to these cases

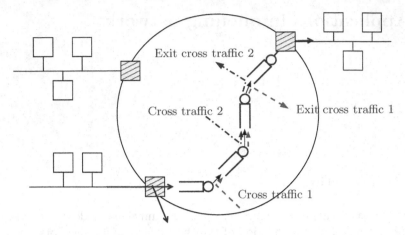

Fig. 5.2. A network with a connection and two cross traffic

5.2 Topological assumptions

As we have seen in Chapter 3, any network (or part of a network) which can be modeled by an event graph fits in the framework. This is the case for G/G/1 queues in tandem as shown in Chapter 3. More generally, this is the case for networks of FIFO mono-server queues with finite or infinite buffer and general blocking which do not contain any routing.

In particular this means that all customers must follow the same route and that no losses are allowed. We shall investigate in a later chapter (Chapter 8), admission control and routing into some particular networks with losses.

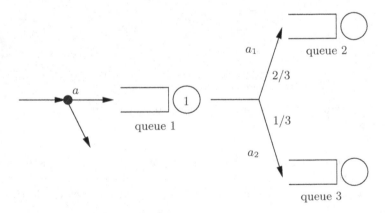

Fig. 5.3. A network with routing of the customers

Imagine a network as in Figure 5.2. The input sequence is periodic with one arrival every time unit. The admission control filters the input with a Sturmian sequence with rate $2/3$. The first queue has a deterministic service rate of one customer per second (therefore the waiting time in this queue is always 0). The stream of customers out of queue one is split into two with $2/3$ of the customers sent to queue two (the rest is sent to queue three). It is rather easy to check that no matter how the splitting is done, the stream in queue 2 is not regular. In particular, this stream will contain customers with inter-arrival time equal to one. However, if the admission control a is not Sturmian but for example $a = (1, 0, 1, 1, 1, 0, 1, 0, 1)^\infty$, then it is possible to split the output of queue one with ratio $(2/3, 1/3)$ and shape the stream of customers in queue two (resp. queue three) such that $a_1 = (1, 0, 1, 0, 1, 0, 1, 0, 0)$ (resp. $a_2 = (0, 0, 0, 1, 0, 0, 0, 0, 1)$) which are Sturmian, and therefore optimal, for the waiting times in queues two and three.

This means that in general, when the network contains several routes for the customers, the Sturmian admission may not be optimal.

Cross traffic as perturbations. When the network is (max,plus) linear, the expected (virtual) traveling time of a customer from the entrance in the network to its destination, or more generally to any node in the network, is multimodular and the general theorems can be used to show that Sturmian admission is optimal.

The example displayed in Figure 5.1 fits in this framework. However the example in Figure 5.2 does not since it contains several classes of customers with different routes. However this case can be approximated by replacing the cross traffic by perturbation of the service in all stations shared between the main connection and the cross traffic.

The idea is to replace the cross traffic by perturbed service times for the connection under study: just replace the cross traffic in node i between costumers $n - 1$ and n by a modified service time for costumer n by adding the service time of all the cross traffic to the service time of n. This new service time is called "visible" service time in the rest of this section. If the cross traffic at every node is stationary, then the perturbation at each node will also be stationary. However, the service times now become a functional of the arrival times for the connection under study. It also induces tricky dependences between the services times in the different nodes of the network.

5.3 Stochastic assumptions

The stochastic assumptions used in Theorem 21 are the following ones. The arrival process is stationary as well as the service times in all queues and the arrival process is independent of all the service processes.

First note that the stationary assumptions are rather minimal. In particular, the arrival process does not even need to be ergodic. Without stationarity, the theorem is not true. Just imagine a D/D/1 queue with a non-stationary fluctuating input sequence with inter-arrival times

$$\delta_{6n} = 1,$$
$$\delta_{6n+1} = 1,$$
$$\delta_{6n+2} = 1,$$
$$\delta_{6n+3} = 2,$$
$$\delta_{6n+4} = 2,$$
$$\delta_{6n+5} = 2,$$

and with deterministic service times equal to 2. If one wants to admit optimally a proportion $\alpha = 2/3$ of the packets, the optimal policy is $a = (100111)^{\infty}$ with null waiting time and all Sturmian sequence with density $2/3$ have a positive average waiting time.

Another interesting property of Theorem 21 is that the different queues in the network are not required to have independent service times. For example, a "large" packet may have large service times all through the network. This could be used as a simplified model for different classes of customers, all following the same path and distributed randomly in the global input process. It can also be used in order to model the influence of cross traffic for the visible service times in several nodes of the network.

As for the independence assumption between the arrival sequence and the service times, this may seem as a rather strong condition imposed on the system. It reduces the set of queuing networks where the general theory can be applied. For example, the case with cross traffic does induce dependences (as shown in the previous section).

5.3.1 Independence between service times and inter-arrival times.

If we do not have independence between the service times and the inter-arrival time, then Theorem 18 cannot be used because the multimodularity in expectation may not hold anymore. Furthermore, one can easily construct an example where the optimal admission sequence is not Sturmian. Here is an example. Consider a G/G/1 queue. We want to admit half of the customers in the queue (this means $\alpha = 1/2$). We consider a Poisson arrival process of intensity λ. As for the service times, we fix N arbitrarily and we construct the service process as follows. The first N customers have service times coupled with their inter-arrival times such that $\sigma_k = \delta_k$ for all $1 \le k < N$. As for the next N customers, their service times, σ_k is exponentially distributed with parameter λ independent of the inter-arrival times as well as the other service times. The next N customers have arrivals coupled with their service times again and so forth. Thus,

$$\sigma_k = \delta_k \text{ if } k < N \text{ mod } 2N,$$
$$\sigma_k \perp \{\delta_i\}_{i \in \mathbb{N}}, \text{ and } \sigma_k \text{ i.i.d. exponential, otherwise.}$$

In this case, the inter-arrivals are i.i.d. (hence stationary), the service times are i.i.d. (hence stationary) but the two processes are not independent of each other.

We consider the following admission policy with admission rate $\alpha = 1/2$.

$$a_k = 1 \text{ if } k < N \text{ mod } 2N,$$

$$a_k = 0 \text{ otherwise.}$$

If the queue is initially empty, then for all $i \in \mathbb{N}$, the virtual waiting time at the ith slot is null: $W_i(a_1, \cdots a_i) = 0$. Therefore,

$$g(a) = \overline{\lim_{n \to \infty}} \frac{1}{n} \sum_{i=1}^{n} \mathbb{E}_{\sigma, \delta} W_i(a_1, \cdots, a_i) = 0.$$

On the other hand, if one considers the Sturmian admission sequence with rate $1/2$, m, such that $m_{2i} = 1$ and $m_{2i+1} = 0$, then $\mathbb{E}_{\sigma, \delta} W_i(m_1, \cdots, m_i) = 0$ if $i \leq N$ but for $N < i < 2N$, $\mathbb{E}_{\sigma, \delta} W_i(m_1, \cdots, m_i) > 0$ since in this case the queue can be seen as a transient $E_2/M/1$ queue and the expectation of the waiting time is positive. Therefore, the cost for the Sturmian sequence is strictly larger than the cost of the optimal sequence a.

Note that as N increases, the optimal sequence a get further and further away from the Sturmian sequence m.

This shows that the independence assumption is essential for Sturmian sequences to be optimal admission sequences.

5.3.2 Cross traffic

As mentioned before, the presence of cross traffic in the network can be modeled by perturbation of the service times in each queue. Unfortunately, this trick usually introduces dependences between the arrival times and the visible service times in the connection under study and the general optimization theorems do not apply in general. However, there are cases when these dependences disappear.

1. This is the case if cross traffic has a lower priority than the connection under study (and is preemtable). In this case the visible service time is the same as the service time and independence is preserved.
2. Another case where the independence remains is when all connections (cross traffic as well as the main connection) are all deterministic processes, which is rather sensible at the packet level. In this case, the visible service can be rather difficult to describe (but remains stationary). The arrival process of the main connection being deterministic, it is independent of its visible service times.

6 Optimal routing

6.1 Introduction

It is a rather general problem to consider a system with multiple resources and tasks. Tasks can be performed by any resource and arrive in the system sequentially. The problem is to construct a *routing* of the tasks to the resources to minimize a given cost function. Such models are common in multiprocessor systems and communication networks, where the cost function may be the combined load in the resources.

In this chapter, we show that under rather general assumptions, the optimal routing policy in terms of expected average workload in each resource is given by a *balanced sequence*, that is, a sequence in which the option to route towards a given resource, is taken in an evenly distributed fashion. To show the relevance of this type of problem, let us briefly review the recent literature on load balancing. Load balancing in a distributed multiprocessor computer system has become an important issue to improve their performance. Many papers have been devoted to the load balancing problem, and for an overview, we refer to [33, 60]. Let us assume there is a centralized controller , and the information available to the controller determines the type of control which can be used. In dynamic feedback control (or close loop control), the information on the system (*e.g.* queue sizes) increases as time runs. In static or open loop control, the central controller only knows his past actions. Clearly, the closed loop models have a better performance than the open loop ones.

In the dynamic control setting, it is well known that under various assumptions and cost structures, the "join the shortest queue" policy is optimal for homogeneous processors (*i.e.* all processors are stochastically identical). For non-homogeneous processors, an optimal load balancing policy has not been found. Only partial optimality results are available (see [62] for results on shortest queue policies and [14] for results on the monotonicity structure of optimal policies).

For open loop control, probabilistic routing and pattern allocation have been studied. Again, in the case of homogeneous processors, the optimality of equal probabilities for probabilistic routing is established in [40, 79]. The round robin routing was proved optimal in the case of pattern allocation in [82]. For the non-homogeneous case, the problem of finding the optimal pattern allocation is generally considered difficult. Approximations of optimal allocations are found in [26, 43]. For markovian models, an algorithm has been developed for computing nearly-optimal policies in [65, 64].

6.1.1 Organization of the chapter

Here we will show an application of balanced sequences introduced in 2 for routing problems. This uses again results from convex analysis, established in Chapter 1. The main results that are used here are of two different kinds. First, we use the fact that the workload as well as the waiting time of customers entering a (max,+) linear system are multimodular functions, under fairly general assumption (stationarity of the arrival process and of the service times, as we showed in Chapter 4). We pose the problem of routing in terms of time average of multimodular sequences. Then, we develop the required theory for the case where the performance for each subsystem is stochastic but its expectation is multimodular. This extends the deterministic analysis that we did in Chapter 1, that proves that multimodular functions are minimized by bracket sequences. The superposition of several bracket sequences being a balanced sequence, this is the basis of the main result of this chapter.

It is interesting to exhibit this link between balanced sequences and scheduling problems, such as routing among several systems.

Section 6.2 shows the link between the notion of balanced sequences and the optimal scheduling in networks. It is also used to establish the optimality of balanced sequences for routing customers is a multiple queue system. Section 6.3 presents special cases for which the optimal rates can be computed.

6.2 Routing of customers in multiple queues

Here, we present an application of balanced sequences in arbitrary dimensions to scheduling optimization.

We consider a system where a sequence of tasks have to be executed by several processing units. The tasks arrive sequentially and each task can be processed by any server. The *routing* control consists in assigning to each task a server on which it will be processed. The routing is optimal if it minimizes some cost function that measures the performance of the system.

These kinds of models have been used to study load balancing within several processors in parallel processing problems as well as for efficient network utilization in telecommunication systems, as presented in the introduction.

6.2.1 Presentation of the model

In this section we consider a more precise queuing model of the system that we described. Customers enter a multiple queue system composed of K subsystems. Each sub-system is made of several queues, initially empty which form an event graph (see Chapter 3).

The routing of customers to the different sub-systems is controlled by a sequence of vectors $\{a_n\}$, with a_n is in $\{0,1\}^K$ and $a_n^i = 1$ means that the n-th customer is routed to sub-system i. Note that a is a *feasible* admission sequence as long as for all n, $\sum_i a_n^i = 1$.

The link between a feasible routing policy and an infinite sequence on a finite alphabet as used in the first part, comes from choosing the alphabet $\mathcal{A}$ composed by the letters

$$\{(1,0,\cdots,0),(0,1,0,\cdots),\cdots,(0,\cdots,0,1)\}.$$

Using this alphabet on K letters, a feasible routing policy can be viewed as an infinite sequence on $\mathcal{A}$.

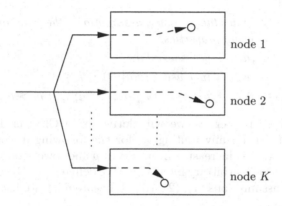

Fig. 6.1. Illustration of the routing of customers in a K node system

Figure 6.1 shows an illustration of the system we are considering.

We denote by T_n the epoch when the n-th customer enters the system. We assume that $T_1 = 0$. The inter-arrival time sequence is $\{\delta_k\} \overset{\text{def}}{=} \{T_{k+1} - T_k\}$. Finally, $\sigma_n^{i,j}$ will denote the service time of the n-*th* customer entering the j-*th* queue in node i.

The sequences $\{\delta_k\}$ and $\{\sigma_k^{i,j}\}$ will be considered as random processes. We also make stochastic assumption on these sequences. The inter-arrival time of the customers and the service times form stationary processes, and we assume that the inter-arrival times are independent of the service times.

6.2.2 Optimal routing sequence

In each sub-system i we pick an arbitrary server s_i (which may be the last server in the sub-system for example). The immediate performance criterion for sub-system i will be the traveling time to server s_i of a virtual

customer that would enter sub-system i at time T_n. Under a given routing policy, this quantity only depends on the values of the n first routing choices. From the routing sequence a, we can isolate the routing decision for node i: if $a_n^i = 1$ then the customer is admitted in node i and if $a_n^i = 0$ then the customer is rejected (for node i). We denote the traveling time at time T_n by $W_n^i(a_1^i, \cdots, a_n^i)$. We will be more particularly interested in the expected value of the traveling time with respect to the service times in all the servers contained in node i and with respect to the inter-arrival times:
$$\overline{W}_n^i(a_1^i, \cdots, a_n^i) \overset{\text{def}}{=} \mathbb{E}_{\sigma,\tau} h(W_n^i(a_1^i, \cdots, a_n^i)), \text{ where } h : \mathbb{R} \to \mathbb{R} \text{ is any convex}$$
increasing function.

A direct application of the results of Chapter 4 to any single node i shows that the function $\overline{W}_n^i(a_1^i, \cdots, a_n^i)$ is *multimodular*.

Theorem 24. *Under the foregoing assumptions, the function $\overline{W}_n^i(a_1^i, \cdots, a_n^i)$ satisfies the following properties.*

(1)- $\overline{W}_n^i(a_1^i, \cdots, a_n^i)$ is multimodular,

(2)- $\overline{W}_n^i(a_1^i, \cdots, a_n^i)$ is increasing in a_k^i, $1 \le k \le n$,

(3)- $\overline{W}_m^i(0, \cdots, 0, a_1^i, \cdots, a_n^i) = \mathbf{W}_n^i(a_1^i, \cdots, a_n^i)$, if $m > n$.

Using these properties, we will derive as in Chapter 1 a lower bound denoted $B_i(\alpha, p)$ for any routing a^i, for the following discounted cost. This function $B_i(\alpha, p)$ is increasing in α and in p and lower-continuous.

Let us fix some arbitrary integer, N. We define $p_\alpha^i = (1-\alpha)\sum_{k=1}^{\infty}\alpha^{k-1}a_k^i$. Now, using assumptions (1), (2), (3) of Theorem 24, we have,

$$\sum_{n=1}^{\infty}(1-\alpha)\alpha^{n-1}\overline{W}_n^i(a_1^i, a_2^i \cdots a_n^i)$$

$$\ge \sum_{n=1}^{N}(1-\alpha)\alpha^{n-1}\overline{W}_N^i(0\cdots, a_1^i, \cdots a_n^i)$$

$$+ \sum_{n=N+1}^{\infty}(1-\alpha)\alpha^{n-1}\overline{W}_N^i(a_{n-N+1}^i \cdots a_n^i)$$

$$= \sum_{n=1}^{N}(1-\alpha)\alpha^{n-1}\widetilde{\overline{W}}_N^i(0\cdots, a_1^i, \cdots a_n^i)$$

$$+ \sum_{n=N+1}^{\infty}(1-\alpha)\alpha^{n-1}\widetilde{\overline{W}}_N^i(a_{n-N+1}^i, \cdots, a_n^i)$$

$$\ge \widetilde{\overline{W}}_N^i\left(\sum_{n=0}^{N-1}(1-\alpha)\alpha^n(0\cdots, a_1^i\cdots a_{n+1}^i)+\sum_{n=N}^{\infty}(1-\alpha)\alpha^n(a_{n-N+2}^i\cdots a_{n+1}^i)\right)$$

$$= \widetilde{\overline{W}}_N^i\left(\alpha^N p_\alpha^i, \alpha^{N-1}p_\alpha^i, \cdots, p_\alpha^i\right),$$

where the last inequality follows from Jensen's inequality, since the function $\widetilde{W}_N^i$ is convex, and since the coefficients $(1-\alpha)\alpha^{n-1}$ are nonnegative and sum to 1. Define

$$B_i(\alpha,p) = \sup_N \widetilde{W}_N^i \left(\alpha^N p, \alpha^{N-1}p, \cdots, p\right). \qquad (6.1)$$

Note that B_i is defined for a fixed sequence $\{a_k^i\}$. Also note that $B_i(\alpha,p)$ is lower semi-continuous in α and in p.

The previous analysis shows that

$$\sum_{n=1}^{\infty}(1-\alpha)\alpha^{n-1}\overline{W}_n^i(a_1^i, a_2^i \cdots, a_n^i) \geq B_i(\alpha, p_\alpha^i). \qquad (6.2)$$

Also, for a given p, we consider the bracket sequence with rate p and arbitrary phase θ,

$$a_n^p(\theta) \stackrel{\text{def}}{=} \lfloor np + \theta \rfloor - \lfloor (n-1)p + \theta \rfloor,$$

(see the definition 5). One can show as in Chapter 1 that $a^p(\theta)$ satisfies

$$\lim_{m\to\infty}\frac{1}{m}\sum_{n=1}^{m}\overline{W}_n^i(a_1^p(\theta), \cdots, a_n^p(\theta)) = B_i(1,p). \qquad (6.3)$$

Here, however, we are interested in the performance of all nodes together. Therefore, we choose as a cost function, the undiscounted average on n of some linear combination of the expected traveling time in all nodes.

Let h be any increasing linear function, $h : \mathbb{R}^K \to \mathbb{R}$. We consider the undiscounted average cost of a feasible routing sequence a,

$$g(a) \stackrel{\text{def}}{=} \underline{\lim}_{N\to\infty}\frac{1}{N}\sum_{n=1}^{N}h(\overline{W}_n^1, \cdots, \overline{W}_n^K). \qquad (6.4)$$

Our objective is to minimize $g(a)$.

Theorem 25. *The following lower bound holds for all policies:*

$$g(a) \geq \inf_{p_1+\cdots+p_K=1} h(B_1(p_1), \cdots, B_K(p_K)).$$

Proof. We adapt the method developed in Chapter 1 for our case. We introduce the following notation.

$$B_i(p_i) \stackrel{\text{def}}{=} \sup_{\alpha\leq 1} B_i(\alpha, p_i).$$

Due to Littlewood's and Jensen's inequalities as well as Equation 6.2, we have

$$\varlimsup_{N \to \infty} \frac{1}{N} \sum_{n=1}^{N} h(\overline{W}_n^1, \cdots, \overline{W}_n^K)$$

$$\geq \varlimsup_{\alpha \to 1} (1 - \alpha) \sum_{n=1}^{\infty} \alpha^{n-1} h(\overline{W}_n^1, \cdots, \overline{W}_n^K)$$

$$\geq \varlimsup_{\alpha \to 1} h \left((1 - \alpha) \sum_{n=1}^{\infty} \alpha^{n-1} \overline{W}_n^1, \cdots, (1 - \alpha) \sum_{n=1}^{\infty} \alpha^{n-1} \overline{W}_n^K \right)$$

$$\geq \varlimsup_{\alpha \to 1} h \left(B_1(\alpha, p_\alpha^1), \cdots, B_K(\alpha, p_\alpha^K) \right) \tag{6.5}$$

By definition, we note that $\sum_{i=1}^{K} p_\alpha^i = 1$. Hence, one may choose a sequence $\alpha_n \uparrow 1$ such that the following limits exist:

$$\lim_{n \to \infty} p_{\alpha_n}^i = p_i, \qquad i = 1, \cdots, K \tag{6.6}$$

and $\sum_{i=1}^{K} p_i = 1$. From the continuity of $B_i(\alpha, p_i)$ in p and α we get from (6.5)

$$g(a) \geq h(B_1(p_1), \cdots, B_K(p_K)) \tag{6.7}$$

$$\geq \inf_{p_1 + \cdots + p_K = 1} h(B_1(p_1), \cdots, B_K(p_K)).$$

$\square$

Note that there exists some p^* that achieves the infimum

$$\inf_{p_1 + \cdots + p_K = 1} h(B_1(p_1), \cdots, B_K(p_K)),$$

since $h(B_1(p_1), \cdots, B_K(p_K))$ is continuous in $p = (p_1, \cdots, p_K)$.

Consider $\theta = (\theta_1, \cdots, \theta_K)$ and the routing policy $a^{p^*}(\theta)$ given for each i by

$$a_k^{i,p^*}(\theta_i) = \lfloor k p_i^* + \theta_i \rfloor - \lfloor (k-1) p_i^* + \theta_i \rfloor. \tag{6.8}$$

There are some p^*'s for which the condition of feasibility of the policy $a^{p^*}(\theta)$ is satisfied, that is, there exists some $\theta = (\theta_1, \cdots, \theta_K)$, such that bracket policy $a^{p^*}(\theta)$ is feasible.

Using the correspondence between a routing policy and a sequence on the alphabet $\mathcal{A}$, these p^*'s correspond precisely to *balanceable* rates.

Theorem 26. *Assume that p^* is balanceable. Then $a^{p^*}(\theta)$ is optimal for the average cost, i.e. it minimizes $g(a)$ over all feasible policies.*

Proof. The proof follows directly from Theorem 25 and Equation (6.3). $\square$

Remark 5. The previous theorem says that bracket sequences are optimal routing policies. As for balanced sequences (which are ultimately bracket, see Theorem 1), they are also optimal if the buffer in each node empties infinitely often. This situation occurs when the system is stable as shown in [22]. In such cases, a finite prefix of the routing sequence does not alter its cost.

6.3 Study of some special cases

The problem which remains to be addressed is to find in which cases, the rate vector p^* is balanceable. We will present several simple examples for which we can make sure that the optimal rate p^* is balanceable.

6.3.1 The case $K = 2$

If $K = 2$, then, the optimal rate vector is of the form $p^* = (p_1^*, 1 - p_1^*)$. Theorem 14 says that p^* is always balanceable and therefore, the optimal routing sequence is given by an associated balanced sequence. Note that this approach does not give any direct way to compute the value of p^*, however, it gives the structure of the optimal policy. The computation of p^* in the deterministic case will be done in the next chapter 7 and uses continuous fraction decomposition of the service times of both queues.

6.3.2 The homogeneous case

Now let K be arbitrary and each node is made of a single server, all servers being identical. This model is displayed in Figure 6.2.

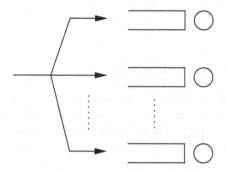

Fig. 6.2. Routing in homogeneous queues.

Also assume that the function h is symmetrical in all coordinates (for example, just the sum of all waiting times) By symmetry and convexity in $(p_1, \cdots, p_K)$ of $h(V_1^1(p_1), \cdots, V_K^1(p_K))$, we get $p^* = (1/K, \cdots, 1/K)$, which is balanceable. The associated balanced sequence is the round robin routing scheme. Applying Theorem 26 yields the following result which is new (to the best of the author's knowledge).

Theorem 27. *The round robin routing to K identical ./G/1 queues, minimizes the total average expected workload of all the queues over all admission sequences with no information on the state of the system.*

In [82], the round robin routing is proved to be optimal in separable-convex increasing order for K identical ./GI/1 queues. Their method uses an intricate coupling argument, whereas our proof is a simple corollary of the general theory on multimodular functions.

To illustrate the advantage of our approach, we further generalize the result to a system composed of K identical (max,+) linear systems with a single entry. In this case, the symmetry argument used in the case of simple queues still holds. Then again, the round robin routing policy minimizes the traveling time in each system. This case includes models such as routing among several identical systems composed of queues in tandem as shown Figure 6.3).

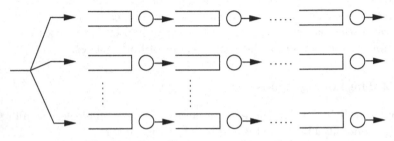

Fig. 6.3. Routing in queues in tandem

6.3.3 Two sets of identical servers

As a consequence of the two previous cases, we can consider a system composed of K_1 identical queues of type 1 and K_2 queues of type 2. Again, assume that h is symmetrical in the K_1 nodes of type 1 and symmetrical in the K_2 nodes of type 2. Then, by symmetry arguments, the optimal rate vector is of the form

$$(\frac{p^*}{K_1}, \cdots, \frac{p^*}{K_1}, \frac{1-p^*}{K_2}, \cdots, \frac{1-p^*}{K_2}),$$

for an appropriate p^*. This rate vector is balanceable indeed. This implies that for the weighted total average expected workload, the optimal routing is of balanced type, if nodes of the same type have the same weight.

Many other examples of this kind can be derived from these examples through similar constructions.

7 Optimal routing in two deterministic queues

7.1 Introduction

The general case of two queues with no state information was studied in Chapter 6 where it was shown that the optimal policy in terms of average waiting time must be a *bracket sequence*. However no explicit computation of the rate p^* of the optimal policy was provided in the heterogeneous case.

In the deterministic case considered here (both the arrivals and the service times are constant deterministic variables), the optimal policy was computed in [112] when the system is fully loaded.

In this chapter, we consider two deterministic FIFO queues with infinite buffers and arbitrary constant arrival and service times (hence the load is arbitrary). As proved in Chapter 6, the optimal policy is a bracket sequence, and we show how to compute explicitly the slope of this optimal sequence, which is a rational number as long as the system is not fully loaded, as in [50]. Hence, the optimal policy is always periodic even when the parameters (service times and arrival times) are irrational numbers.

In order to obtain this result, we start in Section 7.3 by defining a new continued fraction decomposition which helps us for identifying special factors of upper bracket sequences. In Section 7.4, we give an explicit formula for the average waiting time in one deterministic queue when the arrival process is an upper bracket sequence. This function is continuous but not differentiable at certain rational points called *jumps* in the following. It is increasing and concave between jumps. Finally, in Section 7.5, we consider the case of two queues. Using the properties of the average waiting time in one queue, we can show that the optimal routing policy is given by a jump in one of the two queues, and hence is periodic, as long as the system is not fully loaded. An algorithm is provided that computes the optimal jump in finite time (and hence the optimal policy). In the last section, several examples are studied in detail in order to illustrate the strange behavior of the optimal policy in such a simple system (or so it seems). More on this kind of problems (in particular the extension to networks on deterministic queues can be found in [51]. In [68], an algorithm is derived to compute a lower and upper bound for the average waiting time for the optimal routing for $K > 2$ deterministic queues. A key property is the convexity of the optimal average waiting time as a function of the rate. This is shown in [70] for the general stochastic setting od Chapter 4.

7.2 Bracket words

We recall the definition of bracket sequences.

We consider binary sequences over $\mathbb{N}$. In this chapter, to avoid confusion, we will adopt the following notations.

Definition 12 (Bracket sequence). *The upper bracket sequence with slope α is the infinite sequence $\overline{m}_\alpha$ where the nth letter, with $n \geq 0$, is :*

$$\overline{m}_\alpha(n) = \lceil (n+1) \times \alpha \rceil - \lceil n \times \alpha \rceil. \tag{7.1}$$

The bracket sequence with slope α is as usual the infinite sequence $\underline{m}_\alpha$ where the nth letter, with $n \geq 0$, is :

$$\underline{m}_\alpha(n) = \lfloor (n+1) \times \alpha \rfloor - \lfloor n \times \alpha \rfloor. \tag{7.2}$$

The characteristic sequence of slope α is the infinite sequence c_α where the nth letter , with $n \geq 0$, is :

$$c_\alpha(n) = \lfloor (n+2) \times \alpha \rfloor - \lfloor (n+1) \times \alpha \rfloor.$$

In the following, by a slight abuse of notation when an infinite word w is periodic we also denote by w its shortest period.

Example 1 (Graphical interpretation). The terminology comes from the following graphical interpretation. For example let $\alpha = 3/7$, the bracket word is 0010101, the upper bracket word is 1010100 and the characteristic word is 0101010.

Consider the straight line with equation $y = \alpha x$, and consider the points with integer coordinates just below the line : $P_n = (n, \lfloor n\alpha \rfloor)$, the ones just above the line : $P'_n = (n, \lceil n\alpha \rceil)$ and the points $P''_n = (n, \lfloor (n+1)\alpha \rfloor)$. The points P_n form a representation of the bracket word in the following sense : two consecutive points are joined by either an horizontal straight line segment, if $\lfloor n+1\alpha \rfloor - \lfloor n\alpha \rfloor = 0$, or a diagonal, if $\lfloor (n+1)\alpha \rfloor - \lfloor n\alpha \rfloor = 1$. Similarly the points P'_n are a representation of the upper bracket word and the points P''_n are a representation of the characteristic word. In Figure 7.1, the straight line b is the line which equation is $y = \frac{3x}{7}$ while the upper bracket word is represented by a, the bracket word is represented by c and the characteristic word is represented by the dashed curve d.

Remark 6. Since $\lceil n\alpha \rceil = \lfloor n\alpha \rfloor + 1$ except when $n\alpha$ is an integer number, one has $\overline{m}_\alpha(n) = \underline{m}_\alpha(n)$ when $n\alpha$ and $(n+1)\alpha$ are not integer numbers. When α is an irrational number, since $\overline{m}_\alpha(0) = 1$ and $\underline{m}_\alpha(0) = 0$ then $\underline{m}_\alpha = 0c_\alpha$ and $\overline{m}_\alpha = 1c_\alpha$. When α is a rational number $\overline{m}_\alpha$ is a shift of $\underline{m}_\alpha$ (*i.e.* there exists an integer number k such that $\underline{m}_\alpha(n+k) = \overline{m}_\alpha(n)$ for all n).

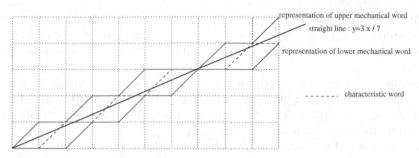

representation of upper mechanical word

straight line : y=3 x / 7

representation of lower mechanical word

------ characteristic word

Fig. 7.1. Mechanical words associated with the line $y = \frac{3}{7}x$.

Lemma 29. *Let α be a real number, then the maximum number of consecutive 0 in $\overline{m}_\alpha$ is $\lceil \alpha^{-1} \rceil - 1$ and the minimum number of consecutive 0 in $\overline{m}_\alpha$ is $\lfloor \alpha^{-1} \rfloor - 1$.*

Proof. The proof can be found in [84]. □

7.3 Expansion in continued fractions

Let ρ, $0 < \rho < 1$ be a real number. The computation of its expansion in a continued fraction is :

$$\rho = 0 + \cfrac{1}{m_1 + \cfrac{1}{m_2 + \dots \cfrac{1}{m_n + \dots}}}.$$

In [84] and [86] one can find an iterative method to compute special factors of the characteristic word of slope ρ and thus build c_ρ, using the coefficients of the expansion in continued fraction of ρ. However, this tight relation between the continued fraction of ρ and the characteristic word c_ρ does not extend to the bracket words. For this reason we will introduce a new expansion in continued fraction (called upper continued fraction expansion). This new continued fraction allows us to find the decomposition in special factors of upper bracket words, decomposition which is necessary for the computations of the average waiting time.

The construction of the upper continued fraction expansion of a number α with $0 < \alpha < 1$ is given by :

$$\left\{ \begin{array}{ll} \alpha = \frac{1}{l_1 + \alpha_1} & , \quad l_1 = \lfloor \alpha^{-1} \rfloor \\ 1 - \alpha_n = \frac{1}{l_{n+1} + \alpha_{n+1}} & , \quad l_{n+1} = \lfloor (1 - \alpha_n)^{-1} \rfloor , \forall n \geq 1 \end{array} \right\} . \quad (7.3)$$

When α is a rational number the construction finishes after a finite number of steps when $1 - \alpha_n = \frac{1}{l_n}$. When α is an irrational number the construction is infinite and we obtain the infinite upper continued fraction expansion of α.

An infinite upper continued fraction expansion is written under the form $\langle l_1, l_2, \ldots, l_{n-1}, l_n, \ldots \rangle$. A finite upper continued fraction is written under the form $\langle l_1, l_2, \ldots, l_{n-1}, l_n \rangle$. A partial upper continued fraction of a number α is written under the form $\langle l_1, l_2, \ldots, l_n + \alpha_n \rangle$.

Let us denote by $\overline{M}$ the set of all upper bracket words and denote by $\overline{M_k}$ the set of upper bracket words with slope α such that $(k+1)^{-1} < \alpha \leq k^{-1}$. According to Lemma 29, all words in $\overline{M_k}$ have k or $k-1$ consecutive zeros between two ones. Let us now introduce the morphism φ_k from $\overline{M_k}$ to $A^* \cup A^{\mathbb{N}}$, which replaces maximal sub-words starting with one and containing only zeros according to the following rules

$$\varphi_k : \begin{array}{c} \overbrace{1\,0\ldots0}^{k-1} \mapsto 1 \\ \underbrace{1\,0\ldots0}_{k} \mapsto 0 \end{array}.$$

For example $\overline{m}_{3/7} = 1010100 \in \overline{M_2}$, therefore $\varphi_2(\overline{m}_{3/7}) = 110$. If $\alpha = k^{-1}$ then the word $\varphi_k(\overline{m}_\alpha)$ is reduced to the letter one.

In order to generalize φ_k let us introduce the morphism

$$\Phi : \begin{array}{ccc} \overline{M} & \to & A^* \cup A^{\mathbb{N}} \\ \overline{m}_\alpha & \mapsto & \varphi_{\lfloor \alpha^{-1} \rfloor}(\overline{m}_\alpha) \end{array}.$$

We now show that Φ has its values in $\overline{M}$.

Lemma 30. *Let $\overline{m}_\alpha$ be the upper bracket word of slope α, then $\Phi(\overline{m}_\alpha)$ is the upper bracket word of slope $1 - (\alpha^{-1} - \lfloor \alpha^{-1} \rfloor)$.*

Proof. If α^{-1} is an integer number $\Phi(\overline{m}_\alpha)$ is reduced to the word "1", then we have the result since $1 - (\alpha^{-1} - \lfloor \alpha^{-1} \rfloor) = 1$.

Recall that $\overline{m}_\alpha(n)$ is defined by Formula (7.1). An integer a_k is the index of the kth occurrence of the letter one in $\overline{m}_\alpha$ if $\overline{m}_\alpha(0) \ldots \overline{m}_\alpha(a_k)$ contains k letters one and $\overline{m}_\alpha(0) \ldots \overline{m}_\alpha(a_k - 1)$ contains $k-1$ letters one. This means

$$\lceil \alpha(a_k + 1) \rceil = k + 1, \quad \lceil \alpha a_k \rceil = k.$$

These equalities imply $a_k = \lfloor k \alpha^{-1} \rfloor$.

Define now the function ψ_k from $\overline{M_k}$ to $A^* \cup A^{\mathbb{N}}$ which replaces maximal sub-words.

$$\psi_k : \begin{array}{c} \overbrace{1\,0\ldots0}^{k-1} \mapsto 0 \\ \underbrace{1\,0\ldots0}_{k} \mapsto 1 \end{array}.$$

Let $l = \lfloor \alpha^{-1} \rfloor$ and let $w = \psi_l(\overline{m}_\alpha)$. The sequence $(w(k))_{k \geq 0}$ of letters of w can be computed by $w(k) = a_{k+1} - a_k - l$. Then $w(k) = \lfloor (k+1)\alpha^{-1} - l(k+1) \rfloor - \lfloor k\alpha^{-1} - kl \rfloor$, hence w is the bracket word of slope $\alpha^{-1} - l$.

Define now the function γ such that

$$\gamma : \begin{array}{c} 1 \mapsto 0 \\ 0 \mapsto 1. \end{array}$$

The function γ transforms a bracket word of slope α into an upper bracket word of slope $1 - \alpha$ and conversely. Let $w' = \gamma(w)$, w' is the upper bracket word of slope $1 - (\alpha^{-1} - l)$.

It can be checked that $\Phi = \gamma \circ \psi$, hence $\Phi(\overline{m}_\alpha)$ is the upper bracket word of slope $1 - (\alpha^{-1} - \lfloor \alpha^{-1} \rfloor)$. $\qquad\square$

The following corollary shows the relation between $\Phi(\overline{m}_\alpha)$ and the upper continued fraction of $\overline{m}_\alpha$.

Corollary 6. *Let* α, $0 < \alpha < 1$ *be any given real number such that* $\alpha = \langle l_1, \ldots, l_{n-1}, l_n + \alpha_n \rangle$, *with* $n \geq 1$. *Let* $w = \Phi(\overline{m}_\alpha)$. *Then* $\alpha(w) = \langle l_2, l_3, \ldots, l_{n-1}, l_n + \alpha_n \rangle$.

Proof. By Equations 7.3, we have $\alpha = (l_1 + \alpha_1)^{-1}$ and also $1 - \alpha_1 = \langle l_2, l_3, \ldots, l_{n-1}, l_n + \alpha_n \rangle$. By Lemma 30, we obtain

$$\alpha(w) = 1 - (\alpha^{-1} - \lfloor \alpha^{-1} \rfloor) = 1 - (\alpha^{-1} - l_1) = 1 - \alpha_1.$$

Hence $\alpha(w) = \langle l_2, l_3, \ldots, l_{n-1}, l_n + \alpha_n \rangle$. $\qquad\square$

This will allow us to use induction on the number of terms in the upper continued fraction in the following.

Theorem 28 ((x,y)-factor decomposition). *Let* $\alpha, 0 < \alpha < 1$ *be any given real number with* $\alpha = \langle l_1, l_2, \ldots, l_n, \ldots \rangle$. *Let us define two sequences* $\{x_i(\alpha)\}_{i \geq 0}$ *and* $\{y_i(\alpha)\}_{i \geq 0}$, *by :*

$$x_0(\alpha) = 1, \quad x_i(\alpha) = x_{i-1}(\alpha)(y_{i-1}(\alpha))^{l_i - 1}, \text{ for } i \geq 1,$$
$$y_0(\alpha) = 0, \quad y_i(\alpha) = x_{i-1}(\alpha)(y_{i-1}(\alpha))^{l_i}, \text{ for } i \geq 1.$$

Then the upper bracket word $\overline{m}_\alpha$ *can be factorized only using the two factors* $x_i(\alpha)$ *and* $y_i(\alpha)$.

These two sequences are called (x-y)-factor decomposition sequences associated with the upper expansion of α.

Proof. We will study finite sequences $\{x_i(\alpha)\}_{0 \leq i \leq n}$, $\{y_i(\alpha)\}_{0 \leq i \leq n}$ associated with the partial upper continued fraction expansion of α and prove the result by induction.

Step 1. We have $\alpha = \langle l_1 + \alpha_1 \rangle$. By Lemma 29, it is immediate to check that $\overline{m}_\alpha$ can be factorized only using x_1 and y_1.

Step n. We have $\alpha = \langle l_1, l_2, \ldots, l_n + \alpha_n \rangle$. Let $w = \Phi(\overline{m}_\alpha)$, by Corollary 6 $\alpha(w) = \langle l_2, l_3, \ldots, l_{n-1}, l_n + \alpha_n \rangle$. Let the two (x-y)-factor decomposition sequences associated with the partial upper expansion of $\alpha(w)$ be $\{x_i(\alpha(w))\}_{0 \leq i \leq n-1}$ and $\{y_i(\alpha(w))\}_{0 \leq i \leq n-1}$. By induction on the number of terms in the partial expansion w can be factorized only using $x_i(\alpha(w))$ and $y_i(\alpha(w))$. Introduce now the sequences x_i' and y_i' such that

$$x_0' = 1, \; y_0' = 0, \; \varphi_{l_1}(x_i') = x_{i-1}(\alpha(w)), \; \varphi_{l_1}(y_i') = y_{i-1}(\alpha(w)), \; \forall i \geq 1.$$

Then $w = \Phi(\overline{m}_\alpha)$ can be factorized only using $\varphi_{l_1}(x_i')$ and $\varphi_{l_1}(y_i')$, $\forall i \geq 0$. Therefore $\overline{m}_\alpha$ can be factorized only using x_i' and y_i', since $\Phi(\overline{m}_\alpha) = \varphi_{l_1}(\overline{m}_\alpha)$. Note now that the sequences x_i', y_i' are the (x-y)-factor decomposition sequences associated with $\langle l_1, l_2, \ldots, l_n + \alpha_n \rangle$. □

Lemma 31. *Let α, $0 < \alpha < 1$ be any given rational number with $\alpha = \langle l_1, \ldots, l_{n-1}, l_n \rangle$. Then $x_n(\alpha) = \overline{m}_\alpha$.*

Proof. We also use here an induction on the number of terms in the partial upper expansion.

Step 1. Considering that $\alpha = \langle l_1 \rangle$ then α is a rational number with $\alpha^{-1} = l_1$. Hence $\overline{m}_\alpha = 10 \ldots 0$, where the number of consecutive 0 is $\alpha^{-1} - 1 = l_1 - 1$.

Step n. We have $\alpha = \langle l_1, l_2, \ldots, l_{n-1}, l_n \rangle$, therefore $w = \Phi(\overline{m}_\alpha)$ has a slope $\alpha(w) = \langle l_2, \ldots, l_n \rangle$. Using the induction hypothesis we obtain $w = x_{n-1}(\alpha(w))$. Since $x_{n-1}(\alpha(w)) = \varphi_{l_1}(x_n(\alpha))$, then $\overline{m}_\alpha = x_n(\alpha)$. □

Remark 7. Since $\Phi^{(n)}(\overline{m}_\alpha) = \varphi_{l_n} \circ \ldots \circ \varphi_{l_2} \circ \varphi_{l_1}(\overline{m}_\alpha)$, then $x_n(\alpha)$ and $y_n(\alpha)$ are the only factors of $\overline{m}_\alpha$, such that $\Phi^{(n)}(x_n(\alpha)) = 1$ and $\Phi^{(n)}(y_n(\alpha)) = 0$.

Remark 8. Since $\forall n \geq 0$, $\Phi^{(n)}(\overline{m}_\alpha)$ is an upper bracket word and since all upper bracket words begin with one, then the sequence $x_i(\alpha)$ is a sequence of prefixes of $\overline{m}_\alpha$ verifying

$$\lim_{i \to \infty} x_i(\alpha) = \overline{m}_\alpha.$$

Example 2. Let $\alpha = \frac{8}{19}$. The upper expansion of α using Equations 7.3 is : $\langle 2, 1, 2, 2 \rangle$, and

$$\overline{m}_{8/19} = 1010100101010010100.$$

Thus $\Phi(\overline{m}_{8/19}) = 11011010$, $\Phi^{(2)}(\overline{m}_{8/19}) = 10100$ and $\Phi^{(3)}(\overline{m}_{8/19}) = 10$. Using the (x-y)-factor decomposition we obtain $x_1 = 10$ and $y_1 = 100$, $x_2 = x_1$ and $y_2 = x_1 y_1$, $x_3 = x_2 y_2$ and $y_3 = x_2(y_2)^2$ with finally $x_4 = x_3 y_3$.

We can check Theorem 28 and Lemma 31 in this case:

$$\overline{m}_{8/19} = \underbrace{\underbrace{\underbrace{10}_{x_1} \; \underbrace{10}_{x_2} \; \underbrace{100}_{y_2}}_{x_3 / x_2} \; \underbrace{\underbrace{10}_{x_1} \; \underbrace{10}_{x_2} \; \underbrace{100}_{y_2} \; \underbrace{10}_{x_1} \; \underbrace{100}_{y_2}}_{y_3}}_{x_4}.$$

Theorem 28 and Lemma 31 show the connection between the construction of factors of an upper bracket word and the upper continued fraction expansion of its slope. This connection could be seen as the analog for upper bracket words of the relation between continued fraction expansion and characteristic words presented in [86, 84].

In the following, the ith factors of the word $\overline{m}_\alpha$ will be denoted x_i and y_i when no confusion is possible.

7.4 Average waiting time in a single queue

The aim of this section is to compute and to study the average waiting time of customers, $\mathcal{W}_S(w)$ in a ./D/1/∞/FIFO queue with a constant service time, S, and constant inter-arrival times before the control. Using an appropriate time scale, the inter-arrival times can be chosen to be equal to one. The input sequence w is such that $w(n)$ is 0 if no customer enter the queue at slot n and $w(n)$ is 1 if one customer enters the queue at slot n. In the following, we will only consider an input sequence which is the upper bracket word of slope α, $\overline{m}_\alpha$.

Let α, $0 \leq \alpha \leq 1$ be the ratio of customers sent in the ./D/1/∞/FIFO queue. The stability condition of the queue is

$$\alpha \leq \frac{1}{S}. \tag{7.4}$$

7.4.1 Jumps

We now show that the computation of the average waiting time $\mathcal{W}_S(\overline{m}_\alpha)$ tightly depends on the factorization of $\overline{m}_\alpha$ in (x-y)-factors of S^{-1}. For that we will exhibit special rational numbers.

Lemma 32. *Let $p = \langle l_1, l_2, \ldots, l_{n-1}, l_n, l_{n+1} \rangle$ be a rational number.*

-i) For all real number α which partial upper expansion is $\langle l_1, l_2, \ldots, l_{n-1}, l_n + \alpha_n \rangle$, such that l_{n+1} is the smallest integer satisfying

$$\alpha_n \leq \frac{l_{n+1} - 1}{l_{n+1}},$$

then $x_{n+1}(\alpha) = x_n(p)(y_n(p))^{l_{n+1}-2}$ and $y_{n+1}(\alpha) = x_{n+1}(p)$.

-ii) For all real number β which partial upper expansion is $\langle l_1, l_2, \ldots, l_{n-1}, l_n + \beta_n \rangle$ with

$$\frac{l_{n+1} - 1}{l_{n+1}} \leq \beta_n < 1,$$

then $x_{n+1}(\beta) = x_n(p)(y_n(p))^k$ and $y_{n+1}(\beta) = x_n(p)(y_n(p))^{k+1}$ with $k \geq l_{n+1} - 1$.

-iii) For any number γ which partial upper expansion is $\langle l_1, \ldots, l_{n-1}, l_n + \gamma_n \rangle$, then $1 - \gamma_n$ is the proportion of $x_n(\beta) = x_n(\alpha) = x_n(p)$ in $\overline{m}_\gamma$ (i.e. the number of $x_n(\beta)$ divided by the number of $x_n(\beta)$ added with the number of $y_n(\beta)$ in $\overline{m}_\alpha$).

Proof. Note that $1 - 1/l_{n+1} = (l_{n+1} - 1)/l_{n+1}$. Hence the Equations 7.3 lead to $\beta \leq p \leq \alpha$.

-i) For all i, $0 \leq i \leq n$, we have $x_i(p) = x_i(\alpha)$ and $y_i(p) = y_i(\alpha)$, since the upper expansions are equal until the nth coefficient. Let us compute l_{n+1} the smallest integer such that

$$\alpha_n \leq \frac{l_{n+1} - 1}{l_{n+1}},$$

that implies

$$l_{n+1} \geq \frac{1}{1 - \alpha_n} \Rightarrow l_{n+1} - 1 = \lfloor \frac{1}{1 - \alpha_n} \rfloor.$$

Hence the upper expansion of α is $\langle l_1, l_2, \ldots, l_n, (l_{n+1} - 1) + \alpha_{n+1} \rangle$. Therefore $x_{n+1}(\alpha) = x_n(p)(y_n(p))^{l_{n+1}-2}$ and $y_{n+1}(\alpha) = x_n(p)(y_n(p))^{l_{n+1}-1} = x_{n+1}(p)$.

-ii) $\frac{l_{n+1}-1}{l_{n+1}} \leq \beta_n$ implies

$$\lfloor \frac{1}{1 - \beta_n} \rfloor \geq l_{n+1},$$

therefore the number of consecutive $y_n(\beta)$ in $x_{n+1}(\beta)$ is larger than $l_{n+1} - 1$. Since $x_n(p) = x_n(\beta)$ and $y_n(p) = y_n(\beta)$ then we proved the result.

-iii) By Remark 7, $x_n(\gamma)$ and $y_n(\gamma)$ are the only factors of the word $\overline{m}_\gamma$, such that $\Phi^{(n)}(x_n(\gamma)) = 1$ and $\Phi^{(n)}(y_n(\gamma)) = 0$. Hence the composition in $x_n(\gamma)$, $y_n(\gamma)$ of $\overline{m}_\gamma$ is the composition in 1 and 0 of $\Phi^{(n)}(\overline{m}_\gamma)$. By Lemma 30 and Corollary 6, $\Phi^{(n)}(\overline{m}_\gamma)$ is an upper bracket word of slope $1 - \gamma_n$. Therefore using the definition of the slope, $1 - \gamma_n$ is the proportion of x_n in $\overline{m}_\gamma$. □

The previous Lemma says that in all bracket words with slope $\alpha \leq p = \langle l_1, l_2, \ldots, l_{n-1}, l_n, l_{n+1} \rangle$, the sequence $x_n(p)y_n(p)^{l_{n+1}-2}$ never appears. But as soon as the slope is larger than p then the sequence $x_n(p)y_n(p)^{l_{n+1}-2}$ appears. Intuitively, when you increase the slope the number of ones increases and the number of zero decreases, hence the minimum number of consecutive y_n decreases until $x_n(p)y_n(p)^{l_{n+1}-2}$ appears.

Definition 13 (jumps). *Let us denote the upper continued fraction expansion of S^{-1} by $\langle l_1, \ldots, l_n, \ldots \rangle$ and let us define the rational number $r_i(S^{-1})$ by $r_i(S^{-1}) = \langle l_1, \ldots, l_{i-1}, l_i + 1 \rangle$. The number $r_i(S^{-1})$ is called the ith jump of $\mathcal{W}_S$. When S is a rational number, then $S^{-1} = \langle l_1, l_2, \ldots, l_N \rangle$ and we define the last jump $r_N(S^{-1})$ by $r_N(S^{-1}) = \langle l_1, l_2, \ldots, l_N \rangle$.*

Note that the jumps form a sequence of rational numbers increasing towards S^{-1}. When S^{-1} is rational, the number of jumps is finite and the last jump equals S^{-1}.

Let us define n_i and q_i to be the two relatively prime integers such that $r_i(S^{-1}) = n_i/q_i$. Therefore by the definition of the slope, $n_i = |\overline{m}_{r_i(S-1)}|_1$ and $q_i = |\overline{m}_{r_i(S-1)}|$. For example, if $r_2 = \langle l_1, l_2 + 1 \rangle$, then it can be checked that $n_2 = l_2 + 1$ and $q_2 = l_1(l_2+1) + l_2$. From now on $r_i(S^{-1})$ will be denoted by r_i when no confusion is possible.

Lemma 33. *The number r_{i+1} is the rational number with the smallest denominator in $]r_i, S^{-1}]$.*

Proof. Let $\alpha = \langle l_1, l_2, \ldots, l_i + \alpha_i \rangle$ be any given rational number in $]r_i, S^{-1}]$. By Equations 7.3 we have $\lfloor (1-\alpha_i)^{-1} \rfloor \geq l_{i+1}$. Hence by Lemma 32, $x_i(y_i)^{l_{i+1}}$ is a factor of $\overline{m}_\alpha$. That means $|\overline{m}_\alpha| \geq |x_i(y_i)^{l_{i+1}}|$. Since $x_i(y_i)^{l_{i+1}}$ is the bracket word $\overline{m}_{r_{i+1}}$ then $|x_i(y_i)^{l_{i+1}}| = q_{i+1}$. Considering that $|\overline{m}_\alpha|$ is the denominator of α ends the proof. $\qquad\square$

Remark 9 (Connection with classical continued fractions). The rational number c_k, obtained by keeping only k terms in a classical expansion in continued fractions is called the kth convergent. The even convergents c_{2n} of a simple continued fraction form an increasing sequence. Hence if $[m_0, m_1, m_2, \ldots]$ denote the classical continued fraction of S^{-1}, then the even convergents satisfy $c_{2n} = [m_0, m_1, \ldots, m_{2n}] \leq S^{-1}$. If $c_{2n} = [m_0, m_1, \ldots, m_{2n}]$ is an even convergent of S^{-1}, then the rational numbers $[m_0, m_1, \ldots, m_{2n-1}, \rho_{2n}]$ with $\rho_{2n} \in \{1, 2, \ldots, m_{2n}\}$ are called intermediate convergents.

Note now that the sequence of intermediate convergents and even convergents is the sequence of jumps $r_i(S^{-1})$, since this sequence increases towards S^{-1} and each term of this sequence is the rational with the smallest denominator in the interval formed by the preceding intermediate convergent and S^{-1} and since $[c_1 + 1] = r_1$.

However it seems hard to use the intermediate convergents to compute the average waiting time as it is shown further.

7.4.2 Formula of the average waiting time

Recall the classical Linley formula for GI/GI/1 queues. Let T^n denote the epoch of the admission of the nth customer and S^n the duration of its service time. The workload w^t which denotes the amount of service (in time units) remaining to be done by the server at the epoch t is given by

$$w^t = \left(w^{T^{n-1}_-} + S^{n-1} - (t - T^{n-1}) \right)^+, \quad \text{for } T^{n-1} \leq t < T^n,$$

where $x_- = \lim_{y \uparrow x} y$ and $(\cdot)^+ = \max(\cdot, 0)$. The waiting time of the nth customer is given by

$$W^n = w^{T^n_-}.$$

The decomposition of $\overline{m}_{r_i}$ in (x-y) factors of S^{-1} allows us to compute recursively the average waiting time of the input sequence $m_{r_i} : W_S(\overline{m}_{r_i})$.

We will see that the factors x_i tend to increase the load in the queue whereas the factors y_i tend to decrease the load. From that, we can compute $W_S(\overline{m}_\alpha)$ for any rational number as well for any irrational number α, $r_i < \alpha < r_{i+1}$.

Let d_i^1 denote the increase of the workload after an input sequence equal to x_i. Let d_i^2 denote the maximal possible decrease of the workload after an input sequence equal to y_i.

Lemma 34. *If the initial workload is equal to zero then using x_i as input sequence, the workload during x_i is never null and remains non negative at the end of the sequence. If the initial workload is equal to zero then using y_i as input sequence the workload is never null during y_i until its last letter (which is always a 0) and is null at the end of the sequence.*

Proof. The proof holds by induction on i.

Step 1. Assume S^{-1} is not an integer, we have $l_1 < S < l_1 + 1$. Therefore $S - l_1 - 1 < 0 < S - l_1$ and the result is proved for step one since the workload after x_1 is equal to $S - l_1$ and the workload after y_1 is $\max(S - l_1 - 1, 0) = 0$. Moreover Equations 7.3 yield $d_1^1 = \alpha_1$ and $d_1^2 = \alpha_1 - 1$. When S^{-1} is an integer, we have $S - l_1 = 0$.

Step 2. By Theorem 28, x_2 is composed by x_1 followed by $(l_2 - 1)$ consecutive y_1 and y_2 is composed by x_1 followed by l_2 consecutive y_1. Hence to prove the result we have to show that $\forall j$, $1 \le j \le \max(l_2 - 1, 1)$:

$$\alpha_1 + (l_2 - j)(\alpha_1 - 1) \ge 0 \,,$$
$$\alpha_1 + l_2(\alpha_1 - 1) + 1 \ge 0 \,,$$
$$\alpha_1 + l_2(\alpha_1 - 1) \le 0 \,,$$

this last inequality implying the nullity of the workload. By Lemma 32, l_2 is the smallest integer satisfying $\alpha_1 \le l_2/(l_2 + 1)$, then for all j such that $1 \le j \le \max(l_2 - 1, 1)$ we have

$$\frac{l_2}{l_2 + 1} \ge \alpha_1 > \frac{l_2 - j}{l_2 + 1 - j} \,.$$

Since $(l_2 - 1)/(l_2) \ge (l_2 - 1)/(l_2 + 1)$, the three inequalities are satisfied.

It can be checked that $\alpha_1 + (l_2 - 1)(\alpha_1 - 1) = (1 - \alpha_1)\alpha_2$ and $\alpha_1 + (l_2)(\alpha_1 - 1) = (1 - \alpha_1)(\alpha_2 - 1)$. Therefore $d_2^1 = (1 - \alpha_1)\alpha_2$ and $d_2^2 = (1 - \alpha_1)(\alpha_2 - 1)$.

Step i+1. Suppose that $d_i^2 = (1 - \alpha_1) \ldots (1 - \alpha_{i-1})(\alpha_i - 1)$ and $d_i^1 = (1 - \alpha_1) \ldots (1 - \alpha_{i-1})\alpha_i$. We have to show that $\forall j$, $1 \le j \le \max(l_{i+1} - 1, 1)$:

$$(1 - \alpha_1) \ldots (1 - \alpha_{i-1}) [\alpha_i + (l_{i+1} - j)(\alpha_i - 1)] > 0 \,, \tag{7.5}$$
$$(1 - \alpha_1) \ldots (1 - \alpha_{i-1}) [\alpha_i + (l_{i+1}(\alpha_i - 1)] \le 0 \,. \tag{7.6}$$

These inequalities are satisfied indeed, since l_{i+1} is the smallest integer such that $\alpha_i \le l_{i+1}/(l_{i+1} + 1)$. Moreover, by the induction hypothesis, during the sequence y_i the workload is positive until the last 0, when the initial workload is null. Consider now the last factor y_i in y_{i+1}, its initial workload is positive by Inequality 7.5. If the workload is never null during an input sequence with an initial workload equal to zero, then the same result holds with a positive initial workload. Hence the workload is never null during the last y_i and also during y_{i+1}, up to its last letter.

We have

$$\alpha_i + (l_{i+1} - 1)(\alpha_i - 1) = (1 - \alpha_i)\alpha_{i+1}, \tag{7.7}$$

$$\alpha_i + (l_{i+1})(\alpha_i - 1) = (1 - \alpha_i)(\alpha_{i+1} - 1), \tag{7.8}$$

then we obtain

$$d_i^2 = (1 - \alpha_1) \ldots (1 - \alpha_{i-1})(\alpha_i - 1) \tag{7.9}$$

$$d_i^1 = (1 - \alpha_1) \ldots (1 - \alpha_{i-1})\alpha_i, \tag{7.10}$$

This finishes the proof. □

Remark 10. When $S = \langle l_1, l_2, \ldots, l_N \rangle$ is a rational number, $1 - \alpha_{N-1} = 1/l_N$ which is equivalent to $\alpha_{N-1} = (l_N - 1)/l_N$. Then $\alpha_{N-1} + (l_N - 1)(\alpha_{N-1} - 1) = 0$, this yields $d_N^1 = 0$. Hence the workload during x_N is positive until the last epoch of the admission sequence $x_N(S)^{-1} = x_N(r_N)$, and null at the end.

We give now a direct consequence of Lemma 34. The workload increase d_i^1 and the maximal workload decrease d_i^2 can be recursively computed for all $i \geq 2$ by :

$$d_i^1 = d_{i-1}^1 + (l_i - 1)d_{i-1}^2 \ , \ d_i^2 = d_{i-1}^1 + (l_i)d_{i-1}^2 \tag{7.11}$$

$$\text{with } d_1^1 = S - l_1 \ , \ d_1^2 = S - l_1 - 1. \tag{7.12}$$

Example 3 (Workload with $S = 51/20$ with $\overline{m}_{5/13}$). Let $S = 51/20 = 2.55$. The partial upper expansion of order three of $20/51$ is $\langle 2, 2, 1 + 2/7 \rangle$. The (x,y)-factor decomposition gives $x_1 = 10$, $y_1 = 100$, $x_2 = 10100$, $y_2 = 10100100$, $x_3 = x_2$ and $y_3 = 1010010100100$. Note that $r_3(20/51) = 5/13$ and that $\overline{m}_{5/13} = y_3$.

The Equation 7.11 leads to

$$d_1^1 = 0.55 \ , \ d_1^2 = -0.45,$$

$$d_2^1 = 0.55 - 0.45 = 0.1 \ , \ d_2^2 = 0.55 - 0.90 = -0.35,$$

$$d_3^1 = d_2^1 = 0.1 \ , \ d_3^2 = 0.1 - 0.35 = -0.25.$$

The figure 7.2 represents the workload during a sequence y_3. The quantity d_1^1 is represented by a, d_1^2 by b, d_2^1 by c, d_2^2 by d and d_3^2 by e.

We have $|\overline{m}_{r_i}|_1 = |x_{i-1}|_1 + l_i|y_{i-1}|_1$, $(|\overline{m}_{r_N}|_1 = |x_{N-1}|_1 + (l_N - 1)|y_{N-1}|_1$ when $S^{-1} = \langle l_1, l_2, l_3, \ldots, l_N \rangle)$ and $|x_{i-1}|_1 = |y_{i-1}|_1 - |y_{i-2}|_1$. Let $S^{-1} = \langle l_1, \ldots, l_i + \alpha_i$, then the recursive computation of $n_i = |\overline{m}_{r_i}|_1$ is given by

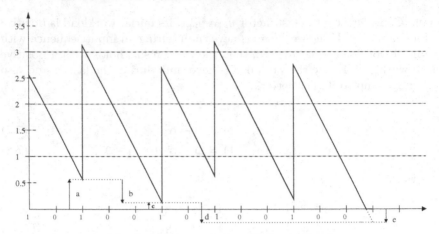

Fig. 7.2. Representation of the workload

$$n_0 = 0, \quad n_1 = 1, \tag{7.13}$$

$$\forall i \geq 2 \qquad n_i = (l_i + 1)n_{i-1} - n_{i-2} \quad \text{if } \alpha_i \neq 0, \tag{7.14}$$

$$n_i = l_i(n_{i-1}) - n_{i-2} \quad \text{if } \alpha_i = 0. \tag{7.15}$$

This allows us to compute the average waiting time of the input sequences $\overline{m}_{r_i}$ recursively.

The sum of the waiting times of customers admitted during the sequence w when the first customer of the sequence waits for a time t is denoted by $K^t(w)$. We also denote the sum of the waiting times of $\overline{m}_{r_i}$ over one period by K_i^t and K_i^0 by K_i. Note first, that we can focus on one period since the workload after $\overline{m}_{r_i}$ is null.

Lemma 35 (Formula of $\mathcal{W}_S(\overline{m}_{r_i})$). *Let* $S^{-1} = \langle l_1, \ldots, l_i + \alpha_i \rangle$. *The average waiting time of* $\overline{m}_{r_i}$ *is :*

$$\mathcal{W}_S(\overline{m}_{r_i}) = \frac{K_i}{n_i},$$

with $K_0 = 0$, $K_1 = 0$ *and* $\forall i \geq 1$

if $\alpha_i \neq 0$

$$K_{i+1} = (l_{i+1} + 1)K_i - K_{i-1} + (n_i l_{i+1} - n_{i-1}) d_i^1 + n_i(l_{i+1} - 1)\frac{l_{i+1}}{2}d_i^2,$$

otherwise

$$K_{i+1} = l_{i+1}K_i - K_{i-1} + (n_i(l_{i+1} - 1) - n_{i-1}) d_i^1 + n_i(l_{i+1} - 2)\frac{l_{i+1} - 1}{2}d_i^2. \tag{7.16}$$

Proof. The minimum number of consecutive zeros in $\overline{m}_{r_1}$ is equal to l_1. This means that the inter-arrival time in the queue is equal to $l_1 + 1$ and since $S < l_1 + 1$ none of the customer has to wait, therefore $K_1 = 0$.

Recall that we have by Definition 13 $\overline{m}_{r_2} = x_1(y_1)^{l_2}$. The waiting time of a customer is the workload at the epoch of its arrival. By Lemma 34 the first customer admitted does not wait, on the other hand the next l_2 customers will have to wait. Moreover the workload during $\overline{m}_{r_2}$ is positive until the last 0 hence the waiting time of the second customer is d_1^1, the one of the third customer is $d_1^1 + d_1^2$, the one of the jth is $d_1^1 + (j-2)d_1^2$ and the one of the last customer, which is the $(l_2+1)^{th}$, is $d_1^1 + (l_2-1)d_1^2$. Summing we obtain $K_2 = l_2 d_1^1 + (l_2-1)(l_2/2)d_1^2$.

Let w be an admission sequence during which the workload is never null if the initial workload is equal to zero. When the first customer of w waits for a time t, the waiting time of a customer admitted during w is its waiting time when the first customer does not wait increased by t. Therefore $K^t(w) = (|w|_1 \times t) + K^0(w)$.

We have $\overline{m}_{r_i} = x_{i-1}(y_{i-1})^{l_i}$. Let us compute the waiting times of the first customers of the l_i consecutive y_{i-1}. The waiting time of the first customer of the first y_{i-1} is d_{i-1}^1. The waiting time of the first customer of the second y_{i-1} is $d_{i-1}^1 + d_{i-1}^2$. The waiting time of the first customer of the jth y_{i-1} is $d_{i-1}^1 + (j-1)d_{i-1}^2$. The waiting time of the first customer of the last y_{i-1} (the l_i^{th}) is $d_{i-1}^1 + (l_i-1)d_{i-1}^2$.

We decompose K_i.

$$K_i = K^0(x_{i-1}) + K_{i-1}^{d_{i-1}^1} + K_{i-1}^{d_{i-1}^1+d_{i-1}^2} + K_{i-1}^{d_{i-1}^1+2d_{i-1}^2} + \ldots + K_{i-1}^{d_{i-1}^1+(l_i-1)d_{i-1}^2},$$

$$= K^0(x_{i-1}) + n_{i-1}d_{i-1}^1 + K_{i-1} + n_{i-1}(d_{i-1}^1 + d_{i-1}^2) + K_{i-1} + \ldots$$
$$+ n_{i-1}(d_{i-1}^1 + (l_i-1)d_{i-1}^2) + K_{i-1},$$

$$= K^0(x_{i-1}) + l_i K_{i-1} + n_{i-1}(l_i d_{i-1}^1 + (l_i-1)\frac{l_i}{2}d_{i-1}^2),$$

since $y_{i-1} = x_{i-1}y_{i-2}$,

$$K_i = K_{i-1} - K_{i-2}^{d_{i-1}^1} + l_i K_{i-1} + n_{i-1}(l_i d_{i-1}^1 + (l_i-1)\frac{l_i}{2}d_{i-1}^2),$$

$$= K_{i-1} - (n_{i-2}d_{i-1}^1 + K_{i-2}) + l_i K_{i-1} + n_{i-1}(l_i d_{i-1}^1 + (l_i-1)\frac{l_i}{2}d_{i-1}^2).$$

When $S^{-1} = \langle l_1, l_2, \ldots, l_N \rangle$, the Nth jump is a special case since $\overline{m}_{r_N}(S^{-1}) = \overline{m}_{S^{-1}}$ is x_N and the number of consecutive y_{N-1} following x_{N-1} is $l_N - 1$. □

We define the average waiting time of a finite word w by

$$\mathcal{W}_S(w) = \frac{K^0(w)}{|w|_1}.$$

We extend this definition by defining the average waiting time of an infinite word w. Let w_n be the prefix of length n of w, then

$$\mathcal{W}_S(w) = \lim_{n \to \infty} \frac{K^0(w_n)}{|w_n|_1},$$

if the limit exists.

Theorem 29 (Formula of $\mathcal{W}_S(\overline{m}_\alpha)$). *Let r_i be the ith jump of $\mathcal{W}_S(\overline{m}_\alpha)$, and r_{i+1} the $i+1^{th}$ jump of $\mathcal{W}_S(\overline{m}_\alpha)$. Let $\alpha = \langle l_1, \ldots, l_i + \alpha_i \rangle$ be any number such that $\alpha \in (r_i, r_{i+1})$. Then the average waiting time of $\overline{m}_\alpha$ is*

$$\mathcal{W}_S(\overline{m}_\alpha) = \frac{(1 - \alpha_i)K_{i+1} + [\alpha_i - l_{i+1}(1 - \alpha_i)] K_i}{(1 - \alpha_i)n_{i+1} + [\alpha_i - l_{i+1}(1 - \alpha_i)] n_i}. \tag{7.17}$$

When α is smaller than r_1, then $\mathcal{W}_S(\overline{m}_\alpha) = 0$.

Proof. The proof is given in Appendix 7.7. ☐

7.4.3 Properties

Here, we will give some properties of the function $\mathcal{W}_S(\overline{m}_\alpha)$. Most of them are required to compute the optimal ratio in the case of two queues.

Lemma 36 (Continuity). *The function $\alpha \mapsto \mathcal{W}_S(\overline{m}_\alpha)$ is continuous.*

Proof. Let $\alpha = \langle l_1, \ldots, l_n + \alpha_n \rangle$ and $\beta = \langle l_1, \ldots, l_n + \beta_n \rangle$ be two points in the interval $[r_n, r_{n+1}]$. If $|\alpha - \beta|$ goes to zero, then $|\alpha_n - \beta_n|$ also goes to zero. By Equation 7.17, $|\mathcal{W}_S(\overline{m}_\alpha) - \mathcal{W}_S(\overline{m}_\beta)|$ goes to zero. ☐

Using similar arguments, one can show that the function $\alpha \to \mathcal{W}_S(\overline{m}_\alpha)$ is infinitely differentiable on each interval $]r_n, r_{n+1}[$, but it is not differentiable at the jumps.

Lemma 37. *The function $i \mapsto \frac{K_i}{n_i}$ is strictly increasing.*

Proof. The proof is given in Appendix 7.8. ☐

Remark 11. As proved in [112] if S is an irrational number then the average waiting time $\mathcal{W}_{S-1}(\overline{m}_{S-1}) = 1/2$. Hence the sequence $\frac{K_i}{n_i}$ goes to $1/2$ when i goes to infinity.

Theorem 30. *If $\alpha \in]r_i, r_{i+1}[$, then*

- *i) the function $\mathcal{W}_S(\overline{m}_\alpha)$ is strictly increasing in α,*
- *ii) the function $\alpha \mapsto \mathcal{W}_S(\overline{m}_\alpha)$ is strictly concave,*
- *iii) the function $\alpha \mapsto \alpha \mathcal{W}_S(\alpha)$ is linear in α.*

Proof. The proof is given in Appendix 7.9. ☐

To illustrate the results obtained in this section let us take an example.

Example 4 (Curve of $\mathcal{W}_{5/4}(\alpha)$). The upper expansion of $4/5$ is $\langle 1,1,1,2 \rangle$ and $\overline{m}_{4/5}$ is equal to 11110. The factor decomposition is : $x_1 = 1$ and $y_1 = 10$, then $x_2 = x_1 = 1$ and $y_2 = x_1 y_1 = 110$, then $x_3 = x_2 = 1$ and $y_3 = x_2 y_2 = 1110$, then $x_4 = x_3 y_3 = 11110 = \overline{m}_{4/5}$.

We have $r_1(4/5) = \langle 2 \rangle = 1/2, r_2(4/5) = \langle 1,2 \rangle = 2/3, r_3(4/5) = \langle 1,1,2 \rangle = 3/4$ and $r_4(4/5) = \langle 1,1,1,2 \rangle = 4/5$, therefore $n_1 = 1$, $n_2 = 2$, $n_3 = 3$ and $n_4 = 4$.

Let us now compute K_i, we obtain $K_1 = 0$, $K_2 = 1/4$, $K_3 = 3/4$ and $K_4 = 3/2$. Hence we have :

$$\frac{K_1}{n_1} = 0, \quad \frac{K_2}{n_2} = \frac{1}{8}, \quad \frac{K_3}{n_3} = \frac{1}{4}, \quad \frac{K_4}{n_4} = \frac{3}{8}.$$

Recall that $\alpha = \langle l_1 + \alpha_1 \rangle = \langle l_1, l_2 + \alpha_2 \rangle = \langle l_1, l_2, l_3 + \alpha_3 \rangle$. Formula 7.17 leads to

$$\mathcal{W}_S(\overline{m}_\alpha) = 0 \text{ , for } \alpha \leq \frac{1}{2}.$$

$$\mathcal{W}_S(\overline{m}_\alpha) = \frac{1-\alpha_1}{4} \text{ , for } \frac{1}{2} \leq \alpha \leq \frac{2}{3} \Leftrightarrow \frac{1}{2} \leq \alpha_1 \leq 1.$$

$$\mathcal{W}_S(\overline{m}_\alpha) = \frac{2-\alpha_2}{4\,(\alpha_2 + 1)} \text{ , for } \frac{2}{3} \leq \alpha \leq \frac{3}{4} \Leftrightarrow \frac{1}{2} \leq \alpha_2 \leq 1.$$

$$\mathcal{W}_S(\overline{m}_\alpha) = \frac{3}{4(2\alpha_3 + 1)} \text{ , for } \frac{3}{4} \leq \alpha \leq \frac{4}{5} \Leftrightarrow \frac{1}{2} \leq \alpha_3 \leq 1.$$

The function $\mathcal{W}_S(\overline{m}_\alpha)$ for all $0 \leq \alpha \leq 4/5$, is shown in Figure 7.3.

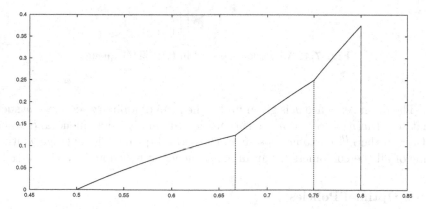

Fig. 7.3. Curve of $\mathcal{W}_{5/4}(\overline{m}_\alpha)$ when α varies from 0.45 to 0.8.

7.5 Average waiting time for two queues

7.5.1 Presentation of the Model

From now on as presented in Figure 7.4, we will study a system made of two ./D/1/∞/FIFO queues, with constant service times S_1 in queue 1 and S_2 in queue 2. It is assumed for convenience in the following that S_2 is the smallest service time. The time unit is chosen such that inter-arrival time slots are constant and equal to one. When they arrive, the customers are routed to a queue where they wait for treatment in a FIFO order. The ratio of customers sent in queue 1 is α and therefore, the ratio of customers sent in queue 2 is $1 - \alpha$ (*i.e.* no customer is lost). Our aim is to find a policy which minimizes the average waiting time of all the customers. The problem consists in finding an optimal allocation pattern and to find the optimal ratio $(\alpha, 1 - \alpha)$. The optimal allocation pattern was found in Chapter 6 but no way to compute the optimal ratio is provided there.

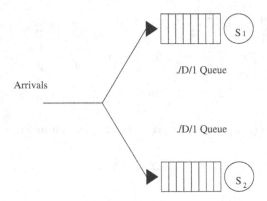

Fig. 7.4. Admission Control in two ./D/1 queues

The input sequence w is given under the form of a binary sequence w such that when $w(n) = 1$ the nth customer is sent in the first queue and when $w(n) = 0$ the nth customer is sent in the second queue. The average waiting time of all the customers for an input sequence w is denoted by $\mathcal{W}_{S_1,S_2}(w)$.

7.5.2 Optimal Policies

To find an optimal allocation pattern we will use results shown in the previous chapters.

Theorem 31. *The average waiting time $\mathcal{W}_{S_1,S_2}(w)$ is minimized for a bracket input sequence $\underline{m}_{\alpha_{opt}}$, with some slope α_{opt}.*

Proof. This is a mere application of the multi-criteria optimization found in Chapter 1 in the deterministic case.   □

Note that the approach proposed in Chapter 6 does not provide any means to compute α_{opt}, which is what we are going to do here for deterministic queues. Moreover, the routing policy described here is optimal among all routing policies, including J.S.Q. (Join the Shortest Queue) and J.S.L (Join the Shortest Load), since the distinction between closed loop and open loop policies is irrelevant for deterministic systems. Note however that it may not be the unique optimal policy.

Remark 12. It can be shown that the average waiting time in a deterministic queue under an arrival process of the form $\underline{m}_\alpha$ is the same as the average waiting time when the arrival process is of the form $\overline{m}_\alpha$. Therefore, Theorem 31 applies with bracket as well as upper bracket sequences. This remark is essential here since when the input sequence in the first queue is an upper bracket word of slope α ($\overline{m}_\alpha$), then the input sequence in the second queue is a bracket word of slope $1 - \alpha$ ($\underline{m}_{1-\alpha}$).

Using the considerations of Remark 12, and conditioning on the queue chosen for each customer the relation between $\mathcal{W}_{S_1,S_2}(\overline{m}_\alpha)$, $\mathcal{W}_{S_1}(\overline{m}_\alpha)$ and $\mathcal{W}_{S_2}(\overline{m}_{1-\alpha})$ is :

$$\mathcal{W}_{S_1,S_2}(\overline{m}_\alpha) = \alpha \cdot \mathcal{W}_{S_1}(\overline{m}_\alpha) + (1 - \alpha) \cdot \mathcal{W}_{S_2}(\overline{m}_{1-\alpha}). \qquad (7.18)$$

7.5.3 Optimal Ratio

We will now compute the optimal ratio α_{opt} associated with a bracket input sequence over all possible stable ratios.

Stability Condition Consider the system of the two ./D/1 queues above, the stability condition of such systems is $\rho \leq 1$, that is

$$\frac{1}{S_2} + \frac{1}{S_1} \geq 1. \qquad (7.19)$$

But the stability of the two queues individually is also necessary. Therefore by Equation 7.4 $\alpha \leq S_1^{-1}$ and $0 \leq 1 - \alpha \leq S_2^{-1}$. Hence for any given number α in the interval $I_s = \left[1 - \frac{1}{S_2}, \frac{1}{S_1}\right] \cap [0,1]$ the input sequence $\overline{m}_\alpha$ is stable. In the following the interval I_s is called the interval of stability.

Special Cases This part considers special degenerated cases where the theory developed in the previous section is not necessary.

Lemma 38. *In the case $2 < S_2 \leq S_1$, no real optimal policy exists.*

Proof. The system is never stable since $S_1^{-1} \leq S_2^{-1} < 1/2$, implying

$$\left(\frac{1}{S_1} + \frac{1}{S_2} \right) < 1 .$$

□

Lemma 39. *In the case* $S_2 \leq 1$, *an optimal policy consists in sending all the customers in the second queue. In this case* $\mathcal{W}_{S_1,S_2}(\overline{m}_0) = 0$.

Proof. When $S_2 \leq 1$, the service time is smaller than the inter-arrival time. Therefore each time a customer arrives the second queue is empty and we have $\mathcal{W}_{S_2}(1) = 0$. Since $\mathcal{W}_{S_1}(0) = 0$ then by Equation 7.18, $\mathcal{W}_{S_1,S_2}(\overline{m}_0) = 0$.

□

Lemma 40. *If* $1 < S_2 \leq S_1 < 2$ *then the round robin policy is an optimal policy.*

Proof. Lemma (35) implies $\mathcal{W}_{S_1}(\overline{m}_{1/2}) = 0$ and $\mathcal{W}_{S_2}(\overline{m}_{1/2}) = 0$. Therefore a possible optimal ratio is $\alpha = 1/2$ and $\overline{m}_{1/2} = \overline{10}$ is an optimal policy.   □

Case $1 < S_2 < 2 < S_1$ This can be considered as the general case.

Characterization of optimal ratios when $\rho < 1$ We are interested to find the optimal ratio α_{opt} defined by :

$$\alpha_{opt} = \arg\min_{\alpha \in I_s} \mathcal{W}_{S_1,S_2}(\overline{m}_\alpha).$$

Let r_i^1 and r_j^2 denote the jumps of $\mathcal{W}_{S_1}$ and $\mathcal{W}_{S_2}$ respectively.

Theorem 32. *For any real service time* S_1 *and* S_2 *the optimal ratio* α_{opt} *is a jump of* $\mathcal{W}_{S_1}$ *or* $\mathcal{W}_{S_2}$. *Hence, it is a rational number and an optimal routing policy is periodic.*

Proof. By Theorem 30, the function $\alpha \mathcal{W}_{S_1}(\overline{m}_\alpha)$ is linear for α in $]r_i^1, r_{i+1}^1[$ and the function $(1 - \alpha)\mathcal{W}_{S_2}(\overline{m}_{1-\alpha})$ is also linear for α in $]1 - r_j^2, 1 - r_{j-1}^2[$. Therefore by Equation 7.18 the function $\mathcal{W}_{S_1,S_2}$ is linear for α in $]r_k^*, r_{k+1}^*[$. Where $\{r_k^*\} = (\{r_i^1\} \cup \{r_j^2\}) \cap I_s$ is the set of jumps of $\mathcal{W}_{S_1,S_2}$.

Hence $\mathcal{W}_{S_1,S_2}$ reaches its minima in the set of all the jumps r_k^* or in S_1^{-1} or in S_2^{-1} when they are irrational. Theorem 42 says that the growing rates of the function $i \to \frac{K_i}{n_i}$ converges to infinity. This allows us to exclude the points S_1^{-1} and S_2^{-1}.

Therefore $\mathcal{W}_{S_1,S_2}$ is minimized by a jump and since all the jumps are rational the optimal ratio α_{opt} is rational. This means that the routing policy $\overline{m}_{\alpha_{opt}}$ is periodic.   □

Lemma 41. *Let* p *be the rational number with the smallest denominator in the interval* I_s. *Then* p *is the smallest jump of* $\mathcal{W}_{S_1}$ *in* I_s *and the largest jump of* $\mathcal{W}_{S_2}$ *in* I_s.

Since p is the unique common jump of $\mathcal{W}_{S_1}$ and $\mathcal{W}_{S_2}$, we call p the double jump.

Proof. Let r_i^1 be the largest jump of $\mathcal{W}_{S_1}(\overline{m}_\alpha)$ such that $r_i^1 < 1 - 1/S_2$. This implies $1 - S_2^{-1} \le r_{i+1}^1 \le 1/S_1$. By Lemma 33 r_{i+1}^1 is the rational number with the smallest denominator in $]r_i^1, 1/S_1]$, then r_{i+1}^1 is the rational number with the smallest denominator in I_s. Let r_j^2 be the largest jump of $\mathcal{W}_{S_2}(1 - \alpha)$ such that $1 - r_j^2 > 1/S_1$. This implies $1 - r_{j+1}^2$ is the largest jump of $\mathcal{W}_{S_2}(1 - \alpha)$ in I_s and the rational number with the smallest denominator in I_s. Then $r_i^1 = 1 - r_{j+1}^2$. $\qquad\square$

The double jump is not always the optimal ratio, as we will see in the example below.

Example 5 (Double jump is not optimal). Let $S_1 = 21/5$ and $S_2 = 6/5$. The upper expansion of $5/21$ is $\langle 4, 1, 1, 1, 2 \rangle$ and the one of $5/6$ is $\langle 1, 1, 1, 1, 2 \rangle$. We have $r_1^1 = 1/5$, $r_2^1 = 2/9$, $r_3^1 = 3/13$, $r_4^1 = 4/17$, $r_5^1 = 1/S_1 = 5/21$ and $1 - r_1^2 = 1/2$, $1 - r_2^2 = 1/3$, $1 - r_3^2 = 1/4$, $1 - r_4^2 = 1/5$, $1 - r_5^2 = 1 - 1/S_2 = 1/6$. The double jump is $1/5$.

The following numerical values have been obtained using exact computations provided by the program presented later (Section 7.5.4 and 7.6).

$$\mathcal{W}_{S_1}(\overline{m}_{1/5}) = 0, \ \mathcal{W}_{S_2}(\overline{m}_{4/5}) = \tfrac{3}{10}, \qquad \mathcal{W}_{S_1,S_2}(\overline{m}_{1/5}) = \tfrac{3}{10} \times \tfrac{4}{5} = \tfrac{6}{25}.$$

$$\mathcal{W}_{S_1}(\overline{m}_{2/9}) = \tfrac{1}{10}, \ \mathcal{W}_{S_2}(\overline{m}_{7/9}) = \tfrac{9}{35}, \ \mathcal{W}_{S_1,S_2}(\overline{m}_{2/9}) = \tfrac{1}{10} \times \tfrac{2}{9} + \tfrac{9}{35} \times \tfrac{7}{9} = \tfrac{2}{9}.$$

Since $\tfrac{2}{9} < \tfrac{6}{25}$, the optimal ratio is not the double jump. It can also be shown that the optimal ratio in this case is $\alpha_{opt} = \tfrac{2}{9}$.

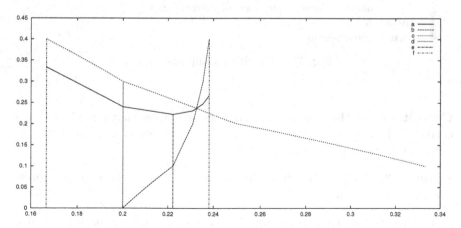

Fig. 7.5. Curves of $\mathcal{W}_{S_1}$, $\mathcal{W}_{S_2}$ and $\mathcal{W}_{S_1,S_2}$ when α varies from $1/6$ to $1/3$.

Figure 7.5 displays the functions $\mathcal{W}_{S_1}(\overline{m}_\alpha)$ (curve b), $\mathcal{W}_{S_2}(\overline{m}_{1-\alpha})$ (curve c) and $\mathcal{W}_{S_1,S_2}(\overline{m}_\alpha)$ (curve a), as well as some jumps of those three curves (the double jump $1/5$ is marked by the dotted line d, and the optimal jump $2/9$ by the dotted line e). The vertical lines f show the interval of stability of the system, $\mathcal{W}_{S_1,S_2}(\overline{m}_\alpha)$ is infinite outside this interval.

Characterization of optimal ratios when $\rho = 1$ In this part, we will assume that the system is fully loaded. When $\rho = 1$ then the stability interval is reduced to a single point. There is just one ratio α satisfying the stability condition, $\alpha = S_1^{-1}$. Therefore the upper bracket word with slope α is optimal. This is the only case where the optimal policy associated with bracket input sequence may be aperiodic.

7.5.4 Algorithm and computational issues

We present in Figure 7.6 an algorithm to compute the optimal ratio α_{opt}, when $\rho \leq 1$.

> Find double jump p
> current-jump $:= p$
> Compute the next-jump-right of p
> **while** $\mathcal{W}_{S_1,S_2}(\overline{m}_{\text{current-jump}}) > \mathcal{W}_{S_1,S_2}(\overline{m}_{\text{next-jump-right}})$ **do**
>   current-jump$:=$ next-jump-right
>   Compute the next-jump-right of current-jump
> **endwhile**
> Compute the next-jump-left of p
> **while** $\mathcal{W}_{S_1,S_2}(\overline{m}_{\text{current-jump}}) > \mathcal{W}_{S_1,S_2}(\overline{m}_{\text{next-jump-left}})$ **do**
>   current-jump$:=$ next-jump-left
>   compute the next-jump-left of current-jump
> **endwhile**
> **return** current-jump

Fig. 7.6. Algorithm computing α_{opt}

Correctness of the algorithm We will now show that the algorithm is correct and converges to the result in a finite number of steps. For that we have to prove two preliminary lemmas.

Lemma 42. *The growth rate of the function $\frac{n_i}{q_i} \mapsto \frac{K_i}{n_i}$ converges to infinity when i goes to infinity.*

Proof. Let $S^{-1} = \langle l_1, \ldots, l_i + \alpha_i \rangle$, the recursive computations of $q_i = |\overline{m}_{r_i}|$ are given as for n_i, by

$$q_0 = 1,$$
$$q_1 = l_1 + 1, \text{ if } \alpha_1 \neq 0, \qquad q_1 = l_1, \text{ if } \alpha_1 = 0,$$
$$\forall i \geq 2 \quad q_i = (l_i + 1)q_{i-1} - q_{i-2}, \text{ if } \alpha_i \neq 0 \tag{7.20}$$
$$q_i = (l_i)q_{i-1} - q_{i-2} \text{ otherwise }.$$

We now show by induction that $\forall i \geq 1$

$$n_{i+1}q_i - n_i q_{i+1} = 1. \tag{7.21}$$

At step one : $n_2 q_1 - n_1 q_2 = (l_2 + 1)(l_1 + 1) - (l_1(l_2 + 1) + l_2) = 1$.
At step $i + 1$: $n_{i+1}q_i - n_i q_{i+1} = (l_{i+1} + 1)q_i n_i - q_i n_{i-1} - ((l_{i+1} + 1)q_i n_i - n_i q_{i-1}) = n_i q_{i-1} - n_{i-1}q_i$.

Hence the growing rate

$$\left(\frac{K_{i+1}}{n_{i+1}} - \frac{K_i}{n_i} \right) \left(\frac{n_{i+1}}{q_{i+1}} - \frac{n_i}{q_i} \right)^{-1} = \frac{(n_i K_{i+1} - n_{i+1}K_i)(q_{i+1}q_i)}{n_{i+1}n_i}$$
$$\geq (n_i K_{i+1} - n_{i+1}K_i)$$
$$\geq (n_{i-1}K_i - n_i K_{i-1}) + n_i(n_i - n_{i-1})d_i^1$$
$$\geq \Sigma_{1 \leq k \leq i} \left(n_k(n_k - n_{k-1})d_k^1 \right).$$

Let us prove now that $\Sigma_{1 \leq k} \left(n_k(n_k - n_{k-1})d_k^1 \right) = +\infty$. We use Equations 7.10 that yields

$$n_k(n_k - n_{k-1})d_k^1 = n_k(n_k - n_{k-1})[(1 - \alpha_1)(1 - \alpha_2) \ldots (1 - \alpha_{k-1})]\alpha_k.$$

By Equation 7.14 we obtain

$$n_i(n_i - n_{i-1}) = [(l_i + 1)n_{i-1} + (n_{i-1} - n_{i-2})][l_i n_{i-1} + (n_{i-1} - n_{i-2})]$$
$$\geq [l_i + 1 + l_i + l_i(l_i + 1)]n_{i-1}(n_{i-1} - n_{i-2}).$$

Hence

$$n_k(n_k - n_{k-1})d_k^1 \geq \frac{((l_k + 1)^2 + l_k)((l_{k-1} + 1)^2 + l_{k-1}) \ldots ((l_2 + 1)^2 + l_2)\alpha_k}{l_2 l_3 \ldots l_{k-1}}$$
$$\geq ((l_k + 3))((l_{k-1} + 3)) \ldots ((l_2 + 3))\alpha_k \geq 4^{k-1}\alpha_k.$$

Two cases may occur, either the sequence $\{\alpha_k\}_{k \geq 0}$ does not converge to zero or it converges to zero. In the first case the sequence $n_k(n_k - n_{k-1})d_k^1$ does not converges to zero and the series $\Sigma_{k \geq 0}n_k(n_k - n_{k-1})d_k^1$ goes to infinity. In the other case there exists a number N such that $\forall k \geq N, \alpha_k < 1/2$. Since $\alpha_k < 1/2$ implies $l_{k+1} = 1$ and $(1 - \alpha_k)\alpha_{k+1} = \alpha_k$ then

$$\frac{n_{k+1}(n_{k+1} - n_k)d_{k+1}^1}{n_k(n_k - n_{k-1})d_k^1} = \frac{n_{k+1}(n_{k+1} - n_k)}{n_k(n_k - n_{k-1})} > 1,$$

which implies that the sequence is strictly increasing and that the series diverges to infinity. $\qquad \square$

Lemma 43. *The function* $i \to \frac{K_i}{q_i}$ *is convex.*

Proof. We have by Equations 7.16 and 7.21

$$\left(\frac{K_{i+1}}{q_{i+1}} - \frac{K_i}{q_i}\right)\left(\frac{n_{i+1}}{q_{i+1}} - \frac{n_i}{q_i}\right)^{-1} - \left(\frac{K_i}{q_i} - \frac{K_{i-1}}{q_{i-1}}\right)\left(\frac{n_i}{q_i} - \frac{n_{i-1}}{q_{i-1}}\right)^{-1} =$$

$$\left(\frac{K_{i+1}}{q_{i+1}} - \frac{K_i}{q_i}\right)(q_{i+1}q_i) - \left(\frac{K_i}{q_i} - \frac{K_{i-1}}{q_{i-1}}\right)(q_i q_{i-1}) =$$

$$q_i K_{i+1} - q_{i+1} K_i - q_{i-1} K_i + q_i K_{i-1}.$$

Using Equation 7.16 leads to

$$q_i K_{i+1} - q_{i+1} K_i = q_i \big((l_{i+1} + 1)K_i - K_{i-1}\big) - q_{i+1} K_i$$

$$+ q_i\left(n_i(l_{i+1} - 1)(d_i^1 + \frac{l_{i+1}}{2}d_i^2) + (n_i - n_{i-1})d_i^1\right)$$

$$= q_i\big((l_{i+1} + 1)K_i - K_{i-1}\big) - (l_{i+1}q_i - q_{i-1})K_i$$

$$+ q_i\left(n_i(l_{i+1} - 1)(d_i^1 + \frac{l_{i+1}}{2}d_i^2) + (n_i - n_{i-1})d_i^1\right)$$

$$= q_{i-1}K_i - q_i K_{i-1} + q_i\left(n_i(l_{i+1} - 1)(d_i^1 + \frac{l_{i+1}}{2}d_i^2) + (n_i - n_{i-1})d_i^1\right).$$

Hence

$$q_i K_{i+1} - q_{i+1} K_i - \big(q_{i-1} K_i - q_i K_{i-1}\big)$$

$$= q_i\left(n_i(l_{i+1} - 1)(d_i^1 + \frac{l_{i+1}}{2}d_i^2) + (n_i - n_{i-1})d_i^1\right).$$

As shown in proof of Lemma 37,

$$n_i(l_{i+1} - 1)(d_i^1 + \frac{l_{i+1}}{2}d_i^2) + (n_i - n_{i-1})d_i^1 \geq 0. \qquad (7.22)$$

This proves the convexity of the function. $\qquad\square$

Lemma 44. *The algorithm is correct and converges in a finite number of steps*

Proof. Correctness. Since by Lemma 43 the function $i \to \frac{K_i}{q_i}$ is convex, then by Equation 7.18 the function $k \to W_{S_1,S_2}(\overline{m}_{r_k^*})$ is also convex. Considering that $\alpha_{opt} \in \{r_k^*\}$ shows the correctness of the algorithm.

Finiteness. Let k, i, j be the integers such that $p = r_k = r_i^1 = 1 - r_j^2$. Let us note that $1 - r_j^2 \leq S_1 < 1 - r_{j-1}^2$. Since $\forall n \geq 1$, $r_{k+n}^* = r_{i+n}^1$, and since by Lemma 42 the growing rate of the function $i \to W_{S_1}(\overline{m}_{r_i^1})$ goes to infinity then by Equation 7.18, $\forall n \geq 1$ the growing rate of the function $k+n \to W_{S_1,S_2}(\overline{m}_{r_{k+n}^*})$ converges to infinity. Therefore the integer $k_0 = k+n_0$ such that $\forall n \geq n_0$

$$\mathcal{W}_{S_1,S_2}(\overline{m}_{r^*_{k+n}}) - \mathcal{W}_{S_1,S_2}(\overline{m}_{r^*_{k+n-1}}) > 0,$$

is finite. Using similar arguments for the jumps smaller than p shows that the algorithm converges in a finite number of steps. □

Concerning this algorithm, which was run hundreds of time with irrational or rational parameters, the maximal number of steps never went over twenty. Nevertheless one can get a large number of steps with well chosen parameters for which the optimal jump is arbitrarily far (in number of jumps) from the double jump.

7.6 Numerical experiments

The algorithm presented above has been implemented in Maple in order to keep exact values for all the rational numbers involved in the computations. This section is dedicated to the presentation of several runs of the program in order to shown how the optimal policy (or equivalently the ratio of the optimal policy) behaves with respect to the parameters of the system, namely S_1 the service time in queue 1, S_2 the service time in queue 2 as well as the inter-arrival time (fixed to one previously, but which can be modified by scaling the time units).

In the first series of computations, we fix $S_1 = 22/5$ and $S_2 = 6/5$. Therefore, the fastest server, S_2 is c times faster than S_1, with $c = 11/3$. One could expect that the optimal routing policy sends c times more customers in the second queue than in the first queue. This policy has a ratio $\alpha = \frac{1}{c+1}$, namely $\alpha = 3/14$. However, as the experiments in Figure 7.7 show, the optimal ratio is $\alpha_{opt} = 1/5$ when the inter-arrival time is one (when $\rho = 66/70 \approx 0.942$).

In Figure 7.7, with $S_1 = 22/5$ and $S_2 = 6/5$, we let the inter-arrival time vary so that the total load ρ goes from 0 to 1. All the results presented in the figure are exact computations and do not suffer from any numerical errors. When ρ is smaller than $11/14$ (dotted line on the left), many ratios are optimal. For instance, for all $\alpha < 1/4$, $\mathcal{W}_{S_1,S_2}(\overline{m}_\alpha) = 0$. This part is not shown in the figure. The main interest lies within the bounds $11/14 \leq \rho \leq 1$ where the optimal ratio is unique. The optimal ratio α_{opt} takes several rational values, ranging from $1/4$ to $1/5$ and ending with the intuitive ratio of the service times, $3/14$. When $\rho = 1$, then $\alpha = 3/14$ is the only point in the stability region. However, for lower values of the load, it is somewhat surprising that the optimal ratio changes with the load and deviates from the ratio of the service times, $3/14$. The changes of the optimal ratio occur according to several reasons. For example, α_{opt} moves away from $1/4$ only when $1/4$ gets out of the stability region (this occurs at $\rho = 6/7$). The same phenomenon occurs when α_{opt} jumps away from $1/5$, which occurs when $\rho = 55/56$. On the other hand, the change from $2/9$ to $1/5$ occurs while $2/9$

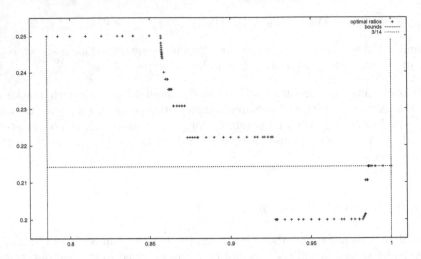

Fig. 7.7. Optimal ratio when the total load varies from $11/14$ to 1

is still a jump in the stability region. In general, the sudden changes of the optimal ratio when the load increases remain mysterious and deserve further studies.

The second set of computations is presented in Figure 7.8. The inter-arrival is fixed to one and we let the inverses of the service times $1/S_1$ and $1/S_2$ vary. We restrict our investigations to the domain of stability that is $\frac{1}{S_1} + \frac{1}{S_2} \geq 1$. The figure displays the points where the value of α_{opt} changes. Therefore, each cell corresponds to couples $(1/S_1, 1/S_2)$ with the same optimal ratio. The larger cell ($\alpha_{opt} = 1/2$) corresponds to the region where the round robin policy is optimal. Using time homogeneity, the optimal ratios along the line $\frac{1}{S_2} = \frac{11}{3}\frac{1}{S_1}$ (bold line in Figure 7.8) are those given in Figure 7.7. We can note that along the line $\frac{1}{S_1} + \frac{1}{S_2} = 1$, the changes of α_{opt} exhibit a fractal behavior which also deserves further investigation.

Figure 7.9 is the same as Figure 7.8 where the cost to be optimized is not the average waiting time of the customers but the average sojourn time of the customers, $S_{S_1,S_2}(\overline{m}_\alpha) = \mathcal{W}_{S_1,S_2}(\overline{m}_\alpha) + \alpha S_1 + (1 - \alpha)S_2$. The optimal fraction of customers sent to queue 1 changes. Roughly speaking, the fastest queue receives more customers when the sojourn time is used as the cost to be minimized instead of the waiting time.

7.7 Appendix: proof of Theorem 29

We will first assume that the number α is a rational number. Thus we work with only finite words.

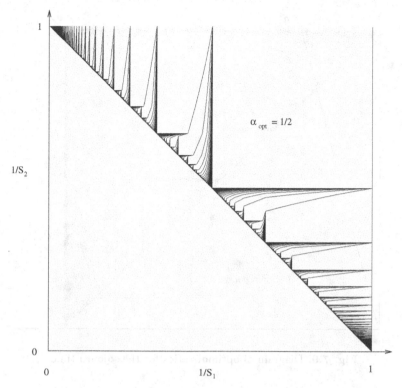

Fig. 7.8. Optimal ratio when $1/S_2$ and $1/S_1$ vary

By Theorem 28, $\overline{m}_\alpha$ can be decomposed only using $x_{i+1}(\alpha)$ and $y_{i+1}(\alpha)$. Since by Equations 7.3

$$\frac{l_{i+1}}{l_{i+1}+1} \leq \alpha_i \leq 1,$$

then we obtain the composition of $x_{i+1}(\alpha)$ and $y_{i+1}(\alpha)$ with Lemma 32:

$$x_{i+1}(\alpha) = x_i(r_{i+1})(y_i(r_{i+1}))^k \text{ and } y_{i+1}(\alpha) = x_i(r_{i+1})(y_i(r_{i+1}))^{k+1}$$

with $k \geq (l_{i+1}+1) - 1$. Let us rewrite these two equalities into :

$$x_{i+1}(\alpha) = x_i(r_{i+1})(y_i(r_{i+1}))^{l_{i+1}}(y_i(r_{i+1}))^{k'},$$

$$y_{i+1}(\alpha) = x_i(r_{i+1})(y_i(r_{i+1}))^{l_{i+1}}(y_i(r_{i+1}))^{k'+1},$$

with $k' \geq 0$. We have $x_{i+1}(\alpha) = \overline{m}_{r_{i+1}}(\overline{m}_{r_i})^{k'}$ and $y_{i+1}(\alpha) = \overline{m}_{r_{i+1}}(\overline{m}_{r_i})^{k'+1}$.

Since by Lemma 34 and Remark 10 the workload after $\overline{m}_{r_i}$ is always null, then the workload after an admission sequence $x_{i+1}(\alpha)$ or $y_{i+1}(\alpha)$ is also null. Therefore the workload after one period of $\overline{m}_\alpha$ is null and we simply have to focus on one period of $\overline{m}_\alpha$. Computing $K^0(x_{i+1}(\alpha))$ and $K^0(y_{i+1}(\alpha))$ gives

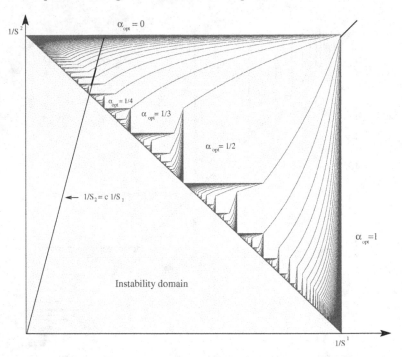

Fig. 7.9. Diagram of optimal ratios for the sojourn time

$$K^0(x_{i+1}(\alpha)) = K_{i+1} + k'K_i \text{ and } K^0(y_{i+1}(\alpha)) = K_{i+1} + (k'+1)K_i.$$

We write $\alpha_i = n/q$, α_i being rational. The number of $\overline{m}_{r_{i+1}}$ in $\overline{m}_\alpha$ is the number of $x_i(S^{-1}) = x_i(r_{i+1})$ in $\overline{m}_\alpha$. By Lemma 32 the total number of $x_i(S^{-1})$ in $\overline{m}_\alpha$ is equal to the total number of ones in $\Phi^{(i)}(\overline{m}_\alpha)$. Since $1 - n/q$ is the slope of $\Phi^{(i)}(\overline{m}_\alpha)$ and since $\overline{m}_\alpha$ and $\Phi^{(i)}(\overline{m}_\alpha)$ are finite words. Then the total number of $x_i(S^{-1})$ in $\overline{m}_\alpha$ is $(1 - \frac{n}{q})q$. Using a similar argument the total number of $\overline{m}_{r_i}$ in $\overline{m}_\alpha$ is $(\frac{n}{q})q$. The number of $\overline{m}_{r_i}$ which do not belong to $\overline{m}_{r_{i+1}}$ is equal to $q(\frac{n}{q}) - l_{i+1}(1 - \frac{n}{q})q$. Therefore

$$K(\overline{m}_\alpha) = q(1 - \frac{n}{q})K_{i+1} + q[\frac{n}{q} - l_{i+1}(1 - \frac{n}{q})]K_i.$$

Compute now n_α the total number of ones in $\overline{m}_\alpha$. This number is the number of $\overline{m}_{r_{i+1}}$ multiplied by the number of ones in $\overline{m}_{r_{i+1}}$ added with the number of $\overline{m}_{r_i}$ which do not belong to $\overline{m}_{r_{i+1}}$ multiplied by the number of ones in $\overline{m}_{r_i}$:

$$n_{\alpha_i} = q(1 - \frac{n}{q})n_{i+1} + q[\frac{n}{q} - l_{i+1}(1 - \frac{n}{q})]n_i.$$

We obtain

$$\mathcal{W}_\alpha(S) = \frac{(1 - \frac{n}{q})K_{i+1} + [\frac{n}{q} - (l_{i+1}(1 - \frac{n}{q}))]K_i}{(1 - \frac{n}{q})n_{i+1} + [\frac{n}{q} - (l_{i+1}(1 - \frac{n}{q}))]n_i}.$$

Suppose now that α is an irrational number, the bracket word is now an infinite word. However we work with finite prefixes which properties are kept when we compute the limit.

Let n be an integer. Let $(\overline{m}_\alpha)_n$ be the prefix of length n of $\overline{m}_\alpha$. Let $(\overline{m}_\alpha)_{s_n}$ be the greatest prefix of $\overline{m}_\alpha$ uniquely composed by factors $x_{i+1}(\alpha)$ and $y_{i+1}(\alpha)$ which length is smaller than n, and let $(\overline{m}_\alpha)_{g_n}$ be the smallest prefix uniquely composed by factors $x_{i+1}(\alpha)$ and $y_{i+1}(\alpha)$ which length is larger than n. We obtain

$$\frac{K^0((\overline{m}_\alpha)_{s_n})}{|(\overline{m}_\alpha)_{g_n}|_1} \le \frac{K^0((\overline{m}_\alpha)_n)}{|(\overline{m}_\alpha)_n|_1} \le \frac{K^0((\overline{m}_\alpha)_{g_n})}{|(\overline{m}_\alpha)_{s_n}|_1}.$$

The prefix $(\overline{m}_\alpha)_{g_n}$ is composed by $(\overline{m}_\alpha)_{s_n}$ followed by either $x_{i+1}(\alpha)$ or $y_{i+1}(\alpha)$. As shown above, the workload after any sequence $x_{i+1}(\alpha)$ and $y_{i+1}(\alpha)$ is null. Hence the workload after $(\overline{m}_\alpha)_{g_n}$ and after $(\overline{m}_\alpha)_{s_n}$ is also null. Since the factor $y_{i+1}(\alpha)$ is longer than the factor $x_{i+1}(\alpha)$, then $|x_{i+1}(\alpha)|_1 \le |y_{i+1}(\alpha)|_1$ and $K^0(x_{i+1}(\alpha)) \le K^0(y_{i+1}(\alpha))$. That yields $K^0((\overline{m}_\alpha)_{g_n}) \le K^0((\overline{m}_\alpha)_{s_n}) + K^0(y_{i+1}(\alpha))$ and also $|(\overline{m}_\alpha)_{g_n}|_1 \le |(\overline{m}_\alpha)_{s_n})|_1 + |y_{i+1}(\alpha)|_1$. Thus

$$\frac{K^0((\overline{m}_\alpha)_{s_n})|(\overline{m}_\alpha)_{s_n}|_1}{|(\overline{m}_\alpha)_{s_n}|_1(|(\overline{m}_\alpha)_{s_n}|_1 + |y_{i+1}(\alpha)|_1)} \le \frac{K^0((\overline{m}_\alpha)_n)}{|(\overline{m}_\alpha)_n|_1} \le \frac{K^0((\overline{m}_\alpha)_{s_n})}{|(\overline{m}_\alpha)_{s_n}|_1} + \frac{K^0(y_{i+1}(\alpha))}{|(\overline{m}_\alpha)_{s_n}|_1}.$$

$$(7.23)$$

Let us focus on $K^0((\overline{m}_\alpha)_{s_n})$. For all $n \ge n_{i+1}$, this sum is strictly positive and only depends on the number of $y_{i+1}(r_{i+1})$ and "remaining" $y_i(r_{i+1})$. The number of $y_{i+1}(r_{i+1})$ and "remaining" $y_i(r_{i+1})$ is given by the slope of $\Phi^{(i)}((\overline{m}_\alpha)_{s_n}))$. Hence

$$K^0((\overline{m}_\alpha)_{s_n}) = \left(|\Phi^{(i)}((\overline{m}_\alpha)_{s_n})| \cdot \alpha(\Phi^{(i)}((\overline{m}_\alpha)_{s_n})) \right) K_{i+1} +$$

$$\left(|\Phi^{(i)}((\overline{m}_\alpha)_{s_n}))| \right) \left[\left(1 - \alpha(\Phi^{(i)}((\overline{m}_\alpha)_{s_n})) \right) - l_{i+1} \cdot \alpha(\Phi^{(i)}((\overline{m}_\alpha)_{s_n})) \right] K_i.$$

and

$$|(\overline{m}_\alpha)_{s_n}|_1 = \left(|\Phi^{(i)}((\overline{m}_\alpha)_{s_n})| \cdot \alpha(\Phi^{(i)}((\overline{m}_\alpha)_{s_n})) \right) n_{i+1} +$$

$$\left(|\Phi^{(i)}((\overline{m}_\alpha)_{s_n}))| \right) \left[\left(1 - \alpha(\Phi^{(i)}((\overline{m}_\alpha)_{s_n})) \right) - l_{i+1} \cdot \alpha(\Phi^{(i)}((\overline{m}_\alpha)_{s_n})) \right] n_i.$$

Finally we have

$$\frac{K^0((\overline{m}_\alpha)_{s_n})}{|(\overline{m}_\alpha)_{s_n}|_1} =$$

$$\frac{\alpha(\Phi^{(i)}((\overline{m}_\alpha)_{s_n})) K_{i+1} + [(1 - \alpha(\Phi^{(i)}((\overline{m}_\alpha)_{s_n}))) - l_{i+1} \cdot \alpha(\Phi^{(i)}((\overline{m}_\alpha)_{s_n}))] K_i}{\alpha(\Phi^{(i)}((\overline{m}_\alpha)_{s_n})) n_{i+1} + [(1 - \alpha(\Phi^{(i)}((\overline{m}_\alpha)_{s_n}))) - l_{i+1} \cdot \alpha(\Phi^{(i)}((\overline{m}_\alpha)_{s_n}))] n_i}.$$

Note that the number $|(\overline{m}_\alpha)_{s_n}|$ depending on n, when $n \to \infty$, $|(\overline{m}_\alpha)_{s_n}| \to \infty$ and also $|(\overline{m}_\alpha)_{s_n}|_1 \to \infty$. Lemma 32 give

$$\lim_{n \to \infty} \alpha(\Phi^{(i)}((\overline{m}_\alpha)_{s_n})) = \lim_{n \to \infty} \alpha(\Phi^{(i)}((\overline{m}_\alpha)_n)) = 1 - \alpha_i .$$

This implies

$$\lim_{n \to \infty} \frac{K^0((\overline{m}_\alpha)_{s_n})}{|(\overline{m}_\alpha)_{s_n}|_1} = \frac{(1 - \alpha_i)K_{i+1} + [\alpha_i - l_{i+1}(1 - \alpha_i)] K_i}{(1 - \alpha_i)n_{i+1} + [\alpha_i - l_{i+1}(1 - \alpha_i)] n_i} . \tag{7.24}$$

Since $|y_{i+1}(r_{i+1})|$ is finite then $|y_{i+1}(r_{i+1})|_1$ and $K(y_{i+1}(r_{i+1}))$ are bounded. It comes

$$\lim_{n \to \infty} \frac{K^0((\overline{m}_\alpha)_{s_n})}{|(\overline{m}_\alpha)_{s_n}|_1} \cdot \frac{|(\overline{m}_\alpha)_{s_n}|_1}{|(\overline{m}_\alpha)_{s_n}|_1 + |y_{i+1}(\alpha)|_1} = \lim_{n \to \infty} \frac{K^0((\overline{m}_\alpha)_{s_n})}{|(\overline{m}_\alpha)_{s_n}|_1},$$

$$\lim_{n \to \infty} \frac{K^0((\overline{m}_\alpha)_{s_n})}{|(\overline{m}_\alpha)_{s_n}|_1} + \frac{K^0(y_{i+1}(\alpha))}{|(\overline{m}_\alpha)_{s_n}|_1} = \lim_{n \to \infty} \frac{K^0((\overline{m}_\alpha)_{s_n})}{|(\overline{m}_\alpha)_{s_n}|_1} .$$

Using Equation 7.23 and Equation 7.24 we obtain

$$\frac{(1 - \alpha_i)K_{i+1} + [\alpha_i - l_{i+1}(1 - \alpha_i)] K_i}{(1 - \alpha_i)n_{i+1} + [\alpha_i - l_{i+1}(1 - \alpha_i)] n_i} \leq \lim_{n \to \infty} \frac{K^0((\overline{m}_\alpha)_n)}{|(\overline{m}_\alpha)_n|_1}$$

$$\leq \frac{(1 - \alpha_i)K_{i+1} + [\alpha_i - l_{i+1}(1 - \alpha_i)] K_i}{(1 - \alpha_i)n_{i+1} + [\alpha_i - l_{i+1}(1 - \alpha_i)] n_i} .$$

This finishes the proof of the formulas.

When $\alpha \leq r_1$, by Lemma 29 the minimum number of consecutive 0 in $\overline{m}_\alpha$ is larger than l_1. Therefore the inter-arrival times of $\overline{m}_\alpha$ are larger than $l_1 + 1$, hence none of the customers have to wait and $\mathcal{W}_S(\overline{m}_\alpha) = 0$.

7.8 Appendix: proof of Lemma 37

We have

$$\frac{K_{i+1}}{n_{i+1}} - \frac{K_i}{n_i} = \frac{n_i \times K_{i+1} - n_{i+1} \times K_i}{n_i \times n_{i+1}} .$$

Since $\forall i \geq 1$, $n_i \times n_{i+1} > 0$ we have to focus on the sign of $n_i K_{i+1} - n_{i+1} K_i$. Using Equations 7.14 and 7.16 yields

$$n_i K_{i+1} - n_{i+1} K_i$$

$$= n_i \left[(l_{i+1} + 1)K_i + (n_i l_{i+1} - n_{i-1}) d_i^1 + n_i(l_{i+1} - 1)\frac{l_{i+1}}{2} d_{i+1}^2 - K_{i-1} \right] - n_{i+1} K_i$$

$$= n_i \left[(l_{i+1} + 1)K_i - K_{i-1} + (n_i l_{i+1} - n_{i-1}) d_i^1 + n_i(l_{i+1} - 1)\frac{l_{i+1}}{2} d_i^2 \right]$$

$$= +(n_{i-1})K_i - n_i(l_{i+1} + 1)K_i$$

$$(n_{i-1} K_i - n_i K_{i-1}) + n_i \left[(n_i l_{i+1} - n_{i-1}) d_i^1 + n_i(l_{i+1} - 1)\frac{l_{i+1}}{2} d_i^2 \right],$$

reordering we obtain

$$n_i K_{i+1} - n_{i+1} K_i$$

$$= (n_{i-1} K_i - n_i K_{i-1}) + n_i^2 (l_{i+1} - 1)(d_i^1 + \frac{l_{i+1}}{2} d_i^2)$$

$$+ n_i(n_i - n_{i-1}) d_i^1 . n_i K_{i+1} - n_{i+1} K_i \qquad (7.25)$$

$$\qquad (7.26)$$

On the first hand, by Lemma 34, we know that d_i^1 is strictly positive and d_{i+1}^1 is non negative and that d_i^2 and d_{i+1}^2 are non positive. On the other hand Equation 7.11 gives us $d_i^1 + (l_{i+1} - 1)d_i^2 = d_{i+1}^1$ and $d_i^1 + l_{i+1} d_i^2 = d_{i+1}^2$. This implies $\forall l_{i+1} \geq 2$,

$$d_i^1 + \frac{l_{i+1}}{2} d_i^2 > 0 .$$

That means $\forall l_{i+1} \geq 1$,

$$(l_{i+1} - 1)(d_i^1 + \frac{l_{i+1}}{2} d_{i+1}^2) \geq 0.$$

Since $K_2 - n_2 K_1 = K_2$, since $K_2 = l_2 \left(d_1^1 + \frac{(l_2-1)}{2} d_1^2 \right)$ and since $d_1^1 + \frac{(l_2-1)}{2} d_1^2 > d_2^1 \geq 0$, then $K_2 > 0$, and by induction we have $\forall i \geq 0$:

$$n_i K_{i+1} - n_{i+1} K_i > 0 . \qquad (7.27)$$

Therefore $\forall i \geq 0$,

$$\frac{K_{i+1}}{n_{i+1}} - \frac{K_i}{n_i} > 0.$$

When $i + 1$ is the last step of the expansion the proof holds replacing l_{i+1} by $l_{i+1} - 1$.

7.9 Appendix 7.9 : proof of Theorem 30

i) Monotonicity. First we will show by induction on i that the derivative function $\frac{\partial \alpha_i}{\partial \alpha}$ is non positive.
Step 1. By Equations 7.3 we get : $\alpha_1 = \alpha^{-1} - l_1$, hence

$$\frac{\partial \alpha_1}{\partial \alpha} < 0 .$$

Step i. By Equations (7.3) we get $\alpha_i = (1 - \alpha_{i-1})^{-1} - l_i$, therefore

$$\frac{\partial \alpha_i}{\partial \alpha_{i-1}} > 0 , \qquad (7.28)$$

and using the induction hypothesis

$$\frac{\partial \alpha_i}{\partial \alpha} = \frac{\partial \alpha_i}{\partial \alpha_{i-1}} \cdot \frac{\partial \alpha_{i-1}}{\partial \alpha} < 0. \tag{7.29}$$

Second, using Equations 7.17 and 7.27, we have

$$\frac{\partial \mathcal{W}_S(\overline{m}_\alpha)}{\partial \alpha_i} = \frac{-(n_i K_{i+1} - n_{i+1} K_i)}{[(1-\alpha_i) n_{i+1} + (\alpha_i - l_{i+1}(1-\alpha_i)) n_i]^2} < 0. \tag{7.30}$$

Since

$$\frac{\partial \mathcal{W}_S(\overline{m}_\alpha)}{\partial \alpha} = \frac{\partial \mathcal{W}_S(\overline{m}_\alpha)}{\partial \alpha_i} \cdot \frac{\partial \alpha_i}{\partial \alpha},$$

then

$$\frac{\partial \mathcal{W}_S(\overline{m}_\alpha)}{\partial \alpha} > 0.$$

ii) Concavity. Using formulas 7.17 and 7.30 it comes

$$\frac{\partial^2 \mathcal{W}_S(\overline{m}_\alpha)}{\partial \alpha_i^2} = \frac{2(n_i K_{i+1} - n_{i+1} K_i)(-n_{i+1} + n_i(1 - l_{i+1}))}{[(1-\alpha_i) n_{i+1} + (\alpha_i - l_{i+1}(1-\alpha_i)) n_i]^3}.$$

Hence we get

$$\frac{\partial^2 \mathcal{W}_S(\overline{m}_\alpha)}{\partial \alpha^2} = \frac{\partial^2 \alpha_i}{\partial \alpha^2} \cdot \frac{\partial \mathcal{W}_S(\overline{m}_\alpha)}{\partial \alpha_i} + \frac{\partial^2 \mathcal{W}_S(\overline{m}_\alpha)}{\partial \alpha_i^2} \cdot \left(\frac{\partial \alpha_i}{\partial \alpha} \right)^2,$$

is equivalent to

$$\frac{\partial^2 \mathcal{W}_S(\overline{m}_\alpha)}{\partial \alpha^2} = -\frac{\partial \mathcal{W}_S(\overline{m}_\alpha)}{\partial \alpha_i} \left[\frac{2n_{i-1} \left(\frac{\partial \alpha_i}{\partial \alpha} \right)^2}{(1-\alpha_i)(n_i - n_{i-1}) + \alpha_i n_i} - \frac{\partial^2 \alpha_i}{\partial \alpha^2} \right]. \tag{7.31}$$

Introducing $\partial \alpha_{i-1}$ and using Formulas 7.3 and 7.14 yield after straightforward computations

$$\frac{2n_{i-1}}{(1-\alpha_i)(n_i - n_{i-1}) + \alpha_i n_i} \cdot \left(\frac{\partial \alpha_i}{\partial \alpha} \right)^2 - \frac{\partial^2 \alpha_i}{\partial \alpha^2} = $$
$$\frac{1}{(1-\alpha_{i-1})^2} \left[\frac{2n_{i-2}(n_{i-1} - n_{i-2})}{(1-\alpha_{i-1}) + \alpha_{i-1} n_{i-1}} \left(\frac{\partial \alpha_{i-1}}{\partial \alpha} \right)^2 - \frac{\partial^2 \alpha_{i-1}}{\partial \alpha^2} \right]. \tag{7.32}$$

At step one we have

$$\frac{2n_0}{(1-\alpha_1)(n_1 - n_0) + \alpha_1 n_1} \left(\frac{\partial \alpha_1}{\partial \alpha} \right)^2 - \frac{\partial^2 \alpha_1}{\partial \alpha^2} = -\frac{2}{(1-\alpha)^3} < 0.$$

Since for all $i \geq 0$, $(1 - \alpha_i) > 0$, it follows from 7.30 and 7.31

$$\frac{\partial^2 \mathcal{W}_S(\overline{m}_\alpha)}{\partial \alpha^2} < 0.$$

iii) Linearity. The function $\alpha \mathcal{W}_S(\overline{m}_\alpha)$ can be computed by replacing n_i by q_i in Formula 7.17. Since the recursive computations of the q_i's are identical for all $i \geq 2$ to the recursive computations of the n_i's, then the Formula 7.32 still holds

$$
\frac{2q_{i-1}}{(1-\alpha_i)(q_i - q_{i-1}) + \alpha_i q_i} \cdot \left(\frac{\partial \alpha_i}{\partial \alpha}\right)^2 - \frac{\partial^2 \alpha_i}{\partial \alpha^2} =
$$
$$
\frac{1}{(1-\alpha_{i-1})^2} \left[\frac{2q_{i-2}\left(\frac{\partial \alpha_{i-1}}{\partial \alpha}\right)^2}{(1-\alpha_{i-1})(q_{i-1} - q_{i-2}) + \alpha_{i-1}q_{i-1}} - \frac{\partial^2 \alpha_{i-1}}{\partial \alpha^2} \right]. \quad (7.33)
$$

From

$$
\frac{2q_0}{(1-\alpha_1)(q_1 - q_0) + \alpha_1 q_1} \left(\frac{\partial \alpha_1}{\partial \alpha}\right)^2 - \frac{\partial^2 \alpha_1}{\partial \alpha^2} = 0,
$$

it comes

$$
\frac{\partial^2 \alpha \mathcal{W}_S(\overline{m}_\alpha)}{\partial \alpha^2} = 0.
$$

Finally, note that the linearity of the function $\alpha \mathcal{W}_S(\overline{m}_\alpha)$ can also be proved directly using Little's law and Farey's intervals as done in [73].

Part III

Several extensions

This part is dedicated to several extensions of the theory developped in the previous parts.

The first extension concerns a system which is not (max,plus) linear but for which the waiting time is still multimodular. This shows that multimodularity is not limited to the class of systems with a (max,plus) linear dynamic and can also be used in systems with losses or dynamic routing.

The second extension shows a model where the control is on the service process and not on the arrival process. This is often called a polling system. By using the duality between the polling system and the routing system, we show that a polling policy based on sturmian sequences is optimal.

The last extension shows that multimodularity is also useful to derive qualitative properties of closed-loop control and not only for open-loop control. In particular, we show monotonicity of the optimal decisions with respect to the state of the system.

8 Networks with no buffers

8.1 Introduction

We begin by considering in this Chapter two types of problems: an admission into a single server with no buffer, and then the extension of this setting to a routing problem. We consider exponential service times and general stationary arrival processes, which include, in particular, the interrupted Poisson process, Markov modulated Poisson Process (MMPP) and Markov arrival process (MAP) . In spite of the generality of the input process, we are able to obtain explicit expressions for the expected cost as a function of the policy. This allows us to derive the multimodularity of two types of cost functions and to conclude that "bracket" policies are optimal as in [1, 2].

Using ideas from [103] we then solve the dual problem [76] which is concerned with the sharing of a single communication link between multiple sources. This is modeled as an optimal assignment problem of a single server to several single buffer queues to which packets arrive according to Poisson processes. The relation to the previous problem is obtained if we take as states in the new problem the vector of free spaces in the buffers (instead of the occupancy of the buffers). The distribution of the process of free spaces in the new model is the same as the distribution of the process of buffer occupancy in the previous problem. This yields the solution of the service assignment problem.

Our original routing problem was also studied recently in [81] (assuming however a much simple i.i.d. arrival process). Using the theory of MDPs with partial information, he established the periodicity of optimal policies and showed for the special case of symmetrical servers and arrival processes that the round-robin policy is optimal. For the case of 2 servers, he showed that a periodic policy whose period has the form $(1, 2, 2, 2, \ldots, 2)$ is optimal, where 2 means sending a packet to the faster server. Similar results have been obtained in the dual model of server assignment in [42]. The approach in [42, 81] heavily depend on the Markovian structure of the arrivals. Surprisingly, the Markovian approach turns out to be useful also in the non Markovian setting (which is due to the general non i.i.d. arrival process that we consider). Indeed, we show here that one can use a simple formulation of the problem as a Markov decision process (MDP) that allows us to conclude in the general setting too, that optimal periodic policies exist. We further relax here the assumption on the distribution of the service times, and allow these to be general independent.

We show in this chapter that policies that are regular in a weaker sense are also optimal. More precisely, we establish that the cost function related to periodic policies are Schur convex. This allows us to conclude that not only regular policies are optimal, but more generally, policies that are regular in the majorization sense. We further show that a policy that majorizes a second policy has a smaller cost.

We conclude this chapter with an application of our framework to the problem of robot scheduling for web search engines.

8.1.1 Organization of the chapter

The Chapter is structured as follows. In the next section we study the admission into a single server, and obtain the optimal policy. We then establish in the following Section properties of the cost and an ordering between the performance of different policies. Section 8.5 deals with the optimal routing problem. The service assignment problem is solved in section 8.6 The technical arguments that establish the multimodularity properties are delayed to Section 8.7. In Section 8.8 we present the MDP formulation of the problem and obtain the optimality of periodic policies . We then study in Section 8.9 the multimodularity of the *global* cost when considering routing into two servers. We conclude with an application to a robot scheduling problem in the Web is discussed in Section 8.11.

8.2 The admission into a single server

Consider a single server with no waiting room. Let T_n be the point process representing the arrival epochs, and assume that service times are i.i.d., independent of the arrival process, and exponentially distributed with parameter μ. We assume that $\delta_n := T_{n+1} - T_n$ is a stationary process (in n). An arrival can be rejected or accepted by an admission controller. An admission policy is a sequence $a = (a_1, a_2, ...)$ where $a_i = 1$ means acceptance of the ith arrival, and $a_i = 0$ means its rejection. The actions are taken without any knowledge of the state of the buffer, and if it is already full when the packet is admitted, then the packet is lost. Let $X_n = X_n(a)$ be the number of packets in the system just after the nth action is taken.

We consider first the following problem:

(P1) Maximize

$$g(a) \stackrel{\text{def}}{=} \lim_{n \to \infty} \frac{1}{n} \sum_{j=1}^{n} \mathbb{E}^a X_n,$$

where $\mathbb{E}^a$ is the conditional expectation given a. The policy a is subject to the constraint that a fraction of no more than a fraction p of the packets is accepted:

$$\overline{\lim_{n\to\infty}} \frac{1}{n} \sum_{j=1}^{n} a_n \le p. \tag{8.1}$$

The reason we consider this reward function for maximization is that maximizing the average number of packets in steady state is related to
- maximizing the throughput (the departure rate), and to
- minimizing the losses (due to both rejection by the controller and to the blocking).

Indeed, every customer has a sojourn time which is exponentially distributed with average μ^{-1}, and therefore the average actual throughput is equal to the product of the long-run average number of packets in the system and μ. The second equivalence follows since the loss rate is the initial given input rate (before the admission control and losses) minus the actual throughput.

Note that the queue size that we maximize in this section is that obtained by averaging over the times T_n. In some cases this will indeed coincide with the time average queue length (for example, if the interarrival times δ_n exponentially distributed). In the case it does not coincide, we propose an alternative approach in Section 8.4.

Another motivation to study this problem follows from an interesting comparison with the infinite queue system. Below we show that for the current system the queue length is *maximized* by assignment sequences as regular as possible; for the infinite buffer system queue lengths are *minimized* by bracket sequences (Chapter 6). Some intuitive understanding can help to explain this at first sight contradictory phenomenon: in an infinite queue system minimizing queue lengths means minimizing waiting by spreading out arrivals; minimizing queue lengths in a system without buffers means maximizing losses by making arrivals bursty.

Before stating the main result, we present a simple coupling property:

Lemma 45. *Fix an arbitrary policy a. Then one can construct a probability space $(\Omega, \mathcal{F}, \mathbb{P})$ such that*
(i) the state trajectories starting from different initial states of the server couple in a time which is finite w.p. 1,
(ii) The following holds:

$$\lim_{n\to\infty} |\mathbb{E}[X_n(a)|X_0 = 1] - \mathbb{E}[X_n(a)|X_0 = 0]| = 0.$$

Proof. We let the interarrival times be the same in both systems. If the policy a never accepts packets or if arrivals never occur, then coupling occurs once the system empties starting at state 1. Otherwise, let T_n be the time at which the first packet is admitted. Note that the admission occurs in both systems. If both systems are empty at time T_n, then coupling occurs at that instant (we take service times to be the same in both systems from time T_n onwards).

Otherwise, system 1 (in which the initial state is of a single packet) has 1 packet at time T_n, and system 0 (which is initially empty) has no packets. Coupling is obtained again at time T_n by assuming that
(a) service times of all packets admitted after time T_n are the same in both systems;
(b) at time T_n, the remaining service time of the packet in service in system 1 equals to the service time of the packet admitted in system 0 at time T_n.

Due to the memoryless property of the exponential distribution and the independence assumption on service times, the above coupling is consistent with the probability distribution of the original state processes. This establishes (i). (ii) follows from the bounded convergence theorem.   □

Theorem 33. *Assume that the system is controlled starting from time T_1. Assume that the inter-arrival times are stationary, and independent of the service times. Assume that the service times are i.i.d. exponentially distributed. Then for any $\theta \in [0,1]$, the bracket policy with rate p and initial phase θ is optimal.*

Proof. Due to Lemma 45, we may assume without loss of generality that the system contains initially one packet at time T_0. $-\mathbb{E}_{T,\sigma} X_n(a)$ is multimodular (we delay the proof of this property to Section 8.7). Here σ denotes the random process governing the service completions. $f_n(a) := -\mathbb{E}_{T,\sigma} X_n(a)$ is clearly monotone decreasing in a. It remains to check conditions $< 2 >$ and $< 3 >$ from Section 1.3.

Recall that Condition $< 3 >$ can be formulated as: the functions $f_n(a)$ must satisfy the following property.
For any sequence $\{a_k\} \exists$ a sequence $\{\beta_k\}$ such that
$\forall k, m, \quad k > m, \ f_k(\beta_1, \cdots, \beta_{k-m}, a_1, ..., a_m) = f_m(a_1, ..., a_m)$.
This clearly holds in our case where $f_n(a) = -\mathbb{E}_{T,\sigma} X_n(a)$ by setting $\beta_i = 1$. (The precise justification of the above is from property (i) which appears in Section 8.7.).

Condition $< 2 >$ states:
$f_k(a_1, ..., a_k) \geq f_{k-1}(a_2, ..., a_k), \ \forall k > 1$;
this holds in our case for $-\mathbb{E}_{T,\sigma} X_n$ due to our assumptions on the initial state.   □

8.3 Properties of the cost and of policies

Lemma 46. $g(a) = g(\vartheta a)$ *for any policy a, where ϑ is the one step shift operator.*

Proof. This follows from Lemma 45. Indeed, with $a = (a_1, a_2, ...)$, we have

$$g(i) = \lim_{n \to \infty} \frac{1}{n} \sum_{j=1}^{n} \mathbb{E}^a X_n = \lim_{n \to \infty} \frac{1}{n-1} \sum_{j=2}^{n} \mathbb{E}^a X_n = \lim_{n \to \infty} \frac{1}{n} \sum_{j=1}^{n} \mathbb{E}^{a'} X_n$$

where $a' = \vartheta a = (a_2, a_3, ...)$.   □

Define for any policy a

$$r_n(a) \stackrel{\text{def}}{=} \min\{m \geq 0 \text{ such that } a_{n-m} = 1\}.$$

Lemma 47. *For any policy, $\mathbb{E}X_n(a)$ is given by*

$$\mathbb{E}X_n(a) = \mathbb{E} \; exp\left(-\mu \sum_{k=1}^{r_n(a)} \delta_k\right).$$

Proof. $X_n(a) = 1$ if and only if since time $T_{n-r_n(a)}$ there has been no departure. Thus $X_n(a) = 1$ if and only if during a time period of duration $\sum_{k=n-r_n(a)}^{n} \delta_k$ there are no end of services. Since δ_k is a stationary sequence, we obtain the result.   □

Assume that a is periodic with period P. Let a' be some shift of a such that $a'_S = 1$. Thus $X_{nP}(a) = 1, n = 0, 1, 2, ...$

$$g(a) = g(a') = \frac{1}{P} \sum_{k=1}^{P} \mathbb{E}(X_k(a))$$

Let $n = \sum_{k=1}^{P} a(k)$. Define

$$\eta(0) = 0; \qquad \eta(j+1) = \min\{l > \eta(j) : a_l = 1\}, j = 0, ..., n-1$$

Note that $\eta(n) = P$. Define

$$d_1 = \eta(1), \qquad d_j = \eta(j+1) - \eta(j), \; j = 2, ..., n-1.$$

Lemma 48. *a (and thus a') is fully determined by the sequence d up to a shift, and the cost g can be expressed as a function of $d(a)$:*

$$g(a) = \frac{1}{P} \sum_{i=1}^{n} \left(\mathbb{E} \sum_{j=0}^{d_i-1} e^{-\mu \sum_{k=0}^{j} \delta_k}\right) \stackrel{\text{def}}{=} g'(d) \qquad (8.2)$$

Proof. By Lemma 47,

$$g(a) = \frac{1}{P} \sum_{i=1}^{n} \sum_{j=0}^{d_i-1} \mathbb{E}X_{\eta(i)-j}(a) = \frac{1}{P} \sum_{i=1}^{n} \sum_{j=0}^{d_i-1} \mathbb{E} \; exp\left(-\mu \sum_{k=0}^{r_{\eta(i)-j}(a)} \delta_k\right)$$

$$= \frac{1}{P} \sum_{i=1}^{n} \sum_{j=0}^{d_i-1} \mathbb{E} \; exp\left(-\mu \sum_{k=0}^{j} \delta_k\right) = \frac{1}{P} \sum_{i=1}^{n} \sum_{j=1}^{d_i} \mathbb{E} \; exp\left(-\mu \delta_1\right)^j$$

     □

We conclude that if a is a periodic policy, then any policy obtained from a by changing the order of the d sequence that characterizes a, achieves the same cost. Thus a non-bracket periodic policy of a period $(1, 0, 1, 0, 1, 0, 0, 1, 0, 0)$ has the same cost as a bracket policy of the form $(1, 0, 1, 0, 0, 1, 0, 1, 0, 0)$.

This property can be seen to be a special corollary of a more general result that will be presented in Theorem 34 below.

Consider two n-dimensional vectors of integers $d(1), d(2)$.

Definition 14. *(Majorization [88])*
We say that $d(2)$ majorizes $d(1)$, which we denote by $d(1) \prec d(2)$, if

$$\begin{cases} \sum_{i=1}^{k} d_{[i]}(1) \le \sum_{i=1}^{k} d_{[i]}(2), \quad k = 1, ..., n-1, \\ \sum_{i=1}^{n} d_{[i]}(1) = \sum_{i=1}^{n} d_{[i]}(2) \end{cases}$$

where $d_{[i]}(j)$ is a permutation of $d_i(j)$ satisfying $d_{[1]}(j) \ge d_{[2]}(j) \ge ... \ge d_{[n]}(j)$, $j = 1, 2$.

A function $f : \mathbb{R}^n \to \mathbb{R}$ is Schur convex if $d(1) \prec d(2)$ implies $f(d(1)) \le f(d(2))$. f is Schur concave is $d(1) \prec d(2)$ implies $f(d(1)) \ge f(d(2))$.

Lemma 49. $g'(d)$, *defined in (8.2) is Schur concave.*

Proof. g' can be written as the sum

$$g'(d) = \frac{1}{P} \sum_{i=1}^{n} \psi_i(d_i))$$

where ψ is the term in brackets in eq. (8.2). Since

$$\psi_i(m+1) - \psi_i(m) = \mathbb{E} e^{-\mu \sum_{k=0}^{m} \delta_k}$$

is monotone *decreasing* in m, it follows that ψ_i are concave in d_i. The proof is then established by using proposition C.1 on p. 64 in [88]. □

Theorem 34. *Assume that a and a' are two periodic policies with the same period P and the same sum*

$$n = \sum_{i=1}^{P} a_i = \sum_{i=1}^{P} a'_i.$$

If $d(a') \prec d(a)$ then $g(a') \ge g(a)$.

Proof. Follows from the Schur concavity of g, see [88] p. 54.   □

The above theorem allows us to "improve" any given periodic policy by replacing it by a more "regular" one, where by more regular we mean a policy whose distance sequences are majorized by the less regular one.

A similar result was obtained in [42] in a related model (which we discuss in Section 8.11 under the assumption that the sequence δ_n is i.i.d.).

8.4 Time averages

In the previous sections we took averages of the costs as seen at times T_n, i.e. at arrival times. The problem with this is, however, that T_n are in fact the times of *potential* rather than actual arrivals. In practice, arrivals occur only at a subsequence of T_n which depend on the policy. In this section we obtain similar results for the actual arrival process.

We consider the same model of the system as well as the statistical assumptions as in Section 8.2.

Instead of describing a policy using a sequence a, it will be more helpful to consider an equivalent description using the distance sequence $d = (d_1, d_2, ...)$. Define $D(n) = \sum_{k=0}^{n} d_k$. The actual arrivals occur at times $T_{D(n)}, n \in \mathbb{N}$. We define the process ξ_n to be the number of packets in the buffer just prior to time $T_{D(n)}$. (If we took, as in the previous sections the time *after* a decision, then the number of packets would always be 1.

The motivation for considering the system at arrival instants is the following. Whenever an actual arrival finds a packet in the system there is a loss. Thus minimizing the average number of packets at actual arrival times will also minimize the fraction of losses and maximize the throughput.

We consider first the following problem:
(Q1) Minimize

$$G(d) \stackrel{\text{def}}{=} \overline{\lim_{n \to \infty}} \frac{1}{n} \sum_{j=1}^{n} \mathbb{E}^a \xi_n$$

subject to the constraint that a fraction of no more than a fraction p of the packets is accepted, or in other words,

$$\lim_{n \to \infty} \frac{1}{n} \sum_{j=1}^{n} d_n \geq 1/p. \tag{8.3}$$

Recall that a policy d is bracket with rate $1/p$ if and only if its related policy a is bracket with rate p, see Lemma 21.

Theorem 35. *Assume that the system is controlled starting from time T_1. Assume that the inter-arrival times δ_n are stationary and independent of the service times. Assume that the service times are i.i.d. exponentially distributed. Then for any $\theta \in [0, 1]$, the bracket policy with rate $1/p$ and initial phase θ is optimal.*

The proof of the Theorem is similar to that of Theorem 33. The multi-modularity of $\mathbb{E}\xi_n(d)$ is established in Section 8.7.

We now present properties of the cost for periodic policies, which are the analogous of Section 8.3. We first define the expected average cost for a periodic sequence $d = (d_1, ..., d_P)$ given by

$$G(d) = \frac{1}{P}\sum_{i=1}^{P}\mathbb{E}\left(-\mu d_i\right) \tag{8.4}$$

when the interarrival times are i.i.d. Since this is a sum of functions that are convex in the d_i's, it is Schur convex in d (proposition C.1 on p. 64 in [88]). We thus conclude the following (see [88] p. 54):

Theorem 36. *Assume that d and d' are two periodic policies with period P and the same sum:*

$$n = \sum_{i=1}^{P} d_i = \sum_{i=1}^{P} d'_i.$$

If $d' \prec d$ then $G(d') \leq G(d)$.

8.5 Routing to several servers

We now consider K servers (all with a single buffer), fed by a stationary input process as in Section 8.2. We make the same probabilistic assumptions as before on the service times in each server. Let X_n^i be the number of packets (0 or 1) being served by server i at the nth time epoch. A routing policy is a sequence $a = (a_1, a_2, ...)$ where $a_i = j$ means routing of the ith arrival to queue j. We consider the following routing problem:
(P2) Maximize

$$g(a) \stackrel{\text{def}}{=} \lim_{n \to \infty} \frac{1}{n}\sum_{j=1}^{n}\sum_{i=1}^{K} h_i \mathbb{E}^a X_j^i,$$

where h_i are some given positive constants.

The following theorems are the result of the properties we established in the previous section, the multimodularity properties (which is established in the next section) as well as the results in Chapters 1 and 4.

Theorem 37. *Consider the symmetric system, i.e., $\mu_1 = \cdots = \mu_K$. Then the round robin policy maximizes $g(a)$ (and hence, the expected average throughput).*

Theorem 38. *Consider the case of two servers. Then there is some p^* such that for any θ, the policy that routes packets to server 1 according to the bracket policy with rate p^* and initial phase θ, and otherwise routes packets to server 2, is optimal.*

Theorem 39. *Consider the case of two types of servers: a set $K_1 \subset \{1, ..., K\}$ of servers with $\mu = \mu_1$, and the remaining set K_2 of servers with $\mu = \mu_2$. Assume that $h_i = h^1$ are the same for all $i \in K_1$ and that $h_i = h^2$ for all $i \in K_2$. Then there exists some p^* such that that for any θ, the following policy is optimal:*
it routes packets to the 1st group server according to the bracket policy with rate p^* and initial phase θ, and otherwise routes packets to the second group of servers. Within each group of servers, the order of service is round-robin.

The proof of all three theorems is based on Proposition 11, which states that if a tuple $(p_1, p_2, ..., p_K)$ is made of less than (or of exactly) two distinct numbers, then it is balanceable. In other words, there exist a policy a such that for each $i = 1, ..., K$, the sequence of indices in a that correspond to routing arrivals to queue i are is a bracket with rate p_i. The optimality of balanceable sequences was established in Chapter 1.

Remark 13. For any a, let $a^i = \{a_n^i, n \in \mathbb{N}\}$ be the binary sequence such that $a_n^i = 1$ if and only if $a_n = i$. Using the result of Section 8.3, and in particular the Schur convexity of g as a function of the d's (see Lemma 49), one can show that if there are two periodic policies a and b with the same period P, such that for any $i = 1, ..., K$,

- $\sum_{j=1}^{P} a_n^i = \sum_{j=1}^{P} b_n^i$,
- $d(b^i) \prec d(a^i)$,

then $g(b) \geq g(a)$.

8.6 The service assignment problem

Consider K Poisson processes with parameters $\mu_1, ..., \mu_K$ respectively, of packet arrivals into K respective single buffer queues. One buffer can obtain transmission opportunity at a time. Let T_n be time at which the nth transmission opportunity occurs. If an arrival occurs to a buffer that is full then it is lost; if the buffer is empty then the arriving packet is stored till a transmission opportunity to that buffer arrives. If a buffer receives a transmission opportunity at time T_n and it has a packet then this packet is transmitted, and immediately after time T_n, a new arriving packet can be stored in this buffer. If there is no packet in the buffer then this transmission opportunity is lost. We assume that $\delta_n := T_{n+1} - T_n$ is a stationary process (in n) and independent on the arrival process.

The role of the controller is to decide to which buffer will the next transmission opportunity be assigned. A service assignment policy is a sequence $a = (a_1, a_2, ...)$ where $a_i = j$ means that the ith transmission opportunity will be to queue j. We assume that the controller has no information about the buffers' contents.

Let Z_n^i be the number of packets (0 or 1) at buffer i just after the nth action is taken. Denote $Y_n^i = 1 - Z_n^i$. It thus corresponds to the 'vacancies' process, as Y_n^i equals one if the ith buffer is empty just after the nth action is taken (i.e. after time T_n). We consider the following problem:

(P3) Minimize

$$g(a) \stackrel{\text{def}}{=} \overline{\lim_{n \to \infty}} \frac{1}{n} \sum_{j=1}^{n} \sum_{i=1}^{K} h_i \mathbb{E}^a Z_j^i,$$

where h_i are some given positive constants.

The above objective corresponds to the minimization of blocking probabilities, since blocking occurs at queue i between T_n and T_{n+1} if and only if $Z_n^i = 1$.

Note that by minimizing blocking probabilities, we maximize the throughput.

We now make the following key observation. Fix an arbitrary sequence a. Then the distribution of the vacancies process $\{Y_n^i\}_{n,i}$ in the service assignment problem is the same as the distribution of the buffer contents process $\{X_n^i\}_{n,i}$ in the routing problem. Hence, using the results of the previous section, we get the following main results.

Theorem 40. *Consider the symmetric system, i.e., $\mu_1 = \cdots = \mu_K$. Then the round robin policy minimizes $g(a)$ (and hence, maximizes the expected average throughput).*

Theorem 41. *Consider the case of two servers. Then there is some p^* such that for any θ, the policy that assigns transmission opportunities to buffer 1 according to the bracket policy with rate p^* and initial phase θ, and otherwise assigns transmission opportunities to server 2, is optimal.*

Similarly, one obtains the dual of Theorem 39.

8.7 The multimodularity

Lemma 50. *The two following statements are true. (i) $\mathbb{E}_\sigma(-X_n(a))$ is multimodular,*
(ii) $\mathbb{E}_\sigma \xi_n(d)$ is multimodular.

Proof. (i) The proof is based on the following useful properties. Let $q(a) \stackrel{\text{def}}{=}$ $\max\{m \le n$ such that $a_m = 1\}$. Consider two policies a, a'. Then,

Property (A): if $q(a) = q(a')$ then $X_n(a) =_{st} X_n(a')$ (this is equivalent to $\mathbb{E}_\sigma(X_n(a)) = \mathbb{E}_\sigma(X_n(a'))$ and to $\mathbb{P}(X_n(a) = 1) = \mathbb{P}(X_n(a') = 1)$).

This is due to the memoryless property of the exponential service times. Thus we can replace our original system by one where at each acceptance, the new packet replaces the one in service, instead of being rejected; the distribution of the process X_n will not change.

Property (B): if $q(a) \leq q(a')$ then $X_n(a) \leq_{st} X_n(a')$ (which is equivalent to $\mathbb{E}_\sigma(X_n(a)) \leq \mathbb{E}_\sigma(X_n(a')))$.

This is obtained by a similar argument.

We shall now check relation (1.1) in the definition of multimodularity (Chapter 1) for any a such that $a, a + w, a + v, a + w + v$ are feasible.

Let $v = e_n$. We have for any $w \in \mathcal{F}$, $w \neq v$,

$$X_n(a + w + v) = X_n(a + v) = X_n(v) = 1,$$

and

$$X_n(a + w) \leq_{st} X_n(a).$$

The first relation follows from property (A) above, and the second from property (B), since

$$q(a + s_i) \leq q(a), i = 2, ..., n \quad \text{and} \quad q(a - e_i) \leq q(a)$$

(recall that s_i corresponds to shifting an arrival to the past). This implies that

$$\mathbb{E}_\sigma(-X_n(a + w + v)) = \mathbb{E}_\sigma(-X_n(a + v)),$$
$$\mathbb{E}_\sigma(-X_n(a + w)) \geq \mathbb{E}_\sigma(-X_n(a)).$$

Hence relation (1.1) is satisfied.

Next, assume $v = -e_1$. We consider $w \neq v$ (and thus restrict to $n \geq 2$). For $a + v$ to be feasible, we must have $a_1 = 1$. For $a + w$ to be feasible, we have $w \neq s_2$, and $q(a) > 1$. It then follows from property (A) that

$$X_n(a) =_{st} X_n(a + v), \qquad X_n(a + w) =_{st} X_n(a + v + w).$$

Hence relation (1.1) holds with equality.

It remains to check $v = s_i, s_j = w$, with $j > i$. Since $a + s_j$ is feasible, it follows that $q(a) \geq j$. Hence, by property (A),

$$X_n(a + s_i) =_{st} X_n(a).$$

Similarly, $q(a + s_j) \geq j - 1$; since it is feasible then $a_{j-1} = 0$, so that $i < j - 1$ (in order for $a + s_i$ to be feasible, we have to exclude $i = j - 1$, since $a_{j-1} = 0$). Hence

$$X_n(a + s_i + s_j) =_{st} X_n(a + s_j).$$

Hence relation (1.1) holds with equality, which concludes the proof of (i).

(ii) Let v be any one of the vectors in the set $\{-e_1, s_2, ..., s_{n-1}\}$. Then due to Property (A), for any $w \neq v$,

$$\xi_n(a) =_{st} \xi_n(a + v), \quad \xi_n(a + w) =_{st} \xi_n(a + v + w).$$

Hence relation (1.1) is satisfied. By symmetry it holds for any w in the set $\{-e_1, s_2, ..., s_{n-1}\}$. It thus remains to check the case $v = e_n, w = s_n$.

From Lemma 47 we have:

$$\mathbb{E}\xi_n(d) = \mathbb{E}\,\exp\left(-\mu\sum_{k=1}^{d_n}\delta_k\right).$$

Let $x_n = m$, and let denote

$$y = \exp\left(-\mu\sum_{k=2}^{m}\delta_k\right).$$

Then

$$\mathbb{E}\xi(x + v + w) = \mathbb{E}\xi(x) = \mathbb{E}ye^{-\mu\delta_{m+1}} = \mathbb{E}ye^{-\mu\delta_1}$$
$$\mathbb{E}\xi(x + v) = \mathbb{E}ye^{-\mu[\delta_{m+1} + \delta_{m+2}]} = \mathbb{E}ye^{-\mu[\delta_1 + \delta_{m+1}]}$$
$$\mathbb{E}\xi(x + w) = \mathbb{E}y.$$

Since the function $f(x) := ye^{-\mu x}$ is convex in x, it follows that $f(\delta_1 + z) - f(z)$ is increasing in z, so that

$$f(\delta_1 + \delta_{m+1}) - f(\delta_{m+1}) \geq f(\delta_1) - f(0).$$

By taking expectations, this implies relation (1.1) for $\mathbb{E}\xi_n(a)$, which concludes the proof.   □

8.8 MDP formulation and new structural results

We reconsider our routing problem into K parallel servers with no waiting room in the framework of Markov Decision Processes. Recall that packets arrive at times $(T_n)_{n\in\mathbb{N}}$. We use the convention that $T_1 = 0$. and we assume that the process $(\delta_n)_{n\in\mathbb{N}}$ of interarrival times is stationary. Upon arrival of a packet a controller must route this packet to one of the K servers in the system.

We allow in this section for more general service times: the service time has a general service distribution G_m when routed to server $m \in \{1, ..., K\}$. If there is still a packet present at the server, where the packet is routed to, then the packet in service is lost (we call this the preemptive discipline). In the special case of an exponential distribution, one can consider instead the non-preemptive discipline in which the arriving packet is lost; the results below will still hold.

We assume that the controller, who wishes to minimize some cost function, has no information on the state of the servers (busy or idle). The only information which the controller possesses is its own previous actions, i.e. to

which server previous packets were routed. We assume that this information does not include the actual time that has elapsed since a packet was
last routed to that server. This assumption enables us to study the embedded Markov chain of the continuous-time process; i.e. we can consider the
discrete-time process embedded at the epochs when packets arrive. Now the
mathematical formulation of this problem is given by the following T-horizon
Markov decision process.

Let $\mathcal{X} = (\mathbb{N} \cup \{\infty\})^K$ be the state space. The mth component x_m of $x \in \mathcal{X}$
denotes the number of arrivals since a packet was last routed to server m.
Let $\mathcal{A} = \{1, \ldots, K\}$ be the action space, where action $a \in \mathcal{A}$ means that a
packet is routed to server a. Since actions are taken before state transitions,
we have the following transition probabilities:

$$p(x' \mid x, a) = \begin{cases} 1, & \text{if } x'_a = 1 \text{ and } x'_m = x_m + 1 \text{ for all } m \neq a \\ 0, & \text{otherwise.} \end{cases}$$

Define the immediate costs by $c_t(x, a) = f_a(x_a)$, which reflects that the costs
only depend on the chosen server and the state of that server for $t < T$. The
terminal costs are given by

$$c_T(x) = \sum_{a \in \mathcal{A}} f_a(x_a) \tag{8.5}$$

(note that the terminal costs use the same functions f_a). Defining these terminal costs will be essential for the mathematical results, as will be illustrated
later in Example 9. It has also a natural physical interpretation, as will be
illustrated in Remark 16.

The set of histories at epoch t of this Markov decision process is defined as
the set $\mathcal{H}_t = (\mathcal{X} \times \mathcal{A})^{t-1} \times \mathcal{X}$. A policy π is a set of decision rules $(\pi_1, \pi_2, \ldots)$
with $\pi_t : \mathcal{H}_t \to \mathcal{A}$. For each fixed policy π and each realization h_t of a
history, the variable A_t is given by $A_t = \pi_t(h_t)$. The variable X_{t+1} takes
values $x_{t+1} \in \mathcal{X}$ with probability[1] $p(x_{t+1} \mid x_t, a_t)$. With these definitions the
expected average cost criterion function $C(\pi)$ is defined by

$$C(\pi) = \limsup_{T \to \infty} \mathbb{E}_x^{\pi} \frac{1}{T} \sum_{t=1}^{T} c_t(X_t, A_t),$$

where $x = (x_1, \ldots, x_K)$ is the initial state. Let Π denote the set of all policies.
The Markov decision problem is to find a policy π^*, if it exists, such that
$C(\pi^*) = \min\{C(\pi) \mid \pi \in \Pi\}$. This Markov decision model is characterized by
the tuple $(\mathcal{X}, \mathcal{A}, p, c)$.

---

[1] In the general MDPs, the variables A_t and X_t are random; here they are deterministic since the transition probabilities take only values of 0's or 1's

Example 6. Suppose that the controller wishes to minimize the number of lost packets (i.e. the number of preempted packets) per unit time. The cost function $f_m(n)$ will typically be a decreasing function in n, because a longer time interval between an assignment to server m results in a smaller probability that a packet, that was previously assigned there, is still in service.

Assume that the arrival process is a Poisson process with rate λ and that services at server i are exponentially distributed with rate μ_i independent of the other servers. Let S_i be a random variable which is exponentially distributed with rate μ_i. Then $f_m(n) = \mathbb{P}(S_m \geq \delta_1 + \cdots + \delta_n) = \left[\lambda/(\lambda + \mu_m)\right]^n$.

In the setting described in Example 6 we obtained a decreasing cost function $f_m(x)$. In Section 8.10 we discuss this application in more detail. In Section 8.11 we describe another application of our model, where the obtained cost function is increasing. In this section, we make some general assumptions on the cost function in order to cover all these different structures. Moreover we investigate structural properties of the optimal policy.

Assume that all the f_m are convex and that one of Conditions (8.6)–(8.8) defined below holds:

$$\lim_{x \to \infty} \left(f_m(x+1) - f_m(x)\right) = \infty \qquad m = 1, \ldots, K. \qquad (8.6)$$

f_m are strictly convex and $\lim_{x \to \infty} \left(f_m(x) - a_m x\right) = C, \qquad m = 1, \ldots, K$
$$\qquad (8.7)$$

(by strictly convex we mean that for all x, $f_m(x+2) - f_m(x+1) > f_m(x+1) - f(x)$).

$$f_m(x) = a_m x + C, \qquad m = 1, \ldots, K, \qquad (8.8)$$

where $a_m \geq 0$ and C are constants (and C does not depend on m). Note that Condition (8.8) is not included in (8.7): it violates its first part. Condition (8.6) covers the case where f_m grows more than linearly, whereas (8.7)–(8.8) cover the case where f_m grows asymptotically linearly. These conditions are complementary to (8.6) since any one of them implies that $\lim_{x \to \infty} \left(f_m(x+1) - f_m(x)\right) < \infty$. In Conditions (8.6) and (8.7), f_m is strictly convex.

Theorem 42. *Assume that one of Conditions (8.6)–(8.8) holds. Then*

(i) *There exists an optimal policy that uses every server infinitely many times, thus* $\sup\{j \mid \pi_j = m\} = \infty$ *for* $m \in \{1, \ldots, K\}$.
(ii) *There exists an periodic optimal policy.*

Before proving the theorem we note that there are certain ways in which the conditions (8.7)-(8.8) cannot be relaxed. We illustrate this below with cost functions for which the above theorem does not hold.

Example 7. Consider the case of two servers with the costs

$$f_i(x) = a_i\, x + \hat{b}_i\, \exp(-\ell_i\, x) + c_i, \quad i = 1, 2,$$

where $c_1 < c_2$, and where $\hat{b}_i \geq 0$ and $\ell_i > 0$ are some constants (as follows from the next remark, the sign of a_i is not important). Then for a sufficiently small value of $\hat{b}_1$, the policy that always routes to server 1 is the only optimal policy for any finite horizon N.

Indeed, assume first $\hat{b}_i = 0$ for all i and let u be any policy that routes at its nth step to server 2. By changing the action at this step into an assignment to server 1 we gain $c_2 - c_1$.

By continuity of the cost in the $\hat{b}_i$'s, we also gain a positive amount using this modified policy if $\hat{b}_i \neq 0$ provided the $\hat{b}_i$'s are sufficiently small. Hence for $\hat{b}_i$ sufficiently small, a policy cannot be optimal if it does not always route to server 1.

When using the average cost, the cost is not affected anymore by any changes in the actions provided that the frequency of such changes converges to zero. Hence for the average cost, there may be other policies that are optimal, but still, any policy for which the fraction of customers routed to server 2 does not converge to 0 cannot be optimal.

We conclude that we cannot relax (8.7) or (8.8) and replace C by C_m.

Remark 14. Note that when the cost $f_i(x)$ contains a linear term $a_i x$ then the total accumulated cost that corresponds to the term $a_i x$ over any horizon of length N is $a_i(N + x_i)$, where $x = (x_1, x_2)$ is the initial state. This part does not depend on the policy. If we use a policy π and then modify it by changing an assignment at time $t < N$ from server i to server $j \neq i$ then the linear part of the cost at time t under the modified policy decreases by $a_i x_i(t) - a_j x_j(t)$, but it then increases by the same amount at time $t + 1$. Thus the accumulated linear cost is independent of the policy. (Note that this argument is valid due to the definition of the cost at time N in (8.5).)

Example 8. Consider the case of two servers with the costs $f_1(x) = a_1 x$ and $f_2(x) = \exp(-\ell_2 x)$. For any finite horizon N and for $\ell_2 > 0$, the only optimal policy is the one that always routes to server 1. Note that the average cost until time N of the policy that always routes to server 1 is

$$\frac{N f_2(1) + f_1(N)}{N} = e^{-\ell_2} + a_1.$$

The average cost of the policy that always routes to server 2 is

$$\frac{N f_1(1) + f_2(N)}{N} = a_1 + \frac{e^{-\ell_2 N}}{N}.$$

Again, for the average cost there are other optimal policies but they have to satisfy the following: the fraction of customers routed to queue 2 by time N should converge to zero as $N \to \infty$.

This illustrates the necessity of the first part of Condition (8.7). For $\ell_2 < 0$, the only optimal policy is the one that always routes to server 2.

Next we present an example to illustrate the importance of the terminal cost.

Example 9. Assume that there are two servers, and that the costs are given by $f_1(x) = x^2$ and $f_2(x) = 2\,x^2$. Assume that the terminal costs $c_N(x)$ were zero, then the policy that always routes to server 1 is optimal.

Proof (**Proof of Theorem 42**). First suppose the cost function satisfies Condition (8.8). Interchanging assignments between any two servers for any finite horizon does not result in changes in cost for that horizon, due to the linearity of the cost function and the terminal cost. Hence any periodic policy that routes packets to all servers is optimal.

We consider next Conditions (8.6) and (8.7). Instead of describing a policy using a sequence π, we use an equivalent description using time distances between packets routed to each server. More precisely, given an initial state x, define the jth instant at which a packet is routed to server m by

$$\eta^m(0) = -x_m$$
$$\text{and } \eta^m(j) = \min\{i \mid \max\left(\eta^m(j-1), 0\right) < i \leq N \text{ and } \pi_i = m\},$$

for $j \in \mathbb{N}$ (the minimum of an empty set is taken to be infinity.) Define the distance sequence $d(m)$ by $d(m) = (d_1(m), d_2(m), \ldots)$, with $d_j(m) = \eta^m(j) - \eta^m(j-1)$, for $j \in \mathbb{N}$. (For simplicity we do not include the N in the notation.)

Let π be an arbitrary policy and m be an arbitrary server (fixed from now on). Assume that the distance sequence $d \stackrel{\text{def}}{=} d(m)$ for this server has the property that $\limsup_{N \to \infty}\{d_j \mid j \in \mathbb{N}\} = \infty$. We shall construct a policy π' with distance sequence $d^{\pi'}$ such that $\limsup_{N \to \infty}\{d_j^{\pi'} \mid j \in \mathbb{N}\}$ is finite and $C(\pi') \leq C(\pi)$.

Assume first that f satisfies Condition (8.6). Choose n_0 such that for all $n > n_0$

$$\min_{1 \leq k \leq 2K+1} \left(f_m(n+k) - f_m(k) - f_m(n) \right)$$

$$> \max_{1 \leq k < 2K+1} \left(f_l(2K+1) - f_l(k) - f_l(2K+1-k) \right),$$

for all l. Since the supremum of the distance sequence is infinity, there is a j (and N) such that $d_j > n + 2K + 1$. Consider the $2K + 1$ consecutive assignments starting at $\eta^m(j-1)$. Since there are K servers, it follows that there is at least one server, to which a packet is assigned three times during that period, say m'. Denote the distance (or interarrival) times of those three assignments to m' by $d_1(m')$ and $d_2(m')$. Replace the second assignment to m' in this sequence by an assignment to server m. Denote the new distance (or interarrival) times to m by d'_j and d''_j (if $\eta(j) = N$ then the distance d''_j is not a real interarrival time). Consider the cost for a horizon of length l where l is an arbitrary integer larger than $\eta(j-1) + n_0 + 2K + 1$.

$$\left[f_m(d_j) + f_{m'}(d_1(m')) + f_{m'}(d_2(m'))\right] -$$
$$\left[f_m(d'_j) + f_m(d''_j) + f_{m'}(d_1(m') + d_2(m'))\right]$$
$$= \left[f_m(d_j) - f_m(d'_j) - f_m(d''_j)\right] -$$
$$\left[f_{m'}(d_1(m') + d_2(m')) - f_{m'}(d_1(m')) - f_{m'}(d_2(m'))\right] > 0,$$

where the last inequality follows from the choice of n_0.

Consider now Condition (8.7). Since by assumption the supremum of the distance sequence $d_l = d_l(m)$, $l = 1, 2, \ldots$ is infinity, there is a j (and N) such that $d_j > 2n + 2K + 1$, for some n. Let $p := \min\{f_m(k) + a_m - f_m(k+1) \mid m = 1, \ldots, K, \; k = 1, \ldots, K\}$. Note that p is positive, since Condition (8.7) implies that $\left(f_l(x) - a_l x - C\right)$ is positive and strictly decreasing (for all l). Now choose n such that $2q = 2\left(f_m(n) - a_m n - C\right) < p$. Note that this is possible, since $f_m(n) - a_m n - C$ goes to zero as n goes to infinity. Consider the $2K + 1$ consecutive assignments starting n units after $\eta(j-1)$. There is at least one server, to which a packet is assigned three times, say m'. Replace the second assignment to m' in this sequence by an assignment to server m.

Define the function $g_i(k) = f_i(k) - a_i k - C$ for all i and consider the cost for a horizon of length l where l is an arbitrary integer larger than $\eta(j-1) + 2n + 2K + 1$. The decrease in cost due to the interchange is

$$\left[f_m(d_j) + f_{m'}(d_1(m')) + f_{m'}(d_2(m'))\right] -$$
$$\left[f_m(d'_j) + f_m(d''_j) + f_{m'}(d_1(m') + d_2(m'))\right]$$
$$= \left[g_m(d_j) + g_{m'}(d_1(m')) + g_{m'}(d_2(m'))\right] -$$
$$\left[g_m(d'_j) + g_m(d''_j) + g_{m'}(d_1(m') + d_2(m'))\right]$$
$$> \left[g_m(d_j) + g_{m'}(d_1(m')) + g_{m'}(d_2(m'))\right] - \left[2g_m(n) + g_{m'}(d_1(m') + 1)\right]$$
$$= g_m(d_j) + g_{m'}(d_2(m')) + \left[g_{m'}(d_1(m')) - g_{m'}(d_1(m') + 1)\right] - \left[2g_m(n)\right]$$
$$> g_m(d_j) + g_{m'}(d_2(m')) + p - 2q > 0,$$

where $d_1(m')$, $d_2(m')$, d'_j and d''_j are defined as before. The first inequality follows from the fact that $n < d'_j$, $n < d''_j$, $d_1(m') + 1 \leq d_1(m') + d_2(m')$ and $f_m(x) - a_m x - C$ is decreasing. The second inequality follows from the definition of p. Since $f_m - a_m - C$ is positive it follows by construction of n that the last inequality holds.

Repeating the interchange procedure for every j for which $d_j > 2n+2K+1$ (when dealing with Condition (8.7)) or for which $d_j > n + 2K + 1$ (when dealing with Condition (8.6)) provides us a policy π' such that $C(\pi') \leq C(\pi)$ and $\sup\{d_{\bar{j}}' \mid j \in \mathbb{N}\} < 2n + 2K + 1$. By repeating this procedure for every server, we get an optimal policy that visits a finite number of states. By Chapter 8 of Puterman [95] we know that the optimal policy can be chosen stationary. It follows that $\pi_n(h_n) = \pi_0(x_n)$. Since the state transitions are deterministic it follows that the optimal policy is periodic.   □

8.9 Multimodularity of the global cost: two servers

We consider in this Section the special case of two servers.

In previous Sections we have already established the multimodularity of the cost related to a single server. This was sufficient to get optimality results for the multidimensional case (the routing problem) since the optimality of bracket policies has been established under the conditions that the cost of each component separately be multimodular (Section 1.4).

In this Section we consider directly the multimodularity of the combined cost of the two queues. Although this is not necessary for obtaining the regularity structure of optimal policies, this is an important property for optimization purposes, as it allows us to use local search procedures (that are based on the fact that local minimum is also a global one, see Corollary 2).

The notation of the distance sequence can be beneficially used to approach the decision problem. After the first assignment to server m, the distance sequence $d(m)$ for server m is periodic, say with period $P(m)$. Therefore in future discussions we will write $\pi = (\pi_1, \ldots, \pi_n)$ for the periodic assignment sequence with period n and with a slight abuse of notation we denote the periodic distance sequence for server m by $d(m) = (d(m)_1, \ldots, d(m)_{P(m)})$.

The periodicity reduces the cost function in complexity. Since we use the expected average cost function, we only have to consider the costs incurred during one period. It would be interesting to establish multimodular properties for any K. Unfortunately it is not clear how even to define multimodularity for $K > 2$. We thus consider below $K = 2$. The expected average cost is given by

$$g(\pi) = \sum_{m=1}^{K} g_m(\pi) = \frac{1}{n} \sum_{m=1}^{K} \sum_{j=1}^{P(m)} f_m(d(m)_j). \qquad (8.9)$$

It is tempting to formulate that $g_m(\pi)$ is multimodular in π for $m = 1, 2$. Note that this is not necessarily true, since an operation $v \in \mathcal{F}$ applied to π leads to different changes in the distance sequences for the different servers.

We shall thus use an alternative description for the average cost through the period of server 1. Define g_m' as follows:

$$g'_m(\pi) = \frac{1}{n} \sum_{j=1}^{d(1)} f_m(d_j(1)).$$ (8.10)

We note that the function $g'_m(\pi)$ only looks at the distance sequence assigned to the first server with respect to π using cost function f_m. By the symmetry between the assignments to the two servers $g(\pi)$ can now be expressed as $g(\pi) = g'_1(\pi) + g'_2(\mathbf{3} - \pi)$. ($\mathbf{3}$ is the vector whose components are all 3.) Note that $d_j(1) = d_j(1)(\pi)$ is a function of π, and we have

$$d_j(1)(\mathbf{3} - \pi) = d_j(2)(\pi).$$

We first prove that $g'_m(\pi)$ is multimodular in π. Then multimodularity of $g(\pi)$ follows as the sum of two multimodular functions.

Lemma 51. *Assume that f_m are convex. Let π be a fixed periodic policy with period n. Let $g'_m(\pi)$ be defined as in (8.10). Then $g'_m(\pi)$ is a multimodular function in π.*

Proof. Since π is a periodic sequence, the distance sequence $d = d(1)$ is also a periodic function, say with period P. Now, define the function h_j for $j = 1, \ldots, P$ by $h_j(\pi) = f_m(d_j)$. The function h_j represents the cost of the $(j+1)^{\text{st}}$ assignment to server m by looking at the jth interarrival time. We will first check the conditions for multimodularity for $\mathcal{V} = \{b_1, \ldots, b_{n-1}\}$, where $b'_i s$ are the elements of $\mathcal{F}$ (i.e. $b_0 = -e_1, b_1 = s_2, \ldots, b_{n-1} = s_n, b_n = e_n$).

Let $v, w \in \mathcal{V}$ with $v \neq w$. If none of these elements changes the length of the jth interarrival time then $h_j(\pi) = h_j(\pi+v) = h_j(\pi+w) = h_j(\pi+v+w)$. Suppose that only one of the elements changes the length of the interarrival time, say v, then $h_j(\pi + v) = h_j(\pi + v + w)$ and $h_j(\pi) = h_j(\pi + w)$. In both cases the function $h_j(\pi)$ satisfies the conditions for multimodularity.

Now suppose that v adds and w decreases the length of the jth interarrival time by one. Then $d_j(\pi + v) - d_j(\pi) = d_j(\pi + v + w) - d_j(\pi + w)$. Since h_j is a convex function, it follows that $h_j(\pi + w) - h_j(\pi + v + w) \geq h_j(\pi) - h_j(\pi + v)$. Now the multimodularity condition in Equation (1.1) directly follows by rearranging the terms. Since $g'_m(\pi)$ is a sum of $h_j(\pi)$ it follows that $g'_m(\pi)$ is multimodular for $\mathcal{V}$.

Now consider the elements b_0 and b_n and note that the application of b_0 and b_n to π splits an interarrival period and merges two interarrival periods respectively. Therefore

$$n\, g'_m(\pi + b_0) = n\, g'_m(\pi) - f_m(d_1) - f_m(d_P) + f_m(d_1 + d_P),$$

$$n\, g'_m(\pi + b_n) = n\, g'_m(\pi) - f_m(d_P) + f_m(d_P - 1) + f_m(1),$$

$$n\, g'_m(\pi + b_0 + b_n) = n\, g'_m(\pi) - f_m(d_1) - f_m(d_P) + f_m(d_1 + 1) + f_m(d_P - 1).$$

Now $n[g'_m(\pi + b_0) + g'_m(\pi + b_n) - g'_m(\pi) - g'_m(\pi + b_0 + b_n)] = [f_m(d_1 + d_P) + f_m(1)] - [f_m(d_1 + 1) + f_m(d_P)]$. Let $k = d_1 + d_P + 1$. Since the function $f_m(x) + f_m(y)$ with $x + y = k$ is a symmetric and convex function, it follows from Proposition C2 of Chapter 3 of Marshall and Olkin [88], that $f_m(x) + f_m(y)$ is also Schur-convex. Since $(d_1 + 1, d_P) \prec (d_1 + d_P, 1)$, the quantity above is non-negative.

In the case that we use $w = b_0$ and $v \in \mathcal{V}$ such that v does not alter d_1, then it follows that $g'_m(\pi + v + w) = g'_m(\pi + v) + g'_m(\pi + w) - g'_m(\pi)$. The same holds for $w = b_n$ and $v \in \mathcal{V}$ such that v does not alter d_P. Suppose that v does alter d_1, then we have $n[g'_m(\pi + b_0) + g'_m(\pi + v) - g'_m(\pi) - g'_m(\pi + b_0 + v)] = [f_m(d_1 + d_P) + f_m(d_1 - 1)] - [f_m(d_1 + d_P - 1) + f_m(d_1)]$. When v alters d_P we have $n[g'_m(\pi + b_n) + g'_m(\pi + v) - g'_m(\pi) - g'_m(\pi + b_n + v)] = [f_m(d_P + 1) + f_m(l)] - [f_m(d_P) + f_m(l + 1)]$ for some $l < d_P$. Now by applying the same argument as in the case of b_0 and b_n we derive multimodularity of $g'_m(\pi)$ for the base $\mathcal{F}$. $\square$

Now we will prove that $g(\pi)$, which is given by $g(\pi) = g'_1(\pi) + g'_2(3 - \pi)$ is multimodular. The proof is based on the fact that if a function is multimodular with respect to $\mathcal{F}$, then it is also multimodular with respect to $\mathcal{G}$ (defined above Definition 1).

Theorem 43. *Let g'_1 and g'_2 be multimodular functions. Then the function $g(\pi)$ given by $g(\pi) = c_1 g'_1(\pi) + c_2 g'_2(3 - \pi)$ for positive constants c_1 and c_2 is multimodular in π.*

Proof. Let $v, w \in \mathcal{F}$, such that $v \neq w$. Then

$$g(\pi + v) + g(\pi + w)$$
$$= c_1 g'_1(\pi + v) + c_2 g'_2(3 - \pi - v) + c_1 g'_1(\pi + w) + c_2 g'_2(3 - \pi - w)$$
$$= c_1 [g'_1(\pi + v) + g'_1(\pi + w)] + c_2 [g'_2(3 - \pi - v) + g'_2(3 - \pi - w)]$$
$$\geq c_1 [g'_1(\pi) + g'_1(\pi + v + w)] + c_2 [g'_2(3 - \pi) + g'_2(3 - \pi - v - w)]$$
$$= c_1 g'_1(\pi) + c_2 g'_2(3 - \pi) + c_1 g'_1(\pi + v + w) + c_2 g'_2(3 - \pi - v - w)$$
$$= g(\pi) + g(\pi + v + w).$$

The inequality in the fourth line holds, since g'_1 is multimodular with respect to $\mathcal{F}$ and g'_2 is multimodular with respect to $\mathcal{G}$. $\square$

8.10 Examples of arrival processes

In today's information and communication systems the traffic pattern may be quite complex, as they may represent a variety of data, such as customer phone calls, compressed video frames and other electronic information. Modern communication systems are designed to accommodate such a heterogeneous input and therefore the arrival process used in mathematical models is of crucial importance to the engineering and performance analysis of these systems. In this section we elaborate on the setting of Example 6 with different arrival processes and derive explicit formulae for the cost function for the corresponding arrival process.

Assume that the controller wishes to minimize the number of lost packets (i.e. the number of preempted packets) per unit time; note that this is equivalent to maximizing the throughput of the system. Furthermore let the services at server i be exponentially distributed with rate μ_i independent of the other servers. Since we know that there exists a periodic optimal policy, we can write the cost induced by using policy π by

$$g(\pi) = \frac{1}{n} \sum_{m=1}^{M} \sum_{j=1}^{P(m)} \left[\frac{\lambda}{\lambda + \mu_m} \right]^{d_j(m)},$$

in case the arrival process is a Poisson process with parameter λ. In Koole [81] it was shown that the optimal policy has a period of the form $(1, 2, \ldots, 2)$, where 2 is the faster server. In Chapter 6 this result was generalized for general stationary arrival processes. Hence suppose that $\lambda = 1$ and $\mu_1 = 1$, then the cost function can be parameterized by the period n and the server speed $\mu_2 \geq \mu_1$ given by

$$g(n, \mu_2) = \frac{1}{n} \left(\frac{1}{2} \right)^n + \frac{1}{n} \left(\frac{1}{1 + \mu_2} \right)^2 + \frac{n-2}{n} \left(\frac{1}{1 + \mu_2} \right).$$

By solving the equations $g(n, \mu_2) = g(n + 1, \mu_2)$ for $n \geq 2$ we can compute the server rates μ_2 for which the optimal policy changes period. For example: the optimal policy changes from $(1, 2)$ to $(1, 2, 2)$ when $\mu_2 \geq 1 + \sqrt{2}$. The results of the computation are depicted in Figure 8.1.

Markov Modulated Poisson Process

An important class of models for arrival processes is given by Markov modulated models. The key idea behind this concept is to use an explicit notion of states of an auxiliary Markov process into the description of the arrival process. The Markov process evolves as time passes and its current state modulates the probability law of the arrival process. The utility of such arrival processes is that they can capture bursty inputs.

The Markov Modulated Poisson Process (MMPP) is the most commonly used Markov modulated model and is constructed by varying the arrival rate of a Poisson process according to an underlying continuous time Markov process, which is independent of the arrival process. Therefore let $\{X_n \,|\, n \geq 0\}$ be a continuous time irreducible Markov process with k states. When the Markov process is in state i, arrivals occur according to a Poisson process with rate λ_i. Let p_{ij} denote the transition probability to go from state i to state j and let Q be the infinitesimal generator of the Markov process. Let $\Lambda = \text{diag}(\lambda_1, \ldots, \lambda_k)$ be the matrix with the arrival rates on the diagonal

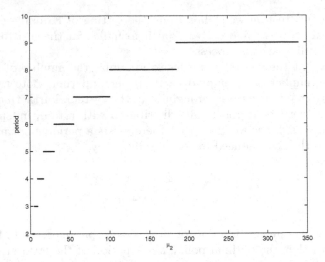

Fig. 8.1. Relationship between n and μ_2

and $\lambda = (\lambda_1, \ldots, \lambda_k)$ the vector of arrival rates. With this notation, we can use the matrix analytic approach to derive a formula for $f_m(n)$.

Theorem 44. *(Section 5.3, [92]) The sequence $\{(X_n, \delta_n) \mid n \geq 0\}$ is a Markov renewal sequence with transition probability matrix $F(t)$ given by*

$$F(t) = \int_0^t e^{(Q-\Lambda)u} \, du \, \Lambda = \left[I - e^{(Q-\Lambda)t}\right] (\Lambda - Q)^{-1} \Lambda.$$

The interpretation of the matrix $F(t)$ is as follows. The elements $F_{ij}(t)$ are given by the conditional probabilities $\mathbb{P}(X_{n+1} = j, \delta_{n+1} \leq t \mid X_n = i)$ for $n \geq 1$. Since $F(t)$ is a transition probability matrix, it follows that $F(\infty)$ given by $(\Lambda - Q)^{-1} \Lambda$ is a stochastic matrix.

The MMPP is fully parameterized by specifying the initial probability vector q, the infinitesimal generator Q of the Markov process and the vector λ of arrival rates. Let the row vector s be the steady state vector of the Markov process. Then s satisfies the equations $sQ = 0$ and $se = 1$, where $e = (1, \ldots, 1)$. Define the row vector $q = s\Lambda/s\lambda$. Then q is the stationary vector of $F(\infty)$ and makes the MMPP interval stationary (see [45]). This is intuitively clear since the stationary vector of $F(\infty)$ means that we obtain the MMPP started at an arbitrary arrival epoch.

In order to find an explicit expression for the cost function, we compute the Laplace-Stieltjes transform $f^*(\mu)$ of the matrix F. Since F is a matrix, we use matrix operations in order to derive $f^*(\mu)$, which will also be a matrix. The interpretation of the elements $f_{ij}^*(\mu)$ are given by $\mathbb{E}\left[e^{-\mu\delta_{n+1}} \mathbf{1}_{\{X_{n+1}=j\}} \mid X_n = i\right]$ for $n \geq 1$, where $\mathbf{1}$ is the indicator function. Let I denote the identity matrix, then $f^*(\mu)$ is given by

$$f^*(\mu) = \int_0^\infty e^{-\mu I t} F(dt) = \int_0^\infty e^{-\mu I t} e^{(Q-\Lambda)t} (\Lambda - Q)(\Lambda - Q)^{-1} \Lambda \, dt$$

$$= \int_0^\infty e^{-(\mu I - Q + \Lambda)t} \, dt \, \Lambda = (\mu I - Q + \Lambda)^{-1} \Lambda.$$

Now we can compute $f_m(n) = \mathbb{P}(S_m \geq \delta_1 + \cdots + \delta_n) = \mathbb{E}[-\mu \sum_{k=1}^n \delta_k]$. The next lemma shows that this is simply given by the product of $f^*(\mu)$ with itself. Note that we do not need the assumption of independence of the interarrival times to derive this result.

Lemma 52. *Let $f^*(\mu)$ be the Laplace-Stieltjes transform of F, where F is a transition probability matrix of a stationary arrival process. Then*

$$\mathbb{E} \, exp\left(-\mu \sum_{k=1}^n \delta_k\right) = q \left[f^*(\mu)\right]^n e.$$

Proof. Define a matrix Q_n with entries $Q_n(i, j)$ given by

$$Q_n(i, j) = \mathbb{E}\left[\exp\left(-\mu \sum_{k=1}^n \delta_k\right) 1_{\{X_n = j\}} \,\Big|\, X_0 = i\right].$$

Note that Q_1 is given by $f^*(\mu)$. By using the stationarity of the arrival process it follows that $Q_n(i, j)$ is recursively defined by

$$Q_n(i, j) = \sum_{l=1}^m Q_{n-1}(i, l) \, \mathbb{E}\left[\exp(-\mu \delta_n) 1_{\{X_n = j\}} \,|\, X_{n-1} = l\right]$$

$$= \sum_{l=1}^m Q_{n-1}(i, l) \, \mathbb{E}\left[\exp(-\mu \delta_1) 1_{\{X_1 = j\}} \,|\, X_0 = l\right]$$

$$= \sum_{l=1}^m Q_{n-1}(i, l) \cdot Q_1(l, j).$$

Note that the last line exactly denotes the matrix product, thus $Q_n = Q_{n-1} \cdot Q_1$. By induction it follows that $Q_n = (Q_1)^n$. Then it follows that

$$\mathbb{E} \, exp\left(-\mu \sum_{k=1}^n \delta_k\right) = \sum_{i=1}^m \sum_{j=1}^m \mathbb{P}(X_0 = i) \, Q_n(i, j) = q \left[f^*(\mu)\right]^n e.$$

The last equation holds since the row vector q is the initial state of the Markov process and summing over all j is the same as right multiplying by e. $\square$

Hence $g(\pi)$ is given by

$$g(\pi) = \frac{1}{n} \sum_{m=1}^{K} \sum_{j=1}^{P(m)} q\big[(\mu_m I - Q + \Lambda)^{-1}\Lambda\big]^{d_j(m)} e.$$

Note that although in case of two servers we know the structure of the optimal policy, it is not intuitively clear that it is optimal in the case of the MMPP. The following argument will clarify this statement. Suppose that one has an MMPP with two states. Choose the rates λ_1 and λ_2 of the Poisson processes such that the policies would have period 2 and 3 respectively if the MMPP is not allowed to change state. One could expect that if the transition probabilities to go to another state are very small, the optimal policy should be a mixture of both policies. But this is not the case.

Markovian Arrival Process

The Markovian arrival process model (MAP) is a broad subclass of models for arrival processes. It has the special property that every marked point process is the weak limit of a sequence of Markovian arrival processes (see Asmussen and Koole [18]). In practice this means that very general point processes can be approximated by appropriate MAP's. The utility of the MAP follows from the fact that it is a versatile, yet tractable, family of models, which captures both bursty inputs and regular inputs.

The MAP can be described as follows. Let $\{X_n \,|\, n \geq 0\}$ be a continuous time irreducible Markov process with k states. Assume that the Markov process is in state i. The sojourn time in this state is exponentially distributed with parameter γ_i. After this time has elapsed, there are two transition possibilities. Either the Markov process moves to state j with probability p_{ij} with generating an arrival or the process moves to state $j \neq i$ with probability q_{ij} without generating an arrival.

This definition also gives rise to a natural description of the model in terms of matrix algebra. Define the matrix C with elements $C_{ij} = \gamma_i q_{ij}$ for $i \neq j$. Set the elements C_{ii} equal to $-\gamma_i$. Define the matrix B with elements $B_{ij} = \gamma_i p_{ij}$. The interpretation of these matrices is given as follows. The elementary probability that there is no arrival in an infinitesimal interval of length dt when the Markov process moves from state i to state j is given by $C_{ij}\, dt$. A similar interpretation holds for B, but in this case it represents the elementary probability that an arrival occurs. The infinitesimal generator of the Markov process is then given by $C + B$. Note that a MMPP can be derived by choosing $C = Q - \Lambda$ and $B = \Lambda$.

In order to derive an explicit expression for the cost function, we use the same approach as in the case of the MMPP. The transition probability matrix $F(t)$ of the Markov renewal process $\{(X_n, \delta_n) \,|\, n \geq 0\}$ is of the form (see [87])

$$F(t) = \int_0^t e^{Cu}\, du B = \big[I - e^{Ct}\big]\big(-C^{-1}B\big).$$

Again the elements of the matrix $F(t)$ have the interpretation that $F_{ij}(t)$ is given by $\mathbb{P}(X_{n+1} = j, \delta_{n+1} \leq t \mid X_n = i)$ for $n \geq 1$. It also follows that $F(\infty)$ defined by $-C^{-1}B$ is a stochastic matrix. Let the row vector s be the steady state vector of the Markov process. Then s satisfies the equations $s(C + B) = 0$ and $se = 1$. Define the vector row vector $q = sB/sBe$. Then q is the stationary vector of $F(\infty)$. This fact can be easily seen upon noting that $sB = s(C + B - C) = s(C + B) - sC = -sC$. With this observation it follows that $q F(\infty) = (sBe)^{-1} s CC^{-1}B = q$. The MAP defined by q, C and B has stationary interarrival times.

The Laplace-Stieltjes transform $f^*(\mu)$ of the matrix F is given by

$$f^*(\mu) = \int_0^\infty e^{-\mu I t} F(dt) = \int_0^\infty e^{-\mu I t} e^{Ct} (-C)(-C^{-1}) B \, dt$$

$$= \int_0^\infty e^{-(\mu I - C)t} \, dt \, B = (\mu I - C)^{-1} B.$$

The interpretation of f^* is given by the elements $f_{ij}^*(\mu)$, which represent the value of the expectation $\mathbb{E}\left[e^{\mu \delta_{n+1}} \mathbf{1}_{\{X_{n+1}=j\}} \mid X_n = i\right]$. By Lemma 52 we know that $f_m(n)$ is given by the product of f^*. Therefore the cost function, when using the MAP as arrival process, is given by

$$g(\pi) = \frac{1}{n} \sum_{m=1}^K \sum_{j=1}^{P(m)} q\left[(\mu_m I - C)^{-1}B\right]^{d_j(m)} e.$$

8.11 Robot scheduling for web search engines

In [10] we specified a routing problem where the expected average weighted loss rate was to be minimized (or equivalently, the average weighted throughput or average weighted number of packets at service was to be maximized). This gave rise (Lemma 47) to an immediate cost of the form:

$$c(x, a) = \mathbb{E} \exp\left[-\mu_a \sum_{i=1}^{x_a} \delta_i\right].$$

Due to stationarity of the interarrival times, this cost function satisfies Condition (8.7) (with $a_m = C = 0$). We assume of course that δ_i are not all zero, which then implies the strict convexity of f_m. Indeed, denote

$$y = \exp\left[-\mu_a \sum_{k=2}^m \delta_k\right].$$

Let x be a state such that $x_a = m > 0$ for a particular action a. Since the interarrival times are a stationary sequence,

$$c(x, a) = \mathbb{E}\, y e^{-\mu_a \delta_{m+1}} = \mathbb{E}\, y e^{-\mu_a \delta_1},$$
$$c(x + e_a, a) = \mathbb{E}\, y e^{-\mu_a [\delta_{m+1} + \delta_{m+2}]} = \mathbb{E}\, y e^{-\mu_a [\delta_1 + \delta_{m+1}]},$$
$$c(x - e_a, a) = \mathbb{E}\, y.$$

Since the function $r(x) := y e^{-\mu_a x}$ is convex in x, it follows that $r(\delta_1 + z) - r(z)$ is increasing in z, so that

$$r(\delta_1 + \delta_{m+1}) - r(\delta_{m+1}) \geq r(\delta_1) - r(0).$$

By taking expectations, this implies the convexity. Thus the results of the previous sections apply. In this section we present another application studied in Coffman, Liu and Weber [42] under assumptions that are more restrictive than ours.

The World Wide Web offers search engines, such as Altavista, Lycos, Infoseek and Yahoo, that serve as a database that allow to search information on the web. These search engines often use robots that periodically traverse a part of the web structure in order to keep the database up-to-date.

We consider a problem where we assume that there is a fixed number of K web-pages. The contents of page i is modified at time epochs that follow a Poisson process with parameter μ_i. The time a page is considered up-to-date by the search engine is the time since the last visit by the robot until the next time instant of modification; at this point the web-page is considered out-of-date until the next time it is visited by the robot. The times between updates by the robot are given by a sequence δ_n. In Coffman, Liu and Weber [42] these times are assumed to be i.i.d., but in our framework we may allow them to form a general stationary sequence.

Let o_i denote the obsolescence rate of page i, i.e. the fraction of time that page i is out-of-date. Then the problem is to find an optimal visiting schedule such that the sum of the obsolescence rates o_i weighted by specified constants c_i is minimized. A reasonable choice for the weights c_i would be the customer page-access frequency, because the total cost then represents the customer total error rate. The case where the customer access frequency $c_i = k \mu_i$ is proportional to the page-change rate μ_i is reasonable under this interpretation, since the greater the interest for a particular page is, the more likely the frequency of page modification is.

We now show that this problem is equivalent to the problem studied in Section 8.8. Indeed, the robot can be considered as the controller in the previous problem. An assignment of the nth packet to server i in the original problem corresponds with sending the robot to page i and requires δ_n time in order to complete the update. The lengths of the periods that page i is up-to-date corresponds to the service times of packets before server i in the original problem. Page i is considered to be out-of-date when server i is idle.

Let S_i be an exponential random variable with parameter μ_i. The cost which one incurs when sending a packet to server a should reflect the expected obsolescence time. Some straightforward computations yield

$$c(x,a) = k\,\mu_a \mathbb{E}\left[\sum_{i=1}^{x_a}\delta_i - S_a\right]^+$$

$$= k\,\mu_a\,\mathbb{E}\left[\sum_{i=1}^{x_a}\delta_i - S_a\right] + k\,\mu_a\,\mathbb{E}\left[S_a - \sum_{i=1}^{x_a}\delta_i\right]^+$$

$$= k\,\mu_a\,x_a\mathbb{E}\,\delta_1 + k\left[\mathbb{E}\,\exp\left(-\mu_a\sum_{i=1}^{x_a}\delta_i\right) - 1\right].$$

This cost function clearly satisfies Condition (8.7). Hence the theorems from the previous sections can indeed be applied to this scheduling problem.

Remark 15. The assumption that the weights c_i are proportional to the page-change rates μ_i is essential in this problem. The cost function for general c_i is given by

$$c(x,a) = c_a\,x_a\mathbb{E}\,\delta_1 + \frac{c_a}{\mu_a}\left[\mathbb{E}\,\exp\left(-\mu_a\sum_{i=1}^{x_a}\delta_i\right) - 1\right].$$

When the c_i are not proportional to the page-change rates, then the cost function is of the type mentioned in Example 7. Therefore if for some i, $\frac{c_i}{\mu_i}$ is sufficiently large (in comparison to others) then it becomes optimal never to update page i. This is an undesirable situation and it shows that the problem is not always well posed when the costs are not proportional to the page-change rates μ_i.

Remark 16. This problem of robot scheduling for web search engines illustrates the importance of including the terminal cost c_T in terms of modeling. Indeed, for any finite horizon T, if we wish that the cost represents indeed the obsolescence time of a page, we have to make sure that if this page is never updated (or at least it stops to be updated after some time), this will be reflected in the cost. It is easy to see that the terminal cost defined in Section 8.8 indeed takes care of that.

9 Vacancies, service allocation and polling

9.1 Introduction

We consider in this chapter the control of vacations in several queuing settings. This chapter is mainly based on [7]. Vacations are time periods during which the server does not serve customers, even when there are some in the system.

We consider three types of vacation models:
(i) vacations driven by service completions,
(ii) vacations driven (triggered) by arrivals, in which both the beginning as well as the end of a vacation are related to arrivals instants, and
(iii) the potential vacation times form a renewal process and the arrival epochs are stationary subsequence of this renewal process.

We shall consider two types of problems. In both cases, we consider open-loop control where the controller has no information on the state of the system.

In the first problem, there is a single infinite FIFO queue. Some *vacation opportunities* are presented (depending on the type of vacations, these opportunities are triggered by service, or arrivals or by some other mechanism). The server should go on vacation during a fraction of at least

p of these opportunities. The goal of the control is to minimize the average workload or waiting time (or any nondecreasing convex functions of these).

The second problem concerns a polling model. There are several infinite queues; when serving one queue, the server is unavailable for other queues. We wish to minimize some linear combination of the average workloads in the different queues (or of waiting times, or of nondecreasing convex functions of these).

9.1.1 Organization of the chapter

The structure of the Chapter is as follows. In Section 9.2 we formulate the two types of control problems: the one of the single queue (P1), and the one of optimal polling of several queues (P2). We then formulate the four generic type of results obtained for these problems. In the following sections we then present and analyze the different models for the vacations and derive the appropriate results for the control. In Section 9.3 we analyze the case where vacations are triggered by service completion. In Sections 9.4 and 9.5 we consider an arrival driven vacations. Finally, in 9.6 and 9.7 we consider models where the vacations are a renewal process.

9.2 The generic control models and main results

We consider two generic control problems in this paper. We shall formulate these problems in an abstract setting, and then focus in the following sections on specific assumptions.

Constrained model:

- Customers (or some demand for service) arrive to a single G/G/1 queue (or to a network) according to some given stationary process.
- There is a single server at some output of the service facility (queue or network) that may be either active in providing service, or may be absent for vacation periods. Some "vacation opportunities" are presented, in which that server can decide whether to take a vacation or not. These opportunities would depend on the model we consider: they could be related to arrivals, to end of services, or to be a renewal process.
- At the nth vacation opportunity, the controller chooses a control a_n that determines the number $num(a_n)$ of vacations to be taken at the nth vacation opportunity. Let a be the control sequence $(a_1, a_2, ...)$.
- Performance measures and objectives:
  - Let $h : \mathbb{R} \to \mathbb{R}$ be a convex nondecreasing function.
  - Given some fraction p, consider the class $\Pi(p)$ of all policies that satisfy the constraint:

$$\liminf_{s \to \infty} \frac{1}{s} \sum_{n=1}^{s} num(a_n) \geq p. \tag{9.1}$$

Consider the following problems.

(P1) The Vacation control for one queue
(P1a): Let $\mathcal{W}_n$ be the waiting time of the nth arriving customer. Define the average expectation of the function h of the waiting time:

$$g(a) = \limsup_{s \to \infty} \frac{1}{s} \sum_{n=1}^{s} \mathbb{E}h(\mathcal{W}_n(a_1, ..., a_n)), \tag{9.2}$$

The objective is to minimize $g(a)$ over $a \in \Pi(p)$.
(P1b): Let V_n be the workload in the system at some special time instants T_n.
Define the average expectation of the function h of the workload:

$$g'(a) = \limsup_{s \to \infty} \frac{1}{s} \sum_{n=1}^{s} \mathbb{E}h(V_n(a_1, ..., a_n)). \qquad (9.3)$$

The objective is to minimize $g'(a)$ over $a \in \Pi(p)$.

Next we present our first generic result that will be established in the following sections for different models. To that end we introduce some definitions. Let p and θ be two positive reals. We will use the *bracket sequence* $\{a_k^p(\theta)\}$ with rate p and initial phase θ:

$$a_k^p(\theta) = \lfloor kp + \theta \rfloor - \lfloor (k-1)p + \theta \rfloor, \qquad (9.4)$$

where $\lfloor x \rfloor$ is the largest integer smaller than or equal to x.

In the different models that we study in the next sections, we shall show the following for both problems (P1a) and (P1b):

Result 45 *There exists some rate p^* such that for any θ, the sequence $a_k^{p^*}(\theta)$ is optimal.*

In Section 9.3, we shall establish Result 45 and show that $p^* = p$, where p is given in constraint (9.1). In all other sections where (P1) is considered, we shall have $p^* = 1 - p$. The difference is simply due to different definitions of the control in different models that we study.

Next we describe the nonconstrained problem (P2):

(P2) Polling of several queues

- There are K queues to which a server is allocated. When serving one queue, the server is unavailable for other queues.
- Again, some "vacation opportunities" (or "switching opportunities") are presented, in which the server can decide to stop serving one queue and start serving another one.
- At the nth vacation opportunity, the controller chooses a control $a_n = (a_n^1, ..., a_n^K)$; for each n, all components of a_n are 0 except one component that may be 1 or 0, $a_n^i = 1$ will mean that the server is assigned to queue i at the nth opportunity instant. Performance measures and objectives: Let $h_i : \mathbb{R} \to \mathbb{R}$ be a convex nondecreasing function, $i = 1, ..., K$.

(P2a): Let W_n^i be the waiting time of the nth arrival to queue i. Define

$$g(a) \overset{\text{def}}{=} \limsup_{N \to \infty} \frac{1}{N} \sum_{i=1}^{K} \left(\sum_{n=1}^{N} f_n^i(a^i) \right),$$

where $f_n^i(a^i) \overset{\text{def}}{=} \mathbb{E}h_i(W_n^i(a))$ and W_n^i depends on a only through a^i. The objective is to minimize $g(a)$.

(P2b): Let V_n^i be the workload in the ith queue at some special time instants T_n. Define

$$g'(a) \stackrel{\text{def}}{=} \limsup_{N \to \infty} \frac{1}{N} \sum_{i=1}^{K} \left(\sum_{n=1}^{N} \tilde{f}_n^i(a^i) \right),$$

where $\tilde{f}_n^i(a^i) \stackrel{\text{def}}{=} \mathbb{E}h_i(V_n^i(a))$.
The objective is to minimize $g'(a)$.

Next we present generic results for problem (P2). To that end we introduce some further definitions. For any vector $\theta \in \mathbb{R}^K$ (which is called a phase vector) and a rate vector $p \in [0,1]^K$, we define the vector valued sequence $a(p, \theta)$ by

$$a_n^i(p, \theta) = \lfloor np_i + \theta_i \rfloor - \lfloor (n-1)p_i + \theta_i \rfloor. \tag{9.5}$$

Note that $a(p, \theta)$ need not correspond to a policy since it may have more than one component that equals to 1 for the same n. In that case we say that it is not feasible, if it defines a policy, we say that it is feasible.

The following results for problems (P2a) and (P2b) follow from the results in Chapter 6.

Result 46 *Assume that $K = 2$. There exist some p^* and θ such that $a(p^*, \theta)$ is a feasible policy and is optimal.*

Result 47 *Consider an arbitrary K. Suppose costs and service disciplines are symmetric for all queues. Then the round robin policy is optimal for (P2a) (resp. for (P2b)).*

And more generally, for $K > 2$ we have:

Result 48 The sequence of functions f_n allows one to construct some convex function $\overline{f} : \mathbb{R}^K \to \mathbb{R}$ as in Chapter 1 with the following property. Let p^* be the vector that minimizes this function. Assume that there is a sequence of numbers $\{i_n\}_n$, where $i_n \in \{1, ..., K\}$ such that for every $k \in 1, ..., K$ the sequence $a_n^k = 1\{i_n = k\}$ is bracket with rate p_k^* (for some θ that may depend on k). Then $\{a_n^k\}$ are optimal for (P2a) and (P2b).

The main tool for obtaining the above results is again by establishing the multimodularity of some sequence of functions $f_n : \overline{\mathbb{Z}}^n \to \mathbb{R}$, where $\overline{\mathbb{Z}}^n$ is some convex subset of $\mathbb{Z}^n$, the set of n-dimensional vectors of integers.

9.3 A single queue with service driven vacations

Consider a single G/G/1 queue (Problem (P1)). The nth customer arrives at time T_n, bringing a workload of σ_n to the system. Customers are served according to the FIFO order. The arrival process will be assumed to be a point process throughout the paper, unless otherwise stated, and we assume that $T_0 < 0 \leq T_1 \leq T_2 \leq ....$

Let $\tau_n = T_{n+1} - T_n$ denote the inter-arrival times. When a service of a customer is completed, the server is allowed to go on vacation. We consider the so-called repeated vacation model, where on each completion of a vacation, another vacation can be initiated.

In this model, "vacation opportunities" are thus triggered by the end of a service or of a vacation.

Let $a = (a_1, a_2...)$ be the server's policy, where $a_i \in \mathbb{N}$ has the interpretation of the number of vacations to be taken after the ith service time completion. (In terms of the notation of Section 9.2, we have $num(a_n) = a_n$.)

Let $v_n, n = 1, 2, ...$ be the duration of the nth vacation period. Let $m(n)$ denote the number of vacations that occur till the $n + 1$st service starts. We set $m(0) = 0$. Denote by

$$S_n \stackrel{\text{def}}{=} \sigma_n + \sum_{j=m(n-1)+1}^{m(n)} v_j \qquad (9.6)$$

the total delay related to the nth customer. It is the sum of its service time, plus the vacations that will take place after its service. The waiting time $\mathcal{W}_n$ of the nth customer is given recursively by

$$\mathcal{W}_{n+1} = (\mathcal{W}_n - \tau_n + S_n)^+. \qquad (9.7)$$

In particular, assume that the system is initially empty. If no vacations are taken before the service of the second customer then $m(1) = 0$, and the waiting time of the second customer is

$$\mathcal{W}_2 = (\mathcal{W}_1 - \tau_1 + S_1)^+ = (-\tau_1 + \sigma_1)^+.$$

If, instead, the 1st vacation is taken just after the service of the 1st customer, then $m(1) = 1$ and

$$\mathcal{W}_2 = (\mathcal{W}_1 - \tau_1 + S_1)^+ = (-\tau_1 + \sigma_1 + v_1)^+.$$

Let $V_n = V_n(a)$ be the virtual workload in the system immediately after the nth arrival; it is defined to be the total time required by the server to serve all the customers actually in the system (including the one that arrives at time T_n) plus all the vacation times that will elapse from the arrival instant T_n until customer $n + 1$ is served. V_n is given by

$$V_n = \mathcal{W}_n + S_n. \qquad (9.8)$$

Fix some integer N. $W(a) \stackrel{\text{def}}{=} \mathcal{W}_{N+1}(a)$ can be written as

$$W(a) = \max(0, w_1 + x, w_2, ..., w_N), \text{ where } w_i = w_i(a) = \sum_{j=i}^{N}(S_j - \tau_j). \quad (9.9)$$

Here, $x = 0$ is the initial workload in the system. Define $V(a) \overset{\text{def}}{=} V_N(a)$.

Denote $\mathbb{E}_v$ the expectation over the v's (for given random realization of τ and σ). Below we shall establish the multimodularity of $\mathbb{E}_v h(W(a))$, where h is any nondecreasing convex function. The dynamics of the vacation model resemble those of the admission control model in Chapter 4.

Property 4. The following holds for $0 < i < N$. If $a_i \geq 1$ then

$$w_i(a - s_i) = w_i(a) + v_{m(i-1)}, \qquad w_j(a - s_i) = w_i(a) \text{ for } j \neq i.$$

Note that $(-s_i)$ corresponds to adding a vacation after the end of service of the ith customer, and delete the last vacation from the $i - 1$st one.

Property 5. Consider a vacation sequence $v = (v_0, ...)$, and the shifted sequence: $v' = (v'_0, v'_1, ...) = (v_1, v_2, ...)$. Let w'_i be defined as w_i in (9.9) with the sequence v' replacing v. The following holds for $0 < i < N$.

$$w'_j(a + e_i) = w_i(a) \qquad j > i,$$

$$w'_j(a + e_i) = w_i(a) + v_{m(j-1)} \qquad j \leq i.$$

Lemma 53. *Assume that v is a stationary sequence. Let $h : \mathbb{R} \to \mathbb{R}$ be a nondecreasing convex function. Then $\mathbb{E}_v h(W(a))$ and $\mathbb{E}_v h(V(a))$ are multimodular in a.*

Proof. We consider the basis $\mathcal{F} = (e_1, -s_2, ..., -s_{N-1}, -e_m)$ and check the condition $h(W(a - v)) + h(W(a - w)) \geq h(W(a)) + h(W(a - v - w))$, $v \neq w$.

Case 1: we check for $s_i, s_j, i \neq j$.

$$W(a - s_i) = \max(W(a), w_i + v_{m(i-1)}),$$

$$W(a - s_j) = \max(W(a), w_i + v_{m(j-1)}),$$

$$W(a - s_i - s_j) = \max(W(a), w_i + v_{m(i-1)}, w_j + v_{m(j-1)}).$$

If $W(a)$ is maximizing in the above equation, then

$$h(W(a - s_i)) + h(W(a - s_j)) = 2h(W(a)) = h(W(a)) + h(W(a - s_i - s_j))$$

and the condition is satisfied. It is then also satisfied for $V(a)$ since $V(a) = W_{N-1}(a) + S_N(a)$, and $S_N(a)$ is the same for $a, a - s_i, a - s_j$ and $a - s_i - s_j$. If the maximizer is $w_i + v_{m(i-1)}$, then $W(a - s_i - s_j) = W(a - s_i)$, and the condition follows from the monotonicity of h. By symmetry we obtain the argument for j instead of i. The same argument holds for $V(a)$. Since this inequality holds samplewise, it also holds in expectation.

Case 2: we check for the first term of the basis. It corresponds to adding an additional vacation v_0 after the service of the first customer. In order to check the inequality for the expectation, we consider the following coupling. We consider a second system defined on the same probability space. Quantities in the new system will be denoted with an over-line. We let $\bar{v}_{n+1} = v_n$ for all n.

We compute $W(a)$ and $W(a - s_i)$ in our original system, and compare them to $\overline{W}(a + e_1)$ and $\overline{W}(a - s_i + e_1)$ in our new system.

$$\overline{W}(a + e_1) = \max(W(a), w_1 + v_0),$$

$$\overline{W}(a + e_1 - s_i) = \max(W(a), w_1 + v_0, w_1 + v_0 + v_{m(i-1)}),$$

and $W(a - s_i) = \max(W(a), w_1 + v_{m(i-1)})$. The condition for the multimodularity holds for both $h(W)$ and $h(V)$ by arguments as in Case 1. Since this inequality holds for any sample, it holds in expectation.

Case 3: we check for the last term of the basis. It corresponds to removing the last vacation $v_{m(N)}$.

$$W(a - e_N) = (W(a) - v_{m(N)})^+ \tag{9.10}$$

$$W(a - e_N - s_i) = \max(W(a) - v_{m(N)}, w_i + v_m(i - 1) - v_{m(N)})^+ \tag{9.11}$$

If the argmax of the last maximization is 0, then $W(a - e_N) = W(a - e_N - s_i)$ and the multimodularity condition is seen to hold (since h is nondecreasing). If it is not 0, then $W(a - s_i) - W(a - e_N - s_i) = v_{m(N)}$. Hence

$$W(a) - W(a - e_N) \le W(a - s_i) - W(a - e_N - s_i) = v_{m(N)}.$$

Since h is convex nondecreasing and $W(a - e_N - s_i) \ge W(a - e_N)$,

$$h(W(a - s_i)) - h(W(a - e_N - s_i))$$
$$= h(W(a - e_N - s_i) + v_{m(N)}) - h(W(a - e_N - s_i))$$
$$\ge h(W(a - e_N) + v_{m(N)}) - h(W(a - e_N))$$
$$\ge h(W(a) - h(W(a - e_N)).$$

Next, we check this case for the workload. We have

$$V(a - e_N) = V(a) - v_{m(N)} \tag{9.12}$$
$$V(a - e_N - s_i) = V(a - s_i) - v_{m(N)} \tag{9.13}$$

Since h is convex nondecreasing, and since $V(a - s_i) \ge V(a)$, this implies that

$$h(V(a)) - h(V(a - e_N))$$
$$= h(V(a) + v_{m(N)}) - h(V(a)) \le h(V(a - s_i) + v_{m(N)}) - h(V(a - s_i))$$
$$= h(V(a - s_i)) - h(V(a - e_N - s_i)).$$

Thus the multimodularity condition holds for $h(V)$ as well.

Again, the condition for the multimodularity holds samplepathwise, and thus in expectation. ∎

Theorem 49. *Consider problem (P1), where the expectation* $\mathbb{E}$ *in (9.2) is with respect to the random sequences* σ, τ, v, *and where* $num(a_n) = a_n$. *Assume that*

- *the inter-arrival and service time sequence* (τ_n, σ_n) *is stationary, and is independent of the sequence* v (τ *and* σ *may depend on each other*),
- *the* v *sequence is stationary,*
- *the following stability condition holds: the queue is in a stationary regime at time 0, corresponding to the policy that does not take vacations (see more details in Remark 17 below).*

Then Result 45 holds for (P1a) and (P1b), where $p^* = p$ *is the fraction given in the constraint (9.1).*

Proof. The proof is based on Theorem 6. For any integer n, the function $f_n(a) \overset{\text{def}}{=} \mathbb{E}_{\tau,\sigma,v} h(\mathcal{W}_n(a_1, ..., a_n))$ is nondecreasing in each a_i, $i = 1, ..., n$. Moreover,

$$f_k(a_1, ..., a_k) = f_m(\underbrace{0, ..., 0}_{m-k}, a_1, ..., a_k), \quad k < m, \tag{9.14}$$

(this implies conditions $< 2 >$ and $< 3 >$ in Section 1.3) and it is multimodular (condition $< 1 >$ in 1.3).

The monotonicity condition follows from Property 5, the stationarity of the vacation times, and the fact that the vacations are independent of the interarrival and service times.

The second condition is satisfied due to the stationarity assumptions. Indeed, since the system is assumed to be in a stationary regime at time 0, corresponding to the policy that does not take vacations, the Palm probability P_N^0 (of the process seen at the times T_n) is invariant under the shift ϑ^{τ_1} (see, e.g. [21] p. 19). In particular, if we do not take vacations till time n, then $\mathcal{W}_k(a) = \mathcal{W}_1(a)$, $k < n$ in distribution. Hence, the distribution of $\mathcal{W}_n$ under the policy a is the same as the distribution of $\mathcal{W}_{n+j}$ under the policy $a' = (\underbrace{0, ..., 0}_{j}, a_1, ..., a_k, ...)$, for any $j \geq 0$. This implies (9.14).

The multimodularity condition was established in Lemma 53. ∎

Remark 17. A sufficient condition for the stability condition in Theorem 49 is that

- (τ_n, σ_n) is a stationary ergodic sequence, and
- $\mathbb{E}\sigma_1 < \mathbb{E}\tau_1$

(see [34, 35]). (This sufficient condition also implies coupling to the stationary regime from any initial state, provided that we do not take vacations.)

Remark 18. Throughout, when we say that a sequence is stationary ergodic, then we mean with respect to the 1-step shift, unless otherwise stated. Note that this implies the stationarity under any k-shift but not the ergodicity under that shift. Indeed, define $\alpha_n = (-1)^n$, $\gamma_n = -\alpha_n$, $n \in \mathbb{N}$. Let a be a random sequence with $\mathbb{P}(a = \alpha) = \mathbb{P}(a = \gamma) = 1/2$. Then a is ergodic with respect to the 1-step shift (see [35]). It is a stationary sequence (with respect to the k-shift for any integer k, but it is not ergodic for a shift operator ϑ of two steps, $(\vartheta a)_n \stackrel{\text{def}}{=} a_{n+2}$. In particular, the expectation of a_0 does not equal to the sample average of the sequence a_{2n}.

Extension to an arbitrary network

Let queue i be one of several possible output queues of an arbitrary network. Assume that every customer that is served in that queue leaves the system. Then the waiting time of the nth customer equals to its sojourn time till it arrives to that queue, plus its waiting time in queue i. Since customers served at queue i are not rerouted, the first component of the waiting time does not depend on the polling strategy. The total waiting time of a customer is thus the sum of a part that does not depend on the control, plus the waiting time in a G/G/1 queue which is influenced by the controller of the vacation. It is thus multimodular, due to the results of the first part. Hence the optimality of a bracket policy for the total average waiting time is directly obtained.

9.4 An arrival-driven vacation model

Consider a single G/G/1 queue (problem (P1)). The nth customer arrives at times T_n, bringing a workload of σ_n to the system. Let $\tau_n = T_{n+1} - T_n$ denote the inter-arrival times. Immediately after an arrival occurs, the server may go on vacation, that lasts till the next arrival occurs. Then it may go back serving, or take another vacation, etc. A vacation policy $a = (a_1, a_2, \ldots)$ indicates, for each n, whether the server goes on vacation ($a_n = 0$) or continues serving ($a_n = 1$) immediately after the nth arrival. (In terms of the notation of Section 9.2, we have $num(a_n) = 1 - a_n$.)

We call this system an *arrival driven* vacation model since the beginning and end of vacations are initiated by arrivals.

In this section, and in all following sections that deal with problem (P1), the constraint (9.1) translates to the following one, directly on the rates of a_n:

$$\limsup_{s \to \infty} \frac{1}{s} \sum_{n=1}^{s} a_n \le 1 - p. \tag{9.15}$$

The waiting time $\mathcal{W}_n(a)$ of the nth arriving customer is given recursively by

$$\mathcal{W}_{n+1}(a) = (\mathcal{W}_n(a) + \sigma_n - a_n \tau_n)^+, \tag{9.16}$$

or explicitly by

$$\mathcal{W}_{n+1}(a) = \max(0, w_1, w_2, ..., w_n), \text{ where } w_i = \sum_{j=i}^{n} (\sigma_j - a_j \tau_j). \qquad (9.17)$$

The workload in the system immediately after the nth arrival is given by $V_n = \mathcal{W}_n + \sigma_n$. Note that it satisfies the recursion

$$V_{n+1}(a) = (0, V_n(a) - a_n \tau_n)^+ + \sigma_{n+1}. \qquad (9.18)$$

The equation (9.16) seems dual to the dynamics of the admission control in Chapter 4. Therefore, it seems natural to expect to obtain the same type of multimodularity results, and therefore also the optimization results. In order to obtain multimodularity in Chapter 4, it is necessary to let σ_n, the nth service time, be the service of the nth customer actually accepted. Thus, the service time of a customer that is rejected is not defined. Then stationarity conditions are assumed on this σ_n sequence, rather than on the sequence of service times of all customers (both the ones accepted as well as the ones rejected).

We thus proceed similarly, and define τ'_n to be the duration of the nth slot during which a vacation *was not* taken. These are the *effective interarrival times*, since, as we see in (9.16), only these have influence on the dynamics of $\mathcal{W}_n$ and V_n.

Let $k(n) \stackrel{\text{def}}{=} \sum_{i=1}^{n} a_i$. Then

$$\mathcal{W}_{n+1} = \max(0, \mathcal{W}_n + \sigma_n - a_n \tau'_{k(n)}). \qquad (9.19)$$

To see why (9.19) holds, we first note that it agrees with (9.16) for those n's for which $a_n = 0$. On the other hand, if $a_n = 1$ then $\tau'_{k(n)} = \tau_n$, so (9.19) again agrees with (9.16).

If we now assume that τ'_n (and not τ_n) is a stationary sequence, we could expect to obtain results dual to those of the admission control.

However, this does not seem natural: it would mean that the vacation control decisions influence the actual interarrival times. In the case of i.i.d. interarrival times, however, both τ'_n and τ_n are stationary. We shall hereafter use the i.i.d. assumption.

Let

$$S_n = \sigma_n - a_n \tau'_{k(n)}.$$

The waiting time of the $n + 1$st customer is given by $\mathcal{W}_{n+1} = (\mathcal{W}_n + S_n)^+$. The workload just after this arrival is $V_n = \mathcal{W}_n + \sigma_n$.

Remark 19. Since σ_n does not depend on a, V_n is multimodular if and only if $\mathcal{W}_n$ is.

$W(a) \stackrel{\text{def}}{=} \mathcal{W}_{N+1}(a)$ is given explicitly by

$$W(a) = \max(0, w_1, w_2, ..., w_N), \text{ where } w_i = \sum_{j=i}^{N} S_j. \qquad (9.20)$$

We can now obtain the multimodularity of the expected waiting time as we did in the first section. The corresponding properties are:

Property 6. The following holds for $0 < i < N$. If $a_i \geq 1$ then

$$w_i(a + s_i) = w_i(a) + \tau'_{k(i)}, \qquad w_j(a + s_i) = w_i(a) \text{ for } j \neq i.$$

Note that s_i corresponds to removing a vacation at time T_{i-1} and adding the vacation at time T_i.

Property 7. Consider the sequence of effective interarrival times $\tau' = (\tau'_0, ...)$, and the shifted sequence: $\vartheta\tau' = (\tau'_1, \tau'_2, ...)$. Let w'_i be defined as w_i in (9.20) with the sequence $\vartheta\tau'$ replacing τ. The following holds for $0 < i < N$.

$$w'_j(a - e_i) = w_i(a) \qquad j > i,$$

$$w'_j(a - e_i) = w_i(a) + \tau'_{k(j-1)} \qquad j \leq i.$$

Denote $\mathbb{E}_\tau$ the expectation over the effective interarrival times (for a fixed realization σ).

Lemma 54. *Assume that τ_n are i.i.d. Let $h : \mathbb{R} \to \mathbb{R}$ be a nondecreasing convex function. Then $\mathbb{E}_\tau h(W(a))$ and $\mathbb{E}_\tau h(V(a))$ are multimodular in a.*

Proof. The proof for the expected waiting time is the same as the one of Lemma 53, with $\tau'_{k(i)}$ replacing $v_{m(i)}$. The proof for the expected workload follows from Remark 19. ∎

Theorem 50. *Assume that*
    (i) the service time sequence σ is stationary,
    (ii) the interarrival times (τ_n) are i.i.d. and independent of σ,
    (iii) the queue is initially (at time 0) in a unique stationary regime corresponding to the policy that never takes vacations.
    Then Result 45 holds where $p^* = 1 - p$, and where p is the fraction given in the constraint (9.1) (or (9.15)).

Proof. The proof is based on Theorem 7. We need only check that
    For any integer n, the function

$$f_n(a) \stackrel{\text{def}}{=} \mathbb{E}_{\tau,\sigma} h(\mathcal{W}_n(a_1, ..., a_n)) \text{ is non-increasing in } a_i, \ i = 1, ..., n, \quad (9.21)$$

and it satisfies the following conditions:

− condition $< 2 >$ in Section 1.3:

$$f_k(a_1, ..., a_k) \geq f_{k-1}(a_2, ..., a_k), \forall k > 1 \qquad (9.22)$$

− a sufficient condition for < 3 > in 1.3:

$$f_k(a_1, ..., a_k) = f_m(\underbrace{1, ..., 1}_{m-k}, a_1, ..., a_k), \quad k < m, \tag{9.23}$$

and
− $f_k(a)$ is multimodular (condition < 1 > there).

The monotonicity follows directly from the explicit solution to the Lindley equations (9.17). (9.22) and (9.23) follow from Property 7, the assumption that the initial state is initially in the stationary regime corresponding to no vacations, the fact that τ_n are i.i.d. (and thus stationary), and the fact that they do not depend on σ_n. For (9.22) we also make use of the monotonicity property. ∎

In Theorem 50, we assume that the queue is initially (at time 0) in a unique stationary regime corresponding to the policy that never takes vacations. Let β_0 be the corresponding distribution of the initial waiting time $\mathcal{W}_1$. We next show that the results hold for other distributions β as well.

Lemma 55. *(Relaxing the assumption on the initial condition)*
Consider any initial distribution β of $\mathcal{W}_1$. Assume that conditions (i) and (ii) of Theorem 50 hold and instead of condition (iii), the following is satisfied:

− *β is stochastically larger than the β_0,*
− *(σ, τ) is a stationary ergodic under the k-shift for any integer k,*
− *the stability condition $\mathbb{E}\sigma < (1 - p)\mathbb{E}\tau$ holds.*

Then the results of Theorem 50 still hold.

Proof. For any policy a, both the waiting time as well as the workload at any time n are strictly nondecreasing in $\mathcal{W}_1$ as can be seen from (9.19). By the definition of stochastic ordering, the expectation of any nondecreasing function of $\mathcal{W}_1$ is larger for the initial distribution β, than for β_0. This implies that the average expected costs $g(a)$ and $g'(a)$ are larger for the initial distribution β. In order to establish the theorem it suffices thus to show that for the optimal policy, the average expected costs do not depend on the initial state.

If p (in the constraint (9.1)) is rational, then the (candidate for the) optimal policy $a_k^{p^*}(\theta)$ (with $p^* = 1 - p$) is periodic. In that case, the process corresponding to the optimal policy starting from any two different initial states couple in a time that is finite with probability one. We may couple now the initial states, i.e. construct a common probability space where the initial state corresponding to β is larger than the one corresponding to β_0. It follows that the convergence of the difference between $\mathcal{W}_n$ corresponding to β and to β_0 is *monotone* decreasing. This implies that the difference between $f_n(a) = \mathbb{E}_{\tau, \sigma} h(\mathcal{W}_n(a_1, ..., a_n))$, starting at the different initial distributions of $\mathcal{W}_1$, converges to 0. Hence the expected average cost under the two initial distributions is the same under the optimal policy.

The same convergence (and hence the same conclusion) is obtained for p irrational. Indeed, if p is irrational, then the (candidate for the) optimal policy $a^{p^*}(\theta)$ is aperiodic. This cost obtained by that policy is unchanged if we replace θ by a random variable Θ, uniformly distributed in $[0, 1]$ (this follows from the discussion in the end of Section 1.3.1). The policy $a^{p^*}(\Theta)$ is stationary ergodic with respect to the 1-step shift. Then we can use the theory of (non-controlled) stochastically recursive sequences by Borovkov [34] pp. 260-272 and [35] to obtain the same convergence results as above. ∎

Remark 20. The assumption that β is stochastically larger than the β_0 is not really restrictive. Indeed, if β_t^u is the distribution of the state at time t, then one can show that for any policy u, $\liminf_{t \to \infty} \beta_t^u$ is stochastically larger than the β_0. Hence the assumption is suitable for the case where the system has operated for a sufficiently long time under an arbitrary policy.

9.5 Arrival-driven polling model

We now analyze problem (P2). Consider K queues, each of which behaves like the one in the previous section. The service period for one queue constitutes a vacation for the others.

The nth customer arrives at time T_n, and brings K jobs to the K queues: a workload of σ_n^i arrives to queue i, $i = 1, ..., K$. These components are processed according to the FIFO order in each queue.

Service beginnings and vacations are synchronized with arrivals. More precisely, T_n is also the time at which the nth potential service begins; it may be in any one of the queues. The service time duration τ_n is the difference between consecutive interarrival times: $T_{n+1} = T_n + \tau_n$. If queue i is the nth to be served and is empty then we assume that the server still remains τ_n time at that queue.

Let $\mathcal{W}_n^i$ be the waiting time of the nth job arriving to queue i, and V_n^i denote the workload at queue i just after the arrival of the n customer. The evolution of the waiting time $\mathcal{W}_n^i(a)$ in queue i is given by:

$$\mathcal{W}_{n+1}^i(a) = \max(0, \mathcal{W}_n^i(a) + \sigma_n^i - a_n^i \tau_n),$$

where $a_n^i = 1$ if queue i is served at the nth period, and is, otherwise, zero.

A policy is a sequence $a = (a_1, a_2, ...,)$, where $a_n = \{a_n^k, k = 1, ..., K\}$, as defined in Section 9.2. We adopt the further constraint that for every integer j, only one of the components $a_j^i, i = 1, ..., K$ may be different than 0. Denote $a^i = (a_1^i, a_2^i, ...)$ the actions corresponding to queue i.

The following is a consequence of Section 1.4 and the properties established for a single queue in the previous section.

Theorem 51. *Consider problems (P2a) and (P2b). Assume that*

- *the inter-arrival times are i.i.d., and independent of the service times,*
- *the service time sequence in each queue is stationary,*
- *for each $i = 1, ..., K$, queue i is initially at a unique stationary regime that corresponds to the policy that never takes vacations at that queue.*

Then Results 46, 47 and 48 hold.

Again, we may relax the assumption on the initial distribution, as we did in Lemma 55.

9.6 The potential vacation times are a renewal process

9.6.1 A single queue

Let u_n be an increasing random sequence of potential switching times. Immediately after u_n, the server may decide to go on vacation till the next instant u_{n+1}. As in the previous section, a vacation policy $a = (a_1, a_2, ...)$ indicates, for each n, whether the server goes on vacation ($a_n = 0$) of continues serving ($a_n = 1$) at time u_n.

Let s_n be the sequence of differences between consecutive potential switching times. Thus $u_{n+1} = u_n + s_{n+1}$.

The kth customer arrives at time $T_k = u_{n_k}$, where n_k is some increasing sequence of positive integers. Thus, u_n can be viewed as basic time epochs to which both arrivals and vacations are related. However, unlike the model in the previous section, where arrivals occurred at beginning of each vacation slots, arrivals are only synchronized with u_n, and need not occur at every period. This will allow us to handle dependent arrival times, and more precisely the case of stationary interarrival arrivals.

Customer k brings a workload (request for service time) of σ_k^*. Hence, the amount σ_n of workload that arrives at time u_n is given by

$$\sigma_n = \begin{cases} 0 & \text{if } n \neq n_k, \forall k \\ \sigma_k^* & \text{if for some } k, n = n_k. \end{cases} \tag{9.24}$$

The waiting time $\mathcal{W}_n$ of the (possibly virtual) customer that arrives at time u_n in the system can now be computed using the following recursion:

$$\mathcal{W}_{n+1}(a) = \max(0, \mathcal{W}_n(a) + \sigma_n - a_n s_n).$$

The workload V_n at the nth time epoch (i.e. immediately after u_n) is $\mathcal{W}_n(a) + \sigma_n$. It can also be given recursively as

$$V_{n+1}(a) = \max(0, V_n(a) - a_n s_n) + \sigma_{n+1}. \tag{9.25}$$

We are now back to the model described by (9.16) of section 9.4, and therefore, the multimodularity results in Lemma 54, and the optimality results in Theorem 50 hold.

It is useful to present conditions directly on the original service sequence σ^* (instead of the sequence σ which are used in Theorem 50) for the optimality results.

In order to make general and yet useful probabilistic assumptions on σ_n (i.e. on the marks of the arrival process), we use the stochastic point process formalism. The sequence n_k, which we used in (9.24), defines a *discrete time point process* $(\mathcal{N}, \vartheta, P)$ (where ϑ is the θ_1 of [21] p. 43):

$$\mathcal{N}(\omega, C) = \sum_{k \in \mathbb{Z}} \delta_{n_k(\omega)}(C).$$

Thus,

for $w \in \Omega$, an arrival occurs at time $u_n(\omega)$ if $\mathcal{N}(\omega, \{n\}) = 1$.   (9.26)

We associate to the process $\mathcal{N}$ the marks σ_k^*. We assume that $\mathcal{N}$ is compatible with the ϑ flow, i.e.

$$n_k(\omega) = n_0(\vartheta_k \omega).$$

Assume that the interarrival times are a strictly stationary sequence, i.e.

$$\mathbb{P}_{\mathcal{N}}^0(\vartheta_{n_k} \in \cdot) = \mathbb{P}_{\mathcal{N}}^0(\cdot), \qquad k \in \mathbb{Z},$$ (9.27)

where $\mathbb{P}_{\mathcal{N}}^0$ is the Palm probability related to $\mathcal{N}$. Then there exists a probability measure $\mathbb{P}$ for which $\mathcal{N}$ is stationary (w.r.t. $(\vartheta, \mathbb{P})$). This follows from the inverse construction of Slivnyak (see p. 27 in [21]) in a discrete-time version (which follows from p. 44 in [21]). Define

$$\overline{\sigma}(l) = \sigma_k^* \quad for \quad n_k \le l < n_{k+1}.$$

Then $(\mathcal{N}, \overline{\sigma})$ are jointly stationary (with respect to $(\vartheta, \mathbb{P})$) as follows from the argument in [21] p. 13-14. Since

$$\sigma_n = \overline{\sigma}(n) \times \mathcal{N}(\{n\}),$$

it then follows that σ_n are stationary w.r.t. $(\vartheta, \mathbb{P})$. Indeed,

$$\sigma_n(\omega) = \overline{\sigma}(n, \omega) \times \mathcal{N}(\omega, \{n\}) = \overline{\sigma}(n - 1, \vartheta \omega) \times \mathcal{N}(\vartheta \omega, \{n - 1\}) = \sigma_{n-1}(\vartheta \omega).$$
$$(9.28)$$

We conclude that if we assume that the original process σ_n^* is stationary, then there exists a probability measure under which $\mathcal{N}$ is stationary (w.r.t. $(\vartheta, \mathbb{P})$) (thus in particular, the process σ_n will be stationary). Note that, in general, there may exist other non-stationary processes $\mathcal{N}$ that have a stationary Palm distribution.

If we assume that the service times are independent of the vacation opportunities times s_n and of the sequence n_k, then a simple argument shows that the stationarity of σ_n^* implies that σ_n are stationary too. Indeed, assume as above, that σ_n^* are the marks of the process $\mathcal{N}$, and assume that (9.27) holds. Fix some integer j and let $S_1, S_2, ..., S_j$ be some Borel sets in $\mathbb{R}$. Then

$$\mathbb{P}(\sigma_1^* \in S_1, ...\sigma_j^* \in S_j) = \sum_{k=-\infty}^{\infty} \mathbb{P}(\sigma_1^* \in S_1, ...\sigma_j^* \in S_j | n_1 = k)\mathbb{P}(n_1 = k)$$

$$= \sum_{k=-\infty}^{\infty} \mathbb{P}(\sigma_1^* \in S_1, ...\sigma_j^* \in S_j | n_1 = 0)\mathbb{P}(n_1 = k)$$

$$= \mathbb{P}(\sigma_1^* \in S_1, ...\sigma_j^* \in S_j | n_1 = 0)$$

$$= \mathbb{P}_{\mathcal{N}}^0(\sigma_1^* \in S_1, ...\sigma_j^* \in S_j).$$

Hence, the stationarity of σ_n^* under $\mathbb{P}_{\mathcal{N}}^0$ implies that it is stationary under $\mathbb{P}$, and if u_n are i.i.d. then the process $\mathcal{N}$ is stationary (w.r.t. $(\vartheta, \mathbb{P})$). Thus, as in (9.28), we conclude that σ_n are stationary.

We summarize this in the following Theorem.

Theorem 52. *Assume that*

- *the inter-potential vacation times s_n are i.i.d. and hence u_n is a renewal process,*
- *arrivals occur at u_{n_k}, where n_k defines a point process $\mathcal{N}$,*
- *the service times σ_n^* are marks of the point process $\mathcal{N}$,*
- *σ_n^* is a stationary sequence,*
- *the durations of the potential vacations s_n do not depend on the service durations and on the sequence n_k,*
- *the queue is initially (at time 0) in a unique stationary regime corresponding to the policy that never took vacations.*

Then Result 45 holds where $p^* = 1 - p$, and where p is the fraction given in the constraint (9.1) (or (9.15)).

Proof. We show that Theorem 50 can be applied. As we showed above, we can consider an equivalent model where arrivals occur at each time u_n instead of the original ones. The service time for this new model are σ_n, which are stationary. Due to the independence between s_n, n_k, and the service duration, the interarrival times in the new model are independent of the service times. The conditions of Theorem 50 are thus satisfied. (Note that the fact that in the new model, arrivals occur at times u_n which are independent of other quantities, allows to have dependence between the n_k sequence.) ∎

Note that we may relax the assumption on the initial distribution, as we did in Lemma 55.

9.6.2 The polling control problem

Having seen that the setting described in the previous subsection for a single queue can be embedded into the one in Section 9.5, we can obtain the corresponding results for the optimal control problem (P2) of polling to several queues.

Theorem 53. *Consider problems (P2a) and (P2b). Assume that*

- *the potential switching times u_n are a renewal process,*
- *the service times $\sigma_n^{*,i}$ of the nth customer at queue $i, i = 1, ..., K$ are stationary,*
- *arrivals to queue i occur at times $u_{n_k(i)}, i = 1, ..., K$, where $n_k(i)$ is a stationary point process,*
- *the duration of the slots $s_{n+1} = u_{n+1} - u_n$ do not depend on the service durations and on the sequences $n_k(i)$,*
- *for each $i = 1, ..., K$, queue i is initially at a unique stationary regime that corresponds to the policy that never takes vacations at that queue.*

Then Results 46, 47 and 48 hold.

Note that the fact that service times in different queues were allowed to be dependent in Theorem 51 allows us to have dependence between the $n_k(i), i = 1, ..., K$ sequences in different queues.

We may relax the assumption on the initial distribution, as we did in Lemma 55 and Remark 20.

Remark 21. The assumptions of Theorem 53 contain as a special case the following exponential model. Suppose the arrival process are independent Poisson process with rates λ_j for queue $j = 1, ..., K$. The service time in queue j is exponential with parameter $\mu_j, i = 1, ..., K$. The potential switching times form a Poisson process with parameter $\nu \geq (\lambda_1 + ... + \lambda_K)$, this is a natural assumption on ν as in case we want to uniformize all processes, ν is taken as $\sum_{i=1}^{K}(\lambda_i + \mu_i)$. Also as approximation of continuous-time polling control we may take ν large. Theorem 53 now shows that the optimal polling control is bracket for the exponential model with $K = 2$ and for the symmetrical model with $K > 2$. This shows Property 1 in paper [66], where an algorithm for computing optimal policies is given.

9.7 1-gated service

We describe in this section models that have stationary arrival processes which may be more general than point processes. The vacations opportunities in the following models will be a periodic process, independent of inter-arrival times or service times.

9.7.1 A single queue

We now consider a vacation as in the previous section, but with a "continuous" arrival into a single queue. We assume that the total workload that arrives during the interval $(u_n, u_{n+1}]$ is σ_n. This workload might arrive in a single batch, or continuously, or at several distinct instants in that interval. If the server does not go on vacation on time u_n, then the amount of service given to the queue till $u_{n+1} = u_n + s_{n+1}$ is the minimum between s_{n+1} and V_n (the workload present just before the nth potential switching-interval). Thus, only workload present in the gating epoch u_n is candidate to be served during the interval $(u_n, u_{n+1}]$. We assume that if the server is not on vacation at the beginning of the nth slot, then it remains in that queue till time u_{n+1}, even if there is no workload to be served during a part (or all) the interval.

We shall require that the process σ_n be stationary in n (i.e. w.r.t. the shift ϑ_1). The recursion (9.18) for the workload in the system at gating instants holds in our case, so we could obtain again optimality of the bracket policy (as in Theorem 50).

In order for the conditions of Theorem 50 to hold we need, however, that σ_n be independent of s_n. This is impossible in general, unless s_n are identical, which we shall thus assume below. (For example, assume Poisson arrivals with rate λ, where each customer requires a unit of workload. Then the expected amount of workload arriving during a period v_n, conditioned on s_n, is λs_n. Hence it is not independent of s_n). Note that this problem did not occur in previous sections, since the service time was not related to the arrival instants, but to the order of arrival.

Consider an underlying probability space $(\Omega, \mathcal{F})$. Define { $vartheta_t\}, t \in \mathbb{R}$ to be a measurable flow on $(\Omega, \mathcal{F})$ (see [21] p. 8 and Remark 22 below). We define $\{\vartheta_n\}, n \in \mathbb{Z}$ to be another measurable flow on $(\Omega, \mathcal{F})$ (see [21] p. 44); ϑ_n will be related to shifts in discrete time.

Consider a general random measure Z describing the arriving workload; in particular, if C is an interval in $\mathbb{R}$, then $Z(\omega, C)$ has the interpretation of the amount of workload arriving during that interval for a realization ω. This includes in particular the case where the arrival process is a point process. Assume that Z is stationary with respect to $(\vartheta_t, \mathbb{P})$ (ϑ_t is the continuous time shift). Then the amount of workload σ_n that arrives during the deterministic (constant) periods $s_n(= s_0)$ is stationary in n (i.e. w.r.t to $(\vartheta_1, \mathbb{P})$) due to the stationarity of Z w.r.t. $(\vartheta_t, \mathbb{P})$. Hence the conditions of Theorem 50 hold for any stationary arrival process (not necessarily a point process).

Remark 22. Consider a more general model for the potential vacation process. Let $(N, \vartheta_t, \mathbb{P})$ be a stationary point process corresponding to the potential vacations: associated with N there is given the random sequence u_n, $n = 1, 2, ...$ We have

$$N(\omega, C) = \sum_{n \in \mathbb{Z}} \delta_{u_n(\omega)}(C), \tag{9.29}$$

where we assume $-\infty \leq \dots \leq u_{-1} \leq u_0 \leq 0 \leq u_1 \leq u_2 \dots \leq \infty$ and δ_x is the Dirac measure at x. σ_n can then be considered as marks of the point process N:

Let Z be stationary with respect to $(\vartheta_t, \mathbb{P})$. Then

$$\sigma_n(\omega) \overset{\text{def}}{=} Z(\omega, [u_n, u_{n+1}))$$

satisfies

$$\sigma_n(\omega) = \sigma_0(\vartheta_{u_n}\omega) = \sigma_0(\vartheta_n\omega).$$

Hence, $((N, \sigma), \vartheta_t, \mathbb{P})$ is a stationary marked point process (see [21] p. 10) and σ_n is stationary in n (see Section 1.3.2 in [21]), i.e. w.r.t. ϑ.

To summarize, we have:

Theorem 54. *Assume that*

- *the potential vacation durations (s_n) are constant,*
- *the sequence of workloads σ_n arriving arriving during the nth slot is stationary,*
- *the queue is initially (at time 0) in a unique stationary regime corresponding to the policy that never took vacations.*

Then Result 45 holds where $p^* = 1 - p$, and where p is the fraction given in the constraint (9.1) (or (9.15)).

We may relax the assumption on the initial distribution, as we did in Lemma 55 and Remark 20.

9.7.2 The case of several queues

Consider K queues, each of which behaves like the one in the queue in Subsection 9.7.1. The service period for one queue constitutes a vacation for the others.

More precisely, let u_n, $n = 1, 2, \dots$ be the time at which the nth potential service ends; it may be in either one of the queues. The service time duration s_n is the difference between consecutive inter-switching times: $u_{n+1} = u_n + s_{n+1}$. If queue i is the nth to be served and is empty then we assume that the server still remains s_n time at that queue.

Let σ_n^i be the amount of workload that arrives to queue i during the interval $(u_n, u_{n+1}]$. As in Subsection 9.7.1, we assume a 1-gated regime, where, only workload that is present at time u_n is candidate to being served during the interval $(u_n, u_{n+1}]$, and not workload that arrives during that interval.

The evolution of the waiting time $\mathcal{W}_n^i(a)$ in queue i is given by:

$$\mathcal{W}_{n+1}^i(a) = \max(0, \mathcal{W}_n^i(a) + \sigma_n^i - a_n^i s_n),$$

where $a_n^i = 1$ if queue i is served at the nth period, and is, otherwise, zero. Here, $\mathcal{W}_n^i$ has the interpretation of the waiting time of a customer that would arrive at time u_n, and V_n^i is the workload in the system just after time u_n.

From the discussion in Subsection 9.7.1, we obtain the following results (with notation similar to those in Theorem 51):

Theorem 55. *Assume that*

- *the arriving workloads σ_n^i is a stationary sequence for each i,*
- *s_n are constant,*
- *for each $i = 1, ..., K$, queue i is initially at a unique stationary regime that corresponds to the policy that never takes vacations at that queue.*

Then Results 46, 47 and 48 hold.

We may relax the assumption on the initial distribution, as we did in Lemma 55 and Remark 20.

Remark 23. Note that we allow in this model for different (dependent or independent) arrival streams (and thus interarrival times) to different queues, unlike the model in Section 9.5. The restriction in Section 9.5 to a single sequence T_n that defines the time of arrivals, that occur simultaneously to all queues, was due to the fact that it was the arrival times that triggered the polling (the vacation opportunities). In this section, arrival can be more general. (Note that, even if the arrival is a point process, the interarrival times $\{T_n^i\}$ in queue i do not appear explicitly anymore in the evolution equations, due to the gating.)

We illustrate the usefulness of the previous result in the following optimal scheduling control problem in an telecommunication (ATM) switch.

9.7.3 Application to an ATM switch

We consider an $M \times N$ switch with M inputs and N outputs, as depicted in Fig. 9.1. We assume that there are separate input queues for each output, so we do not have HOL (head-of-line) blocking. Each input is associated with N queues, one for each output. We denote by queue ij the queue for cells arriving to input i and destined for output j. We consider a slotted queuing model where in each time slot at most one cell can be transmitted from each of the M inputs, and at most one cell can be received by each of the N outputs. In ATM (Asynchronous Transfer Mode), indeed packet size are fixed and constant, so it is natural to consider time-slotted models. A scheduling mechanism decides at each time slot, from which inputs and to which output do we send a packet.

The scheduler may send simultaneously packets from different inputs to different outputs, as long as the following constraints are met:

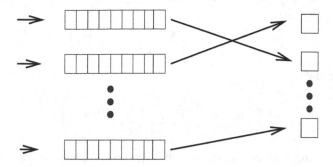

Fig. 9.1. Application to an ATM switch: a feasible schedule

C1: The scheduler cannot send more than one packet from the same input simultaneously, and

C2: it cannot send more than one packet to the same output, simultaneously.

We are interested here in open-loop scheduling policies, i.e. in scheduling that do not rely on queue length information, but only on the input rates (that will be detailed below).

A class of policies have been presented in [13], that achieve 100% throughput of the switch. A natural problem is whether one can obtain a policy that not only achieves the above goal, but also minimizes nondecreasing convex functions of the workload in the system.

Note that a policy that minimizes the workload, maximizes the amount of workload that departs, and therefore, the throughput. Therefore, if any policy achieves maximum throughput (and stability), then so does the policy which minimizes the workload.

For all $1 \leq i \leq M$, $1 \leq j \leq N$, let $A_{ij}(n)$ be the number of cells that arrive at queue ij in time slot n. We assume that the arrival process $\{A_{ij}\}_n$ is stationary with rate λ_{ij} (i.e. the average number of cells arrived in each time slot). The arrival processes may be mutually dependent.

We consider the symmetric case below, i.e. we assume that the λ_{ij} do not depend on i (they may depend on j), and that $N = M$. For any $i = 1, ..., m$, define $j(i, t) = (i+t) mod(N) + 1$. Consider the round-robin scheduling policy u that sends at time t a packet (if there is any) from queue i to queue $j(i, t)$ for each i. This policy clearly meats the constraints C1 and C2 above. For each j, all queues $ij, i = 1, ..., N$ (having j as destination) receive a round-robin service, which, under the conditions of Theorem 55, is optimal among all (open-loop) scheduling policies (in fact, even among those that do not satisfy C1).

A scheduling policy is a sequence $a = (a_1, a_2, ...,)$, where $a_n = \{a_n^{ij}, i, j = 1, ..., N\}$. If the ijth component of a_n is 1, this means that the server polls queue ij at the nth time slot.

Let $h_{ij} = h$ be a convex nondecreasing function. Define for each j,

$$f_n^{ij}(a^{ij}) = \mathbb{E}h(W_n^{ij}(a^{ij})), i, j = 1, ..., N,$$

$$g^j(a) \stackrel{\text{def}}{=} \limsup_{s \to \infty} \frac{1}{s} \sum_{i=1}^{N} \left(\sum_{n=1}^{s} f_n^{ij}(a^{ij}) \right),$$

and further define

$$g(a) \stackrel{\text{def}}{=} \limsup_{s \to \infty} \frac{1}{s} \sum_{i=1}^{N} \sum_{j=1}^{N} \left(\sum_{n=1}^{N} f_n^{ij}(a^{ij}) \right).$$

The following Theorem is then a consequence of Theorem 55:

Theorem 56. *Consider the above $N \times N$ switch. Assume that*

- *the arriving workloads $\sigma_n^{ij} \stackrel{def}{=} A_{ij}(n)$ is a stationary sequence for each i. For each j, the distribution of the processes of arrivals of workload to the queues ij does not depend on i (in other words, $f_n^{ij}(a^{ij}) = f_n^{kj}(a^{kj})$ if $a^{ij} = a^{kj}$),*
- *the time slots s_n are constant,*
- *the workload initially in each queue corresponds to a unique stationary regime that would be obtained if this queue had always been served before.*

Then the round-robin policy u minimizes both $g(a)$ as well as $g^j(a)$, for $j = 1, 2, ..., N$.

Again, we may relax the assumption on the initial distribution, as we did in Lemma 55 and Remark 20.

10 Monotonicity of feedback control

10.1 Introduction

This chapter deals with the problem of closed loop control (unlike the previous ones, which focused on open loop control) for stochastic discrete event systems which are not necessarily Markovian. We present a new approach, based on multimodularity, to show monotonicity properties of the optimal policy which works under various types of the information patterns: total information, sampled information, delayed information.

The concept of multimodularity can be found in applications in feedback control [52, 80, 114, 27] where multimodularity properties of the value functions were formulated in terms of the number of customers in two parallel queues, or in tandem queues. We call this notion multimodularity in "space". It must be distinguished from multimodularity in time, that we are using and that will allow us to obtain monotone properties of the optimal policies which are state dependent.

Moreover, the types of networks for which the multimodularity in space is known (basically Markovian systems) is much more restricted than those for which multimodularity in time holds, as illustrated in this chapter, by some examples. It is thus quite desirable to have a theory that would relate multimodularity in time with the structure of state-dependent optimal policies. As far as the proof techniques are concerned, most of the works on space monotonicity uses a dynamic programming approach and expected costs (see for example [52, 27]). Here, we will primarily use a sample path approach. Some steps in this direction have already been taken in [15], in which some sample-path properties that are related to multimodularity are used to establish the monotonicity of optimal policies in the state (or some partial information that we may have on the state). In this chapter we pursue this goal and construct a general theory for the state-dependent optimal control of discrete event systems with time-multimodular cost.

We consider several types of questions related to different information structures: (i) monotonicity in the initial state, (ii) monotonicity in the current state in the case of full state information, (iii) delayed and sampled information, and others. We show for all these problems that some simple multimodularity properties of the costs imply the monotonicity of optimal policies.

Other general frameworks have been studied in the past for establishing monotonicity of optimal policies, such as [12, 52, 80]. The conditions are related to submodularity, and were typically formulated for Markovian models. In particular, [52] consider only exponential distributions. Our methodology allows for non-Markovian analysis, as multimodularity of workload and waiting times was established for general stationary ergodic sequences of service and interarrival times (see Chapter 4).

10.1.1 Organization of the chapter

The rest of the chapter is organized as follows. In Section 10.2, we define the multimodular ordering on action sequences. In Section 10.3, we introduce the different information patterns that we will consider. In Section 10.4, we investigate the special case of full information whereas the general case is treated in Section 10.5. Finally Section 10.6 shows that the traditional state representation also fits in this framework. The relations between Multimodularity, superconvexity and submodularity is shown in Section 10.7.

10.2 Monotonicity in initial actions

- **Problem P:** Consider a function $f_n : \mathbf{A}^n \to \mathbb{R}$, and assume that f_n is multimodular ($\mathbf{A}^n$ is the action space and will be considered as a bounded convex subspace of $\mathbb{R}^n$). Assume that $a_1, ..., a_{n-1}$ are fixed, and we wish to minimize f_n with respect to a_n.
  We wish to establish properties of optimal a_n as a function of $a_1, ..., a_{n-1}$. In other words, we wish to characterize the impact of some initial choice of actions, which need not be a result of optimization consideration, on the future choice of actions.

In order to establish monotonicity properties, we introduce some ordering on the set of action vectors.

Definition 15 (m-order). *Let $a^1, a^2 \in \mathbf{A}^n$. We say that $a^1 \leq_m a^2$ if there exists a sequence $\gamma_1, ..., \gamma_l$ where $\gamma_i \in \{+e_1, -s_2, -s_3, ..., -s_n\}$, such that $a^2 = a^1 + \gamma_1 + \gamma_2 + ... + \gamma_l$.*

Theorem 57 *Consider problem P. Then there exists optimal control $a_n(\mathbf{a})$ which is monotone nonincreasing in the initial controls $\mathbf{a} = (a_1, ..., a_{n-1})$.*

Proof. Let $a^1 \leq_m a^2$ where $a^i \in \mathbf{A}^{n-1}$. For any $b^i \in \mathbf{A}$, $i = 1, 2$ with $b^1 \geq b^2$ we have

$$f_n(a^2, b^1) - f_n(a^2, b^2) \geq f_n(a^1, b^1) - f_n(a^1, b^2). \tag{10.1}$$

Indeed, if we write $x = (a^1, b^2)$ then the inequality can be rewritten as

$$f_n(x + w - me_n) - f_n(x + w) \geq f_n(x - me_n) - f_n(x) \tag{10.2}$$

where m is some positive integer, and where w is the sum of some elements in $\mathcal{F}$. This inequality then follows from the definition of multimodularity.

Now, assume that b^1 is optimal at a^2 but not at a^1. We shall show that this implies that b^2 is not optimal either at a^1. This will then imply that there exists an optimal policy which is monotone nonincreasing.

Indeed, due to the optimality of b^1 at a^2 we have

$$f_n(a^2, b^1) \le f_n(a^2, b^2).$$

(10.1) then implies that

$$f_n(a^1, b^1) \le f_n(a^1, b^2).$$

Since b^1 is not optimal at a^1, then b^2, which does not perform better, is not optimal either. We conclude that an optimal control at a^1 has to be larger than b^1.

Thus if we define the policy that chooses for a given sequence a^i the maximal element b that is optimal at a^i, then this policy is monotone nonincreasing. This ends the proof.   □

We say that a set $B \subset \mathbf{A}$ is decreasing if $b \in B$ implies that $b' \in B$ for any $b' \le b$. To illustrate the above Theorem, assume that $\mathbf{A} = \{0, 1, \dots, M\}$. Then the theorem implies that there exists an optimal control $a_n(\cdot)$ such that for each $0 < m < M$, there exists some decreasing set $B_m \subset \mathbf{A}^{n-1}$ and such that $\mathbf{a} \in B_m$ if and only if $a_n(\mathbf{a}) \ge m$. B_m is thus the set of initial sequences of length $n - 1$ for which the corresponding optimal action at instant n is greater than or equal to m.

Definition 16. *The maximal elements of B_m defined in the previous paragraph are said to be the mth switching curve that characterizes the optimal action $a_n(\cdot)$.*

We may extend Theorem 57 to the two dimensional case (this extension will be useful in Model 60 in the next section).

Corollary 58 *Consider an action space $\mathbf{A} = \{0, 1\}$. Consider two functions $f_n^i : \mathbf{A}^n \to \mathbb{R}$, $i = 1, 2$, and assume that f_n^i are multimodular. For any given sequence $a^n = (a_1, \dots, a_n) \in \mathbf{A}^n$, define $b^n = (1 - a_1, \dots, 1 - a_n)$. Assume that $a_1, \dots, a_{n-1}$ are fixed, and we wish to minimize the function*

$$g_n(a^n) = f_n^1(a^n) + f_n^2(b^n)$$

with respect to a_n.
Then there exist optimal control $a_n(\mathbf{a})$ which is monotone nonicreasing in the initial controls $\mathbf{a} = (a_1, \dots, a_{n-1})$.

Proof. The proof follows directly from Theorem 57 by noting that f_n^2, and thus g_n are multimodular with respect to a^n.   □

10.3 Exogenous random variables and information patterns

We consider in this section the cost functions that depend not only on the control but also on some random exogenous variables $\{Z_n\}$. More precisely, let $\{Z_n\}$ be a sequence of random variables taking values in some space $\mathbf{Z}$, and let there be a sequence of cost functions: $f_n : \mathbf{A}^n \times \mathbf{Z}^n \to \mathbb{R}$.

We assume that f_n are multimodular in the control actions a for each realization of $z^n \overset{\triangle}{=} \{z_1, \ldots, z_n\}$.

A history of length n has the form $(a_1, \ldots, a_{n-1}; z_1, \ldots, z_n)$, and $\mathbf{H}_n$ is defined to be the set of possible histories of length n. We define the observation function $y_n : \mathbf{H}_n \to \mathbf{Y}_n$, where $\mathbf{Y}_n$ is the observation space available at time n.

A policy $u = (u_1, u_2, \ldots)$ is a sequence of maps where u_n is a function from $\mathbf{Y}_n$ to $\mathbf{A}$. In other words, at time n an action is chosen as a function of the available observation of the history.

We shall consider several information patterns:

1. Full information: $y_n(h_n) = h_n$, so that the whole history is available. This can be seen as the general full information case. (For the standard MDP case, the whole past history can be considered as a state of the system, see Section 10.6.)
2. Sampled information: let k be some fixed integer; we assume that the information on the past actions is always available, but the new information about the values of the variables Z_n becomes known only at sampled times $k, 2k, 3k, \ldots$ In other words, let

$$l(n) \overset{\triangle}{=} k \lfloor \frac{n}{k} \rfloor.$$

Then

$$y_n(a_1, \ldots, a_{n-1}; z_1, \ldots, z_n) \overset{\triangle}{=} (a_1, \ldots, a_{n-1}; z_1, \ldots, z_{l(n)}).$$

Since no new observation on the z sequence arrives during the period $l(n) + 1, \ldots, l(n) + k$, the control under this information structure is equivalent to one that takes decisions only at times $rk, r = 1, 2, \ldots$, where at each time rk the actions $a_{rk+1}, \ldots, a_{(r+1)k}$ are determined. We shall adopt this view below.
3. Delayed information: fix some integer θ. Then the information at time n over the z-variables is delayed by θ units, so that

$$y_n(a_1, \ldots, a_{n-1}; z_1, \ldots, z_n) \overset{\triangle}{=} (a_1, \ldots, a_{n-1}; z_1, \ldots, z_{n-\theta}).$$

Our goal under the different information structures is to minimize

$$J^N(u) = \sum_{n=1}^{N} \mathbb{E}^u f_n(H_n, A_n) + \mathbb{E}^u g_{N+1}(H_{N+1}) \tag{10.3}$$

where N is some given integer and g_{N+1} is some terminal cost. This corresponds to a finite horizon problem where N sequential actions are to be chosen.

In order to illustrate our framework we present three models which will be considered all along the paper.

Model 59 Consider a single queue operating at discrete time. An arrival of a_n packets is generated at the beginning of the nth time unit, where a_n is a control variable. At the end of the time unit, a departure occurs if and only if $z_{n+1} = 1$ (otherwise, $z_{n+1} = 0$). The number x_n of packets in the queue at the end of the nth slot (just after a potential departure), which we call the state of the queue, is given recursively by

$$x_{n+1} = (x_n + a_n - z_{n+1})^+. \tag{10.4}$$

We assume that at $n = 0$ the queue is empty. Using the explicit solution of this Lindley's equation, x_n can be expressed as a function of the history:

$$x_n = \max\left(0, \sum_{k=l}^{n-1}(a_k - z_{k+1}), l = 1, ..., n-1\right). \tag{10.5}$$

We may now choose $f_n(h_n) \overset{\text{def}}{=} r_n(x_n)$ where r_n is any monotone increasing convex function; this will ensure that f_n is indeed multimodular in the a-sequence for any realization of the z-sequence, see [6] and Section 4.3 of [4].

Model 60 Consider a routing problem into two queues: q0, q1. An arrival is generated at the beginning of each time slot. The routing variables a_n, $n \in \mathbb{N}$ take values 0 or 1; the nth packet is routed to queue $b_n = 1 - a_n$. At the end of the time unit, a departure from queue i occurs if and only if $z^i_{n+1} = 1$ (otherwise, $z^i_{n+1} = 0$). The number x^i_n of packets in queue i at the end of the nth slot (just after a potential departure) is given recursively by

$$x^0_{n+1} = (x^0_n + a_n - z^0_{n+1})^+,$$
$$x^1_{n+1} = (x^1_n + b_n - z^1_{n+1})^+.$$

Consider the cost function

$$g_n(x_n) = c_0 r^0_n(x^0_n) + c_1 r^1_n(x^1_n),$$

where $c_i \geq 0$ are some given real numbers and r^i_n are convex increasing functions. It follows from Model 59 that x^0_n is multimodular in $a^{n-1} = (a_1, ..., a_{n-1})$ and x^1_n is multimodular in $b^{n-1} = (b_1, ..., b_{n-1})$ (where $b^{n-1} = (1 - a_1, ..., 1 - a_{n-1})$). Recalling Corollary 58, we conclude that $g_n(x_n)$ is multimodular in a^{n-1}.

Model 61 We consider here (max,+) linear systems. Any system in this class can be modeled by *event graphs*, which are a subset of the Petri nets, as described in Chapter 3.

As in previous chapters, and unsing the notation from Chapter 3 and Chapter 4, let $X_i(n)$ denote the time when transition i initiates its n-th firing, then $X(n) = (X_1(n), \cdots, X_Q(n))^t$ satisfies a linear equation in the (max,+) notation of the form:

$$X(n) = A(n) \otimes X(n-1) \oplus B(n) \otimes T_{\nu(n)}.$$

Define $W_n^q = X_q(n) - T_{\nu(n)}$. It is the *traveling time* for customer n between its entrance in the system and its passage in transition q. Again, we have the Lindley's equation:

$$W_n = A(n) \otimes D(-\tau_n) \otimes W_{n-1} \oplus B_n,$$

where $D(h)$ is the diagonal matrix with h on the diagonal and $-\infty$ everywhere else. The equation is developed into:

$$W_{n+1} = B_n \oplus \bigoplus_{i=1}^{n} C_i, \tag{10.6}$$

with

$$C_i = \bigotimes_{j=i}^{n} (A(j) \otimes D(-\tau_j)) \otimes B(n-i-1), \tag{10.7}$$

and $B(0) = (-\infty, ..., -\infty)$. (see Chapter 3.)

Assume now that σ_n is a stationary sequence, independent of the sequence τ_n. Using (10.7), it was shown in Chapter 4 that for any convex increasing function r and for any realization of the sequence τ_n, $\mathbb{E}_\sigma r(W_n^a)$ (the expectation is taken with respect to the distribution of the σ_n process) are multimodular in a.

In order to be in our general framework, we can identify the sequences σ_n and τ_n as the Z_n variables in the beginning of the section.

Finally, in order to obtain multimodularity of $r(W_n^a)$ for any realization of Z_n (and not just the expectation of $r(W_n^a)$) we shall have to require that the $\sigma_n = \sigma$ do not depend on n (note that they may still be random). The randomness thus exists only in the arrival. We note that such Petri nets with deterministic firing times and stochastic exogenous arrivals have been the subject of several research work, see e.g. [78].

Many other models give rise to recursive equations of a form similar to (10.4), and hence the same type of results that will hold for Models 59 and 60 will also hold for these:

The $G/D/1$ queue: We consider general independent arrivals and a deterministic (constant) service time requirement of one unit time length. Let T_n be the arrival epochs, and let z_{n+1} be the amount of potential service between T_n and T_{n+1}. Hence $z_{n+1} = T_{n+1} - T_n$. At the nth epoch T_n, the number of customers that are accepted is determined by the control and is given by a_n. Then (10.4) holds where x_n denotes the workload at time T_n.

The $D/G/1$ queue: Assume that the time between arrivals is a constant. The waiting time of the nth customer that is admitted at the queue satisfies a recursion similar to (10.4) (this was established in Theorem 6.4 in [4]).

10.4 Monotonicity: full information case

We consider here the the full information case. We consider the cost over a horizon of length N.

Some of the arguments below rely on dynamic programming, which motivates us to define the *cost to go*: having observed the actions a^r and z^r during r periods, we define recursively the cost to go ϕ^{r+1} and the optimal cost to go g_{r+1} as follows. For $r \leq N$ we set $\bar{a}_r \overset{\text{def}}{=} (a_{r+1})$, and define

$$\phi^r(a_r; H_r) = \mathbb{E}^{a_r}\left(f_r(H_r, a_r)|H_r\right) + \mathbb{E}^{a_r}\left(g_{r+1}(H_{r+1})|H_r\right)$$

$$g_r(H_r) = \min_{a_r} \phi^r(a_r; H_r). \tag{10.8}$$

Let U be the set of policies that choose at time r actions achieving the min in (10.8).

Theorem 62 *Consider the control problem under full information. Assume that the cost functions f_n and the terminal cost g_{N+1} are multimodular for any realization of the z sequence.*
Then there exists an optimal policy $u \in U$ such that for each integer $r \leq N$ and each history h_r, the action a_r, is monotone nonincreasing in $\mathbf{a} = (a_1, ..., a_{r-1})$.

Proof. By standard arguments of dynamic programming we know that any policy in U is optimal. The proof will follow directly from Theorem 57 if we show that $\phi^r\left(a_r, H_r\right)$ is multimodular for all integers $r \leq N$.

The multimodularity is established as follows.

– Since f_n and g_{N+1} are multimodular, it follows that the summation

$$f_N(H_N, a_N) + g_{N+1}(H_{N+1})$$

is also multimodular.
– Taking conditional expectation, we obtain the multimodularity of ϕ^{N+1}.

- Using the Key Lemma in the appendix, this implies the multimodularity of g_N.
- Proceeding the same way by induction, we obtain the multimodularity of g_r for all $r \leq N$.

This establishes the proof.   □

10.5 Monotonicity: general case and delayed information

We consider in this section a general information structure. We assume that $y_n(h_n)$ contains the realization of $(a_1, \dots, a_n)$ (i.e. we have full information on prevoius actions). We further assume that it is monotone increasing in n, i.e. y_n is contained in y_{n+1}. Other than that it can be arbitrary. [1]

We note that this general information structure includes both the sampled as well as the delayed information structures. It includes also the noisy-delayed information structure studied in Altman and Koole [11].

In particular, to see that it includes the case of sampled information, we define

$$y_{rk+j}(h_{rk+j}) = (a^{rk+j-1}, z^{rk})$$

where $h_n = (a^{n-1}, z^n)$, $r \in \mathbb{N}$ and $j = 1, \dots, k$. Note that with this interpretation, decisions can be made at each time unit, but no new information on the z's is obtained between information samples.

We use arguments similar to the case of full information. Define (by backward recursion) the cost to go and the optimal cost to go as

$$\phi^n(a_n; Y_n) = \mathbb{E}^{a_n}\left(f_n(H_n, a_n) + g_{n+1}(H_{n+1})|Y_n\right),$$

$$g_n(Y_n) = \min_{a_n} \phi^n(a_n; Y_n), \tag{10.9}$$

for $n = N, N-1, \dots, 1$.

Let U be the set of policies that choose at time n actions achieving the min in (10.9).

Theorem 63 *Consider the control problem under the above information structure. Assume that the cost functions f_n and the terminal cost g_{N+1} are multimodular for any realization of the z sequence.*
Then there exists an optimal policy $u \in U$ such that for each integer $r \leq N$ and each history h_r, u chooses an action a_r which is monotone in $\mathbf{a} = (a_1, \dots, a_{r-1})$.

---

[1] The increasing assumption is necessary for the use of dynamic programming. Indeed, dynamic programming is used to identify an optimal Markov policy in a MDP setting, where the state is the observation. However, if the information is not increasing then given the current observation state y_n, the future need not be independent of the past. Thus the MDP structure is not satisfied and the dynamic programming need not generate optimal Markovian policies.

Let's go back to our models 59, 60 and 61. In all of them we saw that f_n and g_n were indeed multimodular for any realization of the z sequence. We conclude that the monotonicity result of Theorem 63 applies to all these models.

10.6 State representation

The monotonicity results of the first section are in terms of a sequence of actions. However, quite often

- (I) the cost $f_n(H_n, a_n)$ at time n can be expressed as a function of a_n and of some simpler quantity $x_n(H_n)$, called state, which is a function of the history H_n.
- (II) the distribution of the state $x_{n+1}(H_{n+1})$ at time $n + 1$ depends only on the state $x_n(H_n)$ at time n and on a_n, namely:

$$P(x_{n+1}(H_{n+1}) \in \bullet | H_n) = P(x_{n+1}(H_{n+1}) \in \bullet | x_n(H_n), A_n).$$

In other words, we may also say that x_n is a random function of x_{n-1}, a_{n-1} and z_n (z_n represents the randomness).

The advantage in dealing with states is that this allows typically to decrease the dimensionality. For example, the state space is often finite and does not grow with n, whereas the set of histories grows exponentially with n and need not be finite. When such a state representation exists then different histories, possibly with different lengths, may be mapped to the same state.

Model 59 (continued): The same state x may correspond to different histories; in particular, the history

$$a_i = z_{i+1} = 0, i = 1, ..., k - 2, \ z_k = 0, \ a_{k-1} = x$$

implies that $x_k = x$, where k is an arbitrary integer.

Next we present simple conditions under which monotonicity in the actions, as described in Section 10.5, imply monotonicity in the state.

Case 1, an easy case: Assume that

1. the action space is a convex subset of the integers,
2. As in Model 59, we assume that for any history $h_n = (a^n, z^n)$, there exists some action $a = a(h_n)$ such that $x_n(h_n) = x_n((a, 0))$. Consequently, each state can be identified with an action a, and it corresponds to all histories for which

$$x_n(h_n) = x_n((a, 0)). \tag{10.10}$$

The above set of assumptions now implies a natural ordering over the states. Hence, whenever we have monotonicity of the policy in the sequence a in the representation of Section 10.5, this implies monotonicity in the state.

10.6.1 General full information case

Now, consider the standard information structure in MDP, i.e. $y_n(h_n) = (x_1, a_1, \ldots, x_n)$.

Theorem 64 *Consider the control problem under the above information structure. Let $f : X \times A \to \mathbb{R}$ and define $\bar{f}_m : \mathbf{Z}^m \times \mathbb{N}^m \to \mathbb{R}$ such that $\bar{f}_m(z^m, a^m) \triangleq f(x_m(z^m, a^{m-1}), a^m)$. Define similarly for a terminal cost $g_{N+1} : X \to \mathbb{R}$: $\bar{g}_{N+1}(z^{N+1}, a^N) \triangleq g_{N+1}(x_{N+1}(z^{N+1}, a^N)) : \mathbf{Z}^{N+1} \times \mathbb{N}^N \to \mathbb{R}$. Assume that $\bar{f}$ and $\bar{g}$ are multimodular for any z and that the following assumptions hold:*

- *$<1>$: $x_m : \mathbf{H}_m \to \mathbf{X}_m$, where $\mathbf{X}_m$ is some measurable space endowed with some partial order, and $\mathbf{A} \subset \mathbb{N}$;*
- *$<2>$: $x_m(z, a)$ is monotone in a for any z, i.e. $a^{(1)} \leq_m a^{(2)}$ implies that $x_m(z, a^{(1)}) \leq x_m(z, a^{(2)})$. (Equivalently, $x_m(z, a + s_i) \leq x_m(z, a), i = 2, \ldots, m$, and $x_m(z, a - e_1) \leq x_m(z, a)$.)*
- *$<3>$: There exists $\mathbf{a}(x, z) = (a_1, \ldots, a_N)(x, z)$ where a_m is a function of x_m and of $z^m = (z_1, \ldots, z_{m-1})$, $m = 1, \ldots, N$ with the following properties.*
  (i) For any x^m and z^m, if we assume that $x_m = x_m(a^{m-1}(x, z), z^m)$, $\tilde{x}_m = \tilde{x}_m(a^{m-1}(\tilde{x}, z), z^m)$ and $x_m \leq \tilde{x}_m$ then $a^m(x, z) \leq_m \tilde{a}^m(x, z)$.
  (ii) For any $x_m \leq \tilde{x}_m$ there exists some $z^m = (z_1, \ldots, z_m)$ such that $x_m = x_m(a^{m-1}(x, z), z^m)$, $\tilde{x}_m = \tilde{x}_m(a^{m-1}(\tilde{x}, z), z^m)$.

Then there exists an optimal policy $u \in U$ such that for each integer $r < N$ and each history h_r, u chooses an action a_r which is monotone nonincreasing in the current x_r.

Proof. Our goal is to minimize (10.3).

Consider instead the problem of minimizing (10.3) with $\bar{f}_n$ and and $\bar{g}_{N+1}$ replacing f_n and g_{N+1}, where we assume standard information: at time n the controller has the knowledge of all previous actions and realizations of z.

Let $\mathcal{A}_m(a^{m-1}, z^m)$ be the set of optimal actions at time m given the previous actions and z^m, i.e. the set of actions that minimize the sum of immediate cost at time m plus the optimal cost to go as in eq. (10.8). Hence we can apply Theorem 62 and choose a policy $\mathbf{a} = (a_1, \ldots, a_N)$ which is monotone in the history where $a_m \in \mathcal{A}_m$.

Now lets go back to the original problem. Fix some x_m and $\tilde{x}_m$ such that $x_m \leq \tilde{x}_m$. Let z^m and $\mathbf{a}$ be as in $<3>$. Define

$$\bar{a}_m(x_m) = a_m(x, z).$$

Then the monotonicity of $\mathbf{a}$ in its argument implies the monotonicity of $\bar{a}_m$ in x. □

We shall now illustrate the validity of the assumptions used in the Theorem in our Examples.

Lemma 56. *Consider Model 59. If $a^{n-1} \leq_m \tilde{a}^{n-1}$ then for each z^n*

$$x^n(a^{n-1}, z^n) \leq x^n(\tilde{a}^{n-1}, z^n).$$

Hence condition <2> holds.

Proof. ¿From the associative property of the partial ordering $\leq_m$ it follows that it is sufficient to show for all n, a^{n-1} and z^n:

$$x^n(a^{n-1}, z^n) \leq x^n(a^{n-1} + e_1, z^n) \tag{10.11}$$

$$x^n(a^{n-1}, z^n) \leq x^n(a^{n-1} - s_i, z^n), \quad i = 2, 3, ..., n. \tag{10.12}$$

Indeed, it is easy to see that both inequalities follow directly from the representation (10.5). $\qquad \square$

Next, we show that Condition <2> holds for Model 60. To that end, we introduce an ordering on $\mathbf{X} = \mathbb{R}^2$. We say that $x_m \geq \tilde{x}_m$ if $x_m^0 \geq \tilde{x}_m^0$ and $x_m^1 \leq \tilde{x}_m^1$, where $x_m = (x_m^0, x_m^1), \tilde{x}_m = (\tilde{x}_m^0, \tilde{x}_m^1) \in X$.

Lemma 57. *Consider Model 60. If $a^{n-1} \leq_m \tilde{a}^{n-1}$ then for each z^n*

$$x^n(a^{n-1}, z^n) \leq x^n(\tilde{a}^{n-1}, z^n).$$

Hence condition <2> holds.

Proof. Since for a single queue (Model 59) assumption <2> holds, it clearly holds for the component x_{m+1}^0 in Model 60, i.e.

$$a^{(1)} \leq_m a^{(2)} \text{ implies that } x_m^0(z, a^{(1)}) \leq x_m^0(z, a^{(2)}). \tag{10.13}$$

If we expressed the length of queue 1 in terms of the decision variables b^m i.e. $x_m^1(z, a^m) = \tilde{x}_m(z, b^m(a^m))$, then this would also hold for queue 1, i.e. $b^{(1)} \leq_m b^{(2)}$ implies that $\tilde{x}_m^0(z, b^{(1)}) \leq \tilde{x}_m^0(z, b^{(2)})$. However, since $a_i = 1 - b_i$, $a^{(1)} \leq_m a^{(2)}$ implies that $b^{(1)} \geq_m b^{(2)}$ and hence $x_m^1(z, a^{(1)}) \geq x_m^1(z, a^{(2)})$. Combining this with (10.13), we conclude that <2> indeed holds. $\qquad \square$

Lemma 58. *Consider Model 59. If for any n,*

$$x_n = x_n(a^{n-1}, z^n), \quad \tilde{x}_n = x_n(\tilde{a}^{n-1}, z^n),$$

and $x_n \leq \tilde{x}_n$, then there exists $\bar{a}^{n-1}$ such that

$$a^{n-1} \leq_m \bar{a}^{n-1}$$

and

$$\tilde{x}_n = x_n(\bar{a}^{n-1}, z^n).$$

Hence condition <3> holds.

Proof. The proof holds by a rather straightforward induction on n. $\qquad \square$

10.6.2 Monotonicity of the switching curves

Next we illustrate further structural properties of optimal policies. As in Lemma 59, let there be a function $x : \mathbf{H}_m \to \mathbb{N}$, let $f : X \times \mathbf{A} \to \mathbb{R}$ be given and define f' as in Lemma 59. Consider the problem:

$$\min_\alpha f(x, \alpha).$$

We know from the end of Section 10.2 that one may choose the argmin to be of a switching curve structure. In our case it means that there exists some minimizer u (as a function of x) that behaves as follows. There exist thresholds $l_m, m \in \mathbf{A}$ such that $u(x) \leq m$ if and only if $x \geq l_m$. Note that l_m is nonincreasing, i.e.

$$l_{m+1} \leq l_m.$$

Assume that f is superconvex as in property P2, or equivalently

$$f(x, \alpha + 2) - f(x, \alpha + 1) \geq f(x + 1, \alpha + 1) - f(x + 1, \alpha).$$

Then the minimizer u can further be chosen such that the following holds:

$$l_{m+1} \geq l_m - 1.$$

Indeed, assume that at state $x + 1$ it is optimal to use some action m but it is not optimal to use any action $a > m$ at that state. This means that $f(x + 1, m + 1) > f(x + 1, m)$, that $l_m = x + 1$.

Then the superconvexity implies that $f(x, m + 2) > f(x, m + 1)$, which means that it is not optimal to use $m + 2$ at x; thus the smallest state for which there exists some optimal action a among $a \leq m + 1$ is x. Hence $l_{m+1} \geq x = l_m - 1$.

10.6.3 State representation with delayed information structure

We have established in Section 10.5 monotone properties of optimal policies as a function of the previous taken actions for general information structure. In Section 10.6, we considered the monotonicity of optimal policies in case that a state representation exists. The two aspects can now be combined: state representation with more general information patterns.

In case of general information structure, we cannot expect anymore to have monotonicity in the current state, since the current state might not be available to the controller. In order to obtain monotonicity in some quantities other than the previous actions, new objects with the role of a state should be defined.

To illustrate that, consider the case of delayed information on the state. We thus keep the definition of state x in the beginning of this section (requirements (I) and (II) there); we assume that the information available to the controller can be expressed as

$$y_n(h_n) = (a_1, \ldots, a_{n-1}, x_1, \ldots, x_{n-\theta}).$$

We now define an object that will serve as state, and define a new cost function:

$$\xi_n(h_n) = (x_{n-\theta}, a_{n-\theta}, \ldots, a_{n-1}),$$
$$\tilde{f}(\xi_n, a_n) = \mathbb{E}\, f((X_n, a_n)|\xi_n).$$

It is easily seen that this new state and cost are legitimate in the terms of the requirements (I),(II) given in the beginning of the section. In particular, if we note the transition probabilities P_{xaC} for the original state of moving from state x to a subset $C \subset \mathbf{X}$ given that action a is chosen, then for $\mathcal{C} = (C, \alpha_{n-\theta}, \ldots, \alpha_{n-1})$, the transition probabilities for the new state are given by

$$\mathcal{P}_{\tilde{x}, a, \mathcal{C}} = P_{xaC} 1\{\alpha_{n-\theta} = a_{n-\theta+1}, \alpha_{n-\theta+1} = a_{n-\theta+2}, \ldots, \alpha_{n-1} = a\}.$$

Assuming that there exists a partial ordering on x, one can now define the partial order on the new state space as the component-wise ordering: that for x for the first component and the m-order between the second action-vector component.

Now, assume that conditions $< 1 >$-$< 3 >$ of Theorem 64 hold *for the original state* and that $\bar{f}$ and $\bar{g}$ defined there are multimodular. Now, let us consider the delayed case. We define

$$\tilde{f}'(z_1, \ldots, z_m, a_1, \ldots, a_m) \overset{\text{def}}{=} \tilde{f}(\xi(z_1, \ldots, z_m, a_1, \ldots, a_{m-1})), a_m).$$

By definition, we have,

$$\tilde{f}'(z_1, \ldots, z_m, a_1, \ldots, a_m) = \mathbb{E}\, (f(x_m(h_m), a_m)|\xi_m)$$
$$= \mathbb{E}\, (f(x_m(h_m), a_m)|x_{m-\theta}, a_{m-\theta}, \cdots, a_{m-1}).$$

According to the assumption (II) made on $x_m(h_m)$ (in Section 10.6), we know that it is in fact a function of only x_{m-1}, a_{m-1} and z_m. By induction, we may say that $x_m(h_m)$ is a function of $x_{m-\theta}, a_{m-1}, \cdots, a_{m-\theta}$ and $z_{m-\theta+1}, \cdots, z_m$. Therefore, we can write

$$\mathbb{E}\, (f(x_m(h_m), a_m)|x_{m-\theta}, a_{m-\theta}, \cdots, a_{m-1})$$
$$= \mathbb{E}\, (f(x_m(h_m), a_m)|h_{m-\theta}, a_{m-\theta}, \cdots, a_{m-1})$$
$$= \mathbb{E}\, (f'(h_m, a_m)|h_{m-\theta}, a_{m-\theta}, \cdots, a_{m-1})$$
$$= \mathbb{E}\,_{z_{m+1-\theta}, \ldots, z_m} f'(z_1, \ldots, z_m, a_1, \ldots, a_m),$$

and hence $\tilde{f}'$ is also multimodular (we take expectation with respect to the unknown random variables $z_{m+1-\theta}, \ldots, z_{m+1}$). In a similar way, we can show that the terminal cost is also multimodular.

It remains to check that conditions $< 1 >$-$< 3 >$ of Theorem 64 hold for the new state ξ_n. Clearly $< 1 >$-$< 2 >$ hold.

Next we show that $< 3 >$ holds for the new state. Choose some $\xi_m = (x_{m-\theta}, \alpha_{m-\theta}, ..., \alpha_{m-1})$. With the policy $\mathbf{a}$ defined as in $< 3 >$ for the original state, define the new $\bar{a}$ as follows:

$$(\bar{a}_1, ... \bar{a}_{m-1})(\xi_m, z) = ((a_1, ..., a_{m-\theta-1})(x_{m-\theta}, z^{m-\theta-1}), \alpha_{m-\theta}, ..., \alpha_{m-1})$$

With this choice, it is seen that indeed $< 3 >$ holds for the new state. Indeed, let $\xi_m \leq \tilde{\xi}_m$ with $\tilde{\xi}_m = (\tilde{x}_{m-\theta}, \tilde{\alpha}_{m-\theta}, ..., \tilde{\alpha}_{m-1})$. Note that

$$x_{m-\theta} \leq \tilde{x}_{m-\theta}, \qquad (\alpha_{m-\theta}, ..., \alpha_{m-1}) \leq_m (\tilde{\alpha}_{m-\theta}, ..., \tilde{\alpha}_{m-1}).$$

Assumption $< 3 >$ used for the original model implies immediately that $(\alpha_1, ..., \alpha_{m-\theta-1}) \leq_m (\tilde{\alpha}_1, ..., \tilde{\alpha}_{m-\theta-1})$. We conclude that $\bar{\mathbf{a}}(x_m, z) \leq_m \bar{\mathbf{a}}(\tilde{x}_m, z)$ (component-wise), so that $< 3 >$ holds also for the new state.

We can thus extend Theorem 64 to the delayed case:

Theorem 65 *Assume that the conditions of Theorem 64 hold (for the problem with standard information). Then the conditions and the statement of that theorem also hold for the case of delayed information.*

10.7 Relation between multimodularity, superconvexity and submodularity

Lemma 59. *Assume that the action space is a convex subset of the integers and that $<1>$ and $<2>$ from Theorem 64 hold.*
Let $f : X \times \mathbf{A} \to \mathbb{R}$ be given and define

$$f'(z_1, ..., z_{m+1}, a_1, ..., a_m) \triangleq f(x_m(z_1, ..., z_m, a_1, ..., a_m), z_{m+1}, a_{m+1})$$

Then f' is multimodular for any z if and only if f satisfies:
– (P1) it is submodular, i.e., for $\alpha_1 \leq \alpha_2$, $\alpha_i \in \mathbf{A}$, $i = 1, 2$, $\bar{a} \in \mathbf{A}^m$,

$$f(x_m(z, \bar{a}), z_{m+1}, \alpha_1) - f(x_m(z, \bar{a}), z_{m+1}, \alpha_2) \text{ is nonincreasing in } \bar{a},$$

– (P2) it is superconvex (this property is defined e.g. in [80]), i.e. for $\bar{a} = (a_1, ..., a_m)$,

$$\begin{aligned} &f(x_m(z, \bar{a}), z_{m+1}, \alpha + 2) + f(x_m(z, \bar{a} + e_m), z_{m+1}, \alpha) \\ &\geq f(x_m(z, \bar{a} + e_m), z_{m+1}, \alpha + 1) + f(x_m(z, \bar{a}), z_{m+1}, \alpha + 1); \end{aligned}$$

– (P3) for fixed a_{m+1} and z, $f(x_{m+1}(a_1, ..., a_m, z_1, ..., z_{m+1}), a_{m+1}) : \mathbf{A}^m \to \mathbb{N}$ is multimodular.

Proof. Assume $P1, P2, P3$. We have to check that for any z and any $v, w \in \mathcal{F}, v \neq w$,

$$f'(z, a + v) + f'(z, a + w) \geq f'(z, a + v + w) + f'(z, a), \qquad (10.14)$$

where $\mathcal{F} = \{-e_1, s_2, \ldots, s_{m+1}, e_{m+1}\}$. If $v, w \notin \{e_{m+1}, s_{m+1}\}$ then (10.14) holds by property P3.

Let $v = e_{m+1}$, $w \neq s_{m+1}$. Then $x_{m+1}(z, a + w) \leq_m x_{m+1}(z, a)$. Setting $\alpha_1 \stackrel{\text{def}}{=} a_{m+1}$ and $\alpha_2 \stackrel{\text{def}}{=} a_{m+1} + 1$, we see that the submodularity of f implies (10.14):

$$\begin{aligned}
f'(z, a + w) - f'(z, a + w + v) &= \\
f(x_{m+1}(z, a + w), \alpha_1) &- f(x_{m+1}(z, a + w), \alpha_2) \\
\leq f(x_{m+1}(z, a), \alpha_1) &- f(x_{m+1}(z, a), \alpha_2) \\
&= f'(z, a) - f'(z, a + v).
\end{aligned}$$

It now remains to check the case where $v = e_{m+1}$, $w = s_{m+1}$. Let $\bar{a} \stackrel{\text{def}}{=} \{a_1, \ldots, a_m\}$; then

$$\begin{aligned}
f'(z, a + v) + f'(z, a + w) &= \\
f(x_{m+1}(z, \bar{a}), a_{m+1} + 1) &+ f(x_{m+1}(z, \bar{a} + e_m), a_{m+1} - 1) \\
f'(z, a + v + w) &= f(x_{m+1}(z, \bar{a} + e_m), a_{m+1});
\end{aligned}$$

(10.14) now follows from the superconvexity of f, which establishes the multimodularity of f'.

To establish the converse, assume now that f' is multimodular. Then properties P1 and P2 follow directly from Lemma 2.2 (a) in [6], and property P3 follows from (a) and (b.ii) in that Lemma. $\qquad \square$

Now, we give some comments on the generality of assumptions (P1)-(P2)-(P3).

Remark 24. Note that superconvexity is not needed if we restrict to sequences whose elements take only the values 0 and 1 instead of taking values in $\mathbb{N}$. More precisely, if f is defined on $\mathbb{N} \times \{0, 1\} \to \mathbb{R}$ then we do not have to check the case $v = e_{m+1}$, $w = s_{m+1}$, since clearly $a + v$ or $a + w$ are not within the set $\mathbb{N} \times \{0, 1\}$.

Remark 25. Since x_m is assumed to be increasing in $\bar{a}$, a sufficient condition for property (P1) is that $f(x_m, \alpha_1) - f(x_m, \alpha_2)$ is nonincreasing in x_m, for $\alpha_1 \leq \alpha_2$.

Remark 26. Note that the superconvexity and the submodularity imply that f is integer convex in its second argument. Indeed,

$$f(x_m(z, \bar{a} + e_m), \alpha + 1) - f(x_m(z, \bar{a} + e_m), \alpha)$$
$$\le f(x_m(z, \bar{a}), \alpha + 2) - f(x_m(z, \bar{a}), \alpha + 1)$$
$$\le f(x_m(z, \bar{a} + e_m), \alpha + 2) - f(x_m(z, \bar{a} + e_m), \alpha + 1)$$

where the first inequality follows from the superconvexity and the second from the submodularity of f.

Remark 27. The following superconvexity property, which is similar to (P2), is also implied by the multimodularity of a function $f' : \mathbf{Z}^m \times \mathbf{A}^{m+1} \to \mathbb{R}$:

$$f'(z, \bar{a} + 2e_m, \alpha) + f'(z, \bar{a}, \alpha + 1) \ge f'(z, \bar{a} + e_m, \alpha + 1) + f'(z, \bar{a} + e_m, \alpha).$$

Again, this follows directly from Lemma 2.2 in [6].

Simple sufficient conditions can be presented for the assumptions of Lemma 59, which are satisfied by models 59 and 60 respectively.

Indeed, let in <1>, $X = \mathbb{R}$ be the set of real numbers endowed with the standard ordering between numbers. Then (P3) holds if the following conditions hold:
- $f(x_m, a_{m+1})$ is convex increasing in x_m for any fixed value of a_{m+1},
- x_m satisfies the following:

$$x_{m+1}(z, a^m) \vee x_{m+1}(z, a^m - v - w) \le x_{m+1}(z, a^m - v) \vee x_{m+1}(z, a^m - w),$$

for any z, where $v \ne w$, and $v, w \in \mathcal{F}$, see [6]. Note that Model 59 satisfies the above conditions, (see [4]) as well as assumption <2> in Lemma 59.

Next consider $\mathbf{X} = \mathbb{R}^2$, with the ordering introduced just before Lemma 57. Assume that
i) $x^i_{m+1}(z, a^m)$ are multimodular in a^m and satisfy for all $v \ne w, v, w \in \mathcal{F}$;

$$x^i_{m+1}(z, a^m) \vee x^i_{m+1}(z, a^m - v - w) \le x^i_{m+1}(z, a^m - v) \vee x^i_{m+1}(z, a^m - w) \tag{10.15}$$

ii) f is given by

$$f(x_{m+1}, a_{m+1}) = r^0(x^0_{m+1}, a_{m+1}) + r^1(x^1_{m+1}, a_{m+1}),$$

where r^i are convex increasing functions. Then assumption (P3) holds, since (10.15) implies that $r^i(x^i_{m+1}, a_{m+1})$ are multimodular in a^m.

10.8 Appendix: Key Lemma

Lemma 60. *Let $f : A \to \mathbb{R}$ be a convex function where A is a convex set in $\mathbb{R}^n$. If the function g, defined by*

$$g(x_1, x_2, \ldots x_{k-h}) \stackrel{def}{=} \inf_{x_{k-h+1}, \cdots, x_k} f(x_1, x_2, \cdots x_k), \tag{10.16}$$

is finite for all $x_1, x_2, \ldots x_{k-h}$, then it is also convex.

Proof. We decompose a vector x in A into a vector u and a vector v of size h. Therefore, we can rewrite Equation 10.16 as $g(u) = \inf_v f(u, v)$. Let u^1 and u^2 be any two vectors of dimension $k - h$. Since $g(u^1)$ and $g(u^2)$ are both finite, for all $\varepsilon > 0$, there exits v^1 (resp. v^2) such that $(u^1, v^1) \in A$ and $f(u^1, v^1) \le \inf_v f(u^1, v) + \varepsilon$ (resp. $(u^2, v^2) \in A$ and $f(u^2, v^2) = \inf_v f(u^2, v) + \varepsilon$).

By definition of g and by convexity of f, we have

$$
\begin{aligned}
g(\lambda u^1 + (1 - \lambda)u^2) &\le f(\lambda(u^1, v^1) + (1 - \lambda)(u^2, v^2)) \\
&\le \lambda f(u^1, v^1) + (1 - \lambda)f(u^2, v^2), \\
&\le \lambda \inf_v f(u^1, v) + (1 - \lambda)\inf_v f(u^2, v) + 2\varepsilon.
\end{aligned}
$$

This is true for all $\varepsilon > 0$. Therefore, $g(\lambda u^1 + (1-\lambda)u^2) \le \lambda g(u^1) + (1-\lambda)g(u^2)$. $\square$

Lemma 61 (Key Lemma). *Let* $f : \mathbb{N}^k \to \mathbb{R}$ *be a multimodular function. Then* $g : \mathbb{N}^{k-h} \to \mathbb{R}$ *defined by*

$$
g(a_1, a_2, \ldots a_{k-h}) \overset{def}{=} \inf_{a_{k-h+1}, \cdots, a_k} f(a_1, a_2, \cdots a_k)
$$

is also multimodular.

Proof. First, note that it is enough to prove the result for $h = 1$. The proof for $h > 1$ is done by a straightforward induction.

Let $(e_1, e_2, \cdots, e_k)$ be the canonical base of $\mathbb{R}_+^k$. We denote by V_i the vectorial space generated by $(e_1, \cdots, e_i)$. We define the function $\overline{f} : \mathbb{R}^k \to \mathbb{R}$ such that $\overline{f}$ is the linear interpolation of f on the atoms of $\mathbb{R}^k$. Theorem 2.1 in [6] tells us that $\overline{f}$ is convex.

Now, let us define $\tilde{g} : \mathbb{R}_+^{k-1} \to \mathbb{R}$ by:

$$
\tilde{g}(a_1, a_2, \ldots a_{k-1}) \overset{def}{=} \inf_{a_k} \overline{f}(a_1, a_2, \cdots a_k).
$$

By using Theorem 2.1 in [6], we just have to prove that $\tilde{g}$ is convex and linear on all the atoms of $\mathbb{R}_+^{k-1}$.

For convenience, we introduce the following notation: for x in $\mathbb{R}_+^k$ and S a set in $\mathbb{R}_+^k$, $P(x, S)$ is the projection along e_k of x on S. Formally, $P(x, S) \overset{def}{=} x + de_k$ where d is the real number with smallest absolute value such that $x + de_k \in S$.

Choose $m \in \mathbb{N}$. We denote by E_m the set $V_{k-1} \oplus [0, m]e_k$. Note that E_m is a convex union of atoms in $\mathbb{R}_+^k$.

By continuity of $\overline{f}$, for all $u \in \mathbb{R}_+^{k-1}$, there exists $v_m(u) \le m$ such that $g_m(u) \overset{def}{=} \inf_{v \le m} \overline{f}(u, v) = \overline{f}(u, v_m(u))$. By using Lemma 60, then the function g_m is convex. For any $y, z \in E_m$,

$$g_m(\lambda y + (1 - \lambda)z) \leq \lambda g_m(y) + (1 - \lambda)g_m(z). \tag{10.17}$$

In a given atom a of V_{k-1}, pick any two points u^1 and u^2. Let $x \overset{\text{def}}{=}$ $\lambda u^1 + (1 - \lambda)u^2$. The point $(x, v_m(x))$ belongs to an atom, say A of $\mathbb{R}_+^k$, included in E_m.

We consider F^* (resp. F_*), the upper (resp. lower) hyperface of dimension $\mathbb{R}_+^{k-1}$ of A along e_n. We project the point $(x, v_m(x))$ on F_* and F^*.

$$p \overset{\text{def}}{=} P((x, v_m(x)), F_*),$$

$$q \overset{\text{def}}{=} P((x, v_m(x)), F^*).$$

By linearity of $\overline{f}$ over the segment $[p, q]$, $\overline{f}(x, v_m(x))$ is a linear combination of $\overline{f}(p)$ and $\overline{f}(q)$. Moreover, the points p and q belong to E_m. By definition of $v_m(x)$, we also have $\overline{f}(x, v_m(x)) \leq \overline{f}(p)$ and $\overline{f}(x, v_m(x)) \leq \overline{f}(q)$. Therefore, we have $\overline{f}(x, v_m(x)) = \overline{f}(p)$ or $\overline{f}(x, v_m(x)) = \overline{f}(q)$. In the following we will consider the case $\overline{f}(x, v_m(x)) = \overline{f}(p)$. The case $\overline{f}(x, v_m(x)) = \overline{f}(q)$ is similar by using the face F^* instead of F_*.

We project the points $(u^1, v_m(u^1)), (u^2, v_m(u^2))$ on F_*.

$$p^1 \overset{\text{def}}{=} P((u^1, v_m(u^1)), F_*),$$

$$p^2 \overset{\text{def}}{=} P((u^2, v_m(u^2)), F_*).$$

By linearity of $\overline{f}$ over the segment $[p^1, p^2]$, and definition of $v_m(x)$,

$$\overline{f}((x, v_m(x))) = \overline{f}(p), \tag{10.18}$$
$$= \lambda \overline{f}(p^1) + (1 - \lambda)\overline{f}(p^2), \tag{10.19}$$
$$\geq \lambda \overline{f}((u^1, v_m(u^1)) + (1 - \lambda)\overline{f}((u^2, v_m(u^2)), \tag{10.20}$$
$$\geq \overline{f}(\lambda(u^1, v_m(u^1)) + (1 - \lambda)(u^2, v_m(u^2))), \tag{10.21}$$
$$\geq \overline{f}((x, v_m(x))). \tag{10.22}$$

Where Equation 10.19 comes from linearity of $\overline{f}$ over A, Equation 10.20 comes from the definition of $v_m(.)$, Equation 10.21 comes from convexity of $\overline{f}$ and Equation 10.22 comes from the theorem of Thales. Therefore,

$$g_m(\lambda u^1 + (1 - \lambda)u^2) = g_m(x), \tag{10.23}$$
$$= \overline{f}((x, v_m(x))), \tag{10.24}$$
$$= \lambda \overline{f}((u^1, v_m(u^1)) + (1 - \lambda)\overline{f}((u^2, v_m(u^2)), \tag{10.25}$$
$$= \lambda g_m(u^1) + (1 - \lambda)g_m(u^2), \tag{10.26}$$

where Equation 10.25 comes from Equation 10.21, which is an equality.

All these arguments are illustrated in Figure 10.1, which is done with dimension $k = 2$.

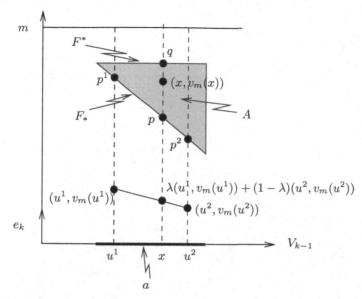

Fig. 10.1. Illustration of the proof of the key lemma.

To finish the proof, it is enough to notice that $\tilde{g}(u) = \lim_{m\to\infty} g_m(u)$, to take the limit in Equation 10.17 to establish the convexity of $\tilde{g}$ and to take the limit in Equation 10.26 to establish the linearity on atom a.

$\square$

Part IV

Comparisons

Part II shows that in general it is very difficult (and sometimes impossible) to construct the optimal policy for routing customers in several parallel systems.

One way to overcome this problem is to replace this ambitious objective by an easier one: improve on one initial policy. To reach this objective, we propose several order relations between policies which all have the following underlying idea. Policy a is smaller than policy b if a is more regular than b.

All the orders introduced in the following three chapters have some relations with each other, but they are indeed different. The utilization of one of them rather than the other depends on the cost function as well as the overall context.

11 Comparison of queues with discrete-time arrival processes

11.1 Introduction

In this chapter, traveling times in a FIFO-stochastic event graph are compared in increasing convex ordering for different arrival processes. As a special case a stochastic lower bound is obtained for the sojourn time in a tandem network of FIFO-queues with a Markov arrival process. A counterexample shows that the extended Ross conjecture is not true for discrete-time arrival processes. This chapter is an adaptation of [61].

We consider an open stochastic queueing network with one input node. The network dynamics are supposed to satisfy a linear recursion in the (max,plus)-algebra on $\mathbb{R}^L$, as described in Chapter 3. It is shown in Chapter 3 that the epoch of the beginning of the n-th firing time of a FIFO-stochastic event graph (FSEG) satisfies such a linear recursion for each transition (see also [25]).

We recall that a special case is a stochastic network of L single-server-FIFO-queues in tandem, with infinite buffer capacity in the first queue and finite buffers with manufacturing blocking or infinite buffers in the other queues. Note that the sojourn time in this tandem network of L single-server-FIFO-queues, is the traveling time to server L plus its service time at L. Hence, the comparison results below hold also for the sojourn time.

Let T_n, $n = 1, 2, \ldots$ be a stationary sequence of potential arrival epochs. The number of arrivals at T_n will be denoted by A_n. In general A_n, $n = 1, 2, \ldots$ may be a stochastic sequence, and $A_n = \ell$ means that ℓ customers arrive at T_n. Note that $A_n = 0$ implies that T_n is not an actual arrival epoch.

Two arrival processes are compared with respect to their implied performance of the stochastic network. We assume that both have the same potential arrival epochs, but different A-sequences, say A_n^1 and A_n^2. Let us call these the admission sequences. Let S_n^j be the firing-time (service-time) of the n-th token (customer) in transition (server) j. We assume that

$$S_n = (S_n^1, \ldots, S_n^L)$$

is a stationary sequence of stochastic vectors. Note that no independence assumption is made on the firing-times, stationarity is sufficient. But, we assume that for $i = 1$ and for $i = 2$ every couple of sequences from $\{A_n^i, T_n, S_n\}$

are stochastically independent. Let $_bW_n^q$ denote the traveling time of the n-th arrival to transition q, i.e. the time between its entrance in the stochastic network and the beginning of its firing time at transition q. Let $_aW_n^q$ be the same time of a potential arrival at T_n. Recall that at T_n there may be no arrival, and the arrival time of the n-th customer is in general not T_n. With Z_n^i we denote the n-th arrival epoch for arrival process $i, i = 1, 2$, i.e.

$$Z_n^i = \min\{k : \sum_{\ell=1}^k A_\ell^i \geq n\}.$$

Then the arrival time of the n-th customer is, $T_{Z_n^i}$. We also need the following notation,

$$B_n^i \overset{\triangle}{=} Z_n^i - Z_{n-1}^i, \quad i = 1, 2, \quad n = 1, 2, \ldots,$$

where we take $Z_0^i = 0$. Note that B_n^i is the n-th inter-arrival length, in general this is not equal to the inter-arrival time. Of course, B_n^i is a function of $A_1^i, A_2^i, \ldots, A_{Z_n^i}^i$; we suppress this in our notation. It is shown in Chapter 4 for a (max, $plus$)-linear system that for any transition q, and $n = 1, 2, \ldots$, $\mathbb{E} \, _bW_n^q$ is a multimodular function of $(B_1, \ldots, B_n)$, and $\mathbb{E} \, _aW_n^q$ is a multimodular function of $(A_1, \ldots, A_n)$, where the expectation is with respect to T_n and S_n, $n \in \mathbb{N}$. These multimodularity properties induce the convexity results which we use to prove our comparison results. The arrival processes in the chapter will be generated by a Markov Arrival Processes (MAP), for which we assume a Markov process on E, a finite state space with intensities λ_{xy}, $x, y \in E$, and an arrival occurs with probability r_{xy} when a transition from state x to state y happens.

In [63] it is explained that a Markov arrival process (MAP) is more general than the Markov-modulated Poisson process (MMPP) or the Phase-type renewal process. In [18] it is shown that any arrival process can be approximated arbitrarily closely by a MAP. Let us mention the stochastic orders we use in this chapter. Random vectors $X^1 = (X_1^1, \ldots, X_n^1)$ and $X^2 = (X_1^2, \ldots, X_n^2)$ are ordered with respect to the convex ordering $(X^1 \leq_{cx} X^2)$ (resp. increasing convex ordering $(X^1 \leq_{icx} X^2)$) if

$$\mathbb{E} \, h(X^1) \leq \mathbb{E} \, h(X^2)$$

for all convex (resp. increasing convex) functions

$$h : \mathbb{R}^n \to \mathbb{R}.$$

Increasing and decreasing are in this chapter always in the non strict sense.

In section 11.2 we give a counterexample which shows that the extension of the Ross-conjecture is not true in our comparison of queues with different admission sequences. In section 11.3 a first comparison lemma is derived for admission (inter-arrival) sequences which are comparable in the convex ordering. It is shown, that the potential (actual) traveling times are ordered in the increasing convex ordering. Similar comparison results hold for the stationary traveling times. As applications of the lemma, we derive the following results:

1. independent sources have a better performance (in increasing convex ordering sense) than coupled sources,
2. fixed batch sizes are better than random batch sizes,
3. fluid scaling improves the performance.

In a second comparison lemma, derived in section 11.4, actual (potential) traveling times are ordered in the icx-ordering for non-integer admission (inter-arrival) sequences. Here a "regularization" procedure is given, which has been used in the theory on balanced sequences and optimal routing presented in Chapters 2 and 6. In section 11.5 we construct the most regular arrival process for a fixed arrival intensity, and we call it the Regular Arrival Process (RAP). We show that the RAP provides a stochastic lower bound for any MAP-source with the same arrival intensity. This result (Theorem 68) can be seen as the 'Ross-conjecture-theorem' in the comparison of discrete-time arrival processes. In the literature on optimal routing to parallel queues, it was claimed (cf. [43]) that good approximations could be obtained through replacing the MAP-arrival process by a renewal process with approximately the same arrival intensity. Theorem 69 and Corollary 9 provide the proof of these claims. Indeed, the performance of a RAP has stochastic lower bounds for arrival processes which are (approximately) renewal processes. For a rational arrival intensity, there is a RAP which is renewal with Erlang distributed inter-arrival times. By a continuity argument we obtain that the renewal-arrival process with constant inter-arrival process gives, for any real stationary arrival probability, a stochastic lower bound on the performance.

11.2 On the Ross conjecture in discrete-time

In his inspiring paper [98] Sheldon Ross conjectured that the mean waiting time in a $\cdot/G/1/\infty$ queue with non-stationary Poisson arrival process is larger than or equal to the mean waiting time of the $M/G/1/\infty$ queue with the same arrival intensity. . This paper of Ross initiated a long sequence of research papers on this and related problems. Rolski proved the Ross conjecture in [97]. Recent publications on this and related topics are [17, 19, 30]. Suppose a $\cdot/G/1/\infty$ queue with potential arrival epochs given by sequence T_n but with different admission sequences A^1 and A^2. Since the potential arrival times are fixed and form a sequence of discrete epochs, we prefer to call this a discrete-time model, although the T_n may have continuous distributions. Suppose A^1 is time-stationary and A^2 not, but they have the same intensity. Is the mean waiting time for the A^1 sequence smaller than or equal to that of A^2? The A-sequence can be seen as a random environment and one may expect that an extended Ross conjecture holds (cf. [21], where the service process has a random environment) which says:

"The $G/G/1/\infty$ queue in a random environment should be bounded below by the corresponding queue where the environment process is 'frozen' to its mean values."

For the setting of this chapter, it is not true as the following counter example shows.

Let $T = \{T_n\}$ be a Poisson process with rate 1. We consider the $\cdot/M/1/\infty$ queue, and we assume that the S_n are i.i.d. with exponential distribution with mean 1. Let A_n^1 be distributed as i.i.d. Bernoulli random variables with mean $1/2$. Let A_n^2 be distributed as independent Bernoulli random variables, let the mean of A_n^2 be p_n. We assume that p_n is random with mean $1/2$. Then A_n^1 is the arrival process for which the random environment of A_n^2 is frozen to its mean. So the extended Ross conjecture would claim that for W^i the stationary waiting time for sequence A^i, $i = 1, 2$

$$\mathbb{E}\, W^1 \leq \mathbb{E}\, W^2. \tag{11.1}$$

Suppose that the sequence $(p_1, p_2, p_3, \ldots)$ is with probability $1/2$ equal to $\frac{1}{2} + \epsilon, \frac{1}{2} - \epsilon, \frac{1}{2} + \epsilon, \ldots$ and with probability $1/2$ equal to $\frac{1}{2} - \epsilon, \frac{1}{2} + \epsilon, \frac{1}{2} - \epsilon, \ldots$, where $0 \leq \epsilon \leq \frac{1}{2}$. Then $\mathbb{E}\, p_n = \frac{1}{2}$ which is the probability in the A_n^1-sequence. It is easily seen that for $\epsilon = 1/2$ we have that W^1 is the stationary waiting time of the $M/M/1/\infty$ queue with traffic intensity $\rho = \frac{1}{2}$, and that W^2 is the stationary waiting time of the $GI/M/1/\infty$ queue with inter-arrival times which have an Erlang distribution with 2 phases of exponential length with mean 1. It is well-known that

$$\mathbb{E}\, W^1 > \mathbb{E}\, W^2,$$

which contradicts relation 11.1. In Section 11.5 we will derive a discrete-time analogue of the Ross-conjecture-theorem. The above example explains why we use regular sequences there.

11.3 A comparison lemma and its applications

In this section we derive a first comparison lemma which is a rather direct consequence of the multimodularity of the traveling times as function of the admission sequence. As we will see, the comparison lemma has some nice implications.

We recall that $_bW_n^i$ $(_aW_n^i)$, $i = 1, 2$ is the traveling time of the n-th arrival (potential arrival at T_n) to a fixed but arbitrarily chosen transition q in the FSEG.

Lemma 62. *The following implications hold for any $n = 1, 2, \ldots$*
1)$(A_1^1 \cdots A_n^1) \leq_{\mathrm{cx}} (A_1^2 \cdots A_n^2) \Rightarrow (_aW_1^1 \cdots _aW_n^1) \leq_{\mathrm{icx}} (_aW_1^2 \cdots _aW_n^2)$.
2)$(B_1^1 \cdots B_n^1) \leq_{\mathrm{cx}} (B_1^2 \cdots B_n^2) \Rightarrow (_bW_1^1 \cdots _bW_n^1) \leq_{\mathrm{icx}} (_bW_1^2 \cdots _bW_n^2)$.

Proof. The proof is given for 1); the proof of 2) goes similarly. Note that $(A_1^i, \ldots, A_n^i) \in \mathbb{N}^n$, $i = 1, 2$, let $h : \mathbb{R} \to \mathbb{R}$ be an increasing convex function. Then, for $i = 1, 2$,

$$W_k(A_1^i, \ldots, A_n^i) \triangleq \mathbb{E}\, h(_a W_k^i(A_1^i, \ldots, A_n^i))$$

is a function from $\mathbb{N}^n$ to $\mathbb{R}$. Theorem 18 (for the proof of part 2, use Theorem 19) shows that $W_k(A_1^i, \ldots, A_k^i)$ is multimodular in $(A_1^i, \ldots, A_k^i)$ as function on $\mathbb{N}^k$ for $k = 1, 2, \ldots$. Since $W_k(A_1^i, \ldots, A_k^i)$ is independent of A_n^i for $n > k$, it trivially follows that $W_k(A_1^i, \ldots, A_n^i)$ is also multimodular in $(A_1^i, \ldots, A_n^i)$ if $n \geq k$. It then follows from Theorem 1 that W_k^i is integer convex on $\mathbb{N}^n$ $i = 1, 2$, $k = 1, \ldots, n$.

The rest of the proof is more or less standard. Since $(A_1^1, \ldots, A_n^1) \leq_{\mathrm{cx}} (A_1^2, \ldots, A_n^2)$ we may by Strassen's representation theorem assume without loss of generality that

$$\mathbb{E}\left((A_1^2, \ldots, A_n^2) \mid (A_1^1, \ldots, A_n^1)\right) = (A_1^1, \ldots, A_n^1).$$

From Jensen's inequality we then have,

$$W_k(A_1^1, \ldots, A_n^1) = W_k(\mathbb{E}\left((A_1^2, \ldots, A_n^2) \mid (A_1^1, \ldots, A_n^1)\right)) \qquad (11.2)$$
$$\leq \mathbb{E}\left(W_k(A_1^2, \ldots, A_n^2) \mid (A_1^1, \ldots, A_n^1)\right). \qquad (11.3)$$

Hence,

$$\mathbb{E}\, W_k(A_1^1, \ldots, A_n^1) \leq \mathbb{E}\, W_k(A_1^2, \ldots, A_n^2).$$

Let $h : \mathbb{R}^n \to \mathbb{R}$ be an increasing convex function. Then

$$h(W_1(A_1^i, \ldots, A_n^i), \ldots, W_n(A_1^i, \ldots, A_n^i))$$

is an increasing convex function of $(A_1^i, \ldots, A_n^i)$ $i = 1, 2$. The first inequality below is now a consequence of (11.3) and the increasingness of h, the second inequality follows from Jensen's inequality

$$h(W_1(A_1^1 \ldots A_n^1) \ldots W_n(A_1^1 \ldots A_n^1))$$
$$\leq h(\mathbb{E}\left(W_1(A_1^2 \ldots A_n^2) \mid (A_1^1 \ldots A_n^1)\right) \ldots \mathbb{E}\left(W_n(A_1^2 \ldots A_n^2) \mid (A_1^1 \ldots A_n^1)\right)$$
$$\leq \mathbb{E}\left(h(W_1(A_1^2 \ldots A_n^2) \ldots W_n(A_1^2 \ldots A_n^2)) \mid (A_1^1 \ldots A_n^1)\right).$$

Hence,

$$\mathbb{E}\, h(W_1(A_1^1 \ldots A_n^1) \ldots W_n(A_1^1 \ldots A_n^1))$$
$$\leq \mathbb{E}\, h(W_1(A_1^2 \ldots A_n^2) \ldots W_n(A_1^2 \ldots A_n^2)).$$

$\square$

Let us assume now the following,

Assumption 66 A_n^i *is a stationary sequence in* $n \in \mathbb{Z}$ *for* $i = 1, 2$.

The following sequences with $i = a$ or b, $j = 1$ or 2 are well-known as the Loynes-sequences (cf.[21]),

$$_i\overline{W}_n^j \triangleq {}_iW_n(A_{-n}^j, A_{-n+1}^j, \ldots, A_{-1}^j).$$

They are monotone increasing in n. Consequently, they have a limit as n tends to infinity, which is possibly ∞. These limits are called Loynes-variables, we denote them as

$$_iW_\infty^j \triangleq \lim_{n \to \infty} {}_i\overline{W}_n^j.$$

It is well-known (cf.[21, 34]) that under strong coupling or renovation it holds that

$$_iW_\infty^j = \lim_{n \to \infty} {}_iW_n^j,$$

i.e. it is the time-forward limit. As an immediate consequence of the comparison Lemma 62 we find,

Corollary 7. *If assumption 66 holds then*
1) $(A_1^1, \ldots, A_n^1) \leq_{cx} (A_1^2, \ldots, A_n^2)$ *for all* $n \geq 1$ $\Rightarrow {}_aW_\infty^1 \leq {}_aW_\infty^2$. *2)*
$(B_1^1, \ldots, B_n^1) \leq_{cx} (B_1^2, \ldots, B_n^2)$ *for all* $n \geq 1$ $\Rightarrow {}_bW_\infty^1 \leq_b W_\infty^2$.

Proof. We prove part 1), the proof of part 2) goes similarly. From assumption 66 we have for all n,

$$(A_{-n}^1, A_{-n+1}^1, \ldots, A_{-1}^1) \overset{\text{dist}}{=} (A_1^1, A_2^1, \ldots, A_n^1)$$
$$\leq_{cx} (A_1^2, A_2^2, \ldots, A_n^2)$$
$$\overset{\text{dist}}{=} (A_{-n}^2, A_{-n+1}^2, \ldots, A_{-1}^2),$$

where $\overset{\text{dist}}{=}$ means equality in distribution.

From the comparison Lemma 62,

$$_a\overline{W}_n^1 \leq_{icx} {}_a\overline{W}_n^2.$$

The monotone convergence theorem then gives,

$$_aW_\infty^1 = \lim_{n \to \infty} {}_a\overline{W}_n^1 \leq_{icx} \lim_{n \to \infty} {}_a\overline{W}_n^2 = {}_aW_\infty^2.$$

$\square$

It is well-known that in case of stability of the stochastic networks the Loynes-variables are a.s. finite and represent the stationary versions of the traveling times. So in case of stability also the stationary versions are icx-ordered. Also a multidimensional marginal distribution of the stationary processes can be shown (also as a consequence of comparison lemma 62) to be ordered in the icx-ordering.

11.3.1 Application 1: two i.i.d. MAP-sources perform better than 2 completely coupled Map-sources

Consider two MAP-arrival processes, say $\text{MAP}^i, i = 1, 2$, which are independent and have the same distribution. Denote by T^i the transition epochs of $\text{MAP}^i, i = 1, 2$, and let $T = T^1 \cup T^2$ be the superposition of T^1 and T^2. Define $A_n^1 = 1$ if $T_n \in T$ is an arrival epoch of MAP^1 or MAP^2. Then $T = \{T_n\}$ are the potential arrival epochs, and the admission sequence A_n^1 generates all arrivals of the two independent MAP's. Consider now two completely coupled MAP-sources, which is equivalent to one MAP-source which generates 2 arrivals at any of its arrival epochs. Say this MAP-source is with probability $1/2$ the MAP^1-source and with probability $1/2$ the MAP^2-source. Define for $i = 1, 2$,

$$E_n^i = \begin{cases} 1 & \text{if } T_n \text{ is arrival epoch of } \text{MAP}^i \\ 0 & \text{otherwise.} \end{cases}$$

Then

$$A_n^1 = E_n^1 + E_n^2,$$

and for A_n^2 the admission sequence of the coupled MAP-sources we have for $1 \leq m \leq n$ that, A_m^2 given $A_1^1, \ldots, A_n^1$ is with probability $1/2$ equal to $2E_m^1$ and with probability $1/2$ equal to $2E_m^2$. Hence,

$$\mathbb{E}\left[A_m^2 \mid A_1^1, \ldots, A_n^1\right] = A_m^1$$

and therefore,

$$(A_1^1, \ldots, A_n^1) \leq_{cx} (A_1^2, \ldots, A_n^2) \quad \text{for all} \quad n.$$

Clearly, the events E_n^i in MAP^i are independent of the transition epochs $T_n^i, i = 1, 2$. This implies that $\{A_n^i\}$ and $\{T_n\}$ are independent for $i = 1, 2$. Hence we can apply comparison lemma 62 and corollary 7, and find that the potential (stationary) traveling times for the i.i.d. MAP-sources are in icx-ordering smaller than those of 2 completely coupled MAP-sources. It is possible to extend this result to: k i.i.d. MAP-sources which generate ℓ customers at each of their arrival epochs perform better in icx-order than $\ell \leq k$ i.i.d. MAP-sources which generate k customers at each of their arrival epochs.

11.3.2 Application 2: a fixed batch size is better than random batch sizes

Consider a MAP-source, and assume that at each of its arrival-epochs a random batch number of customers arrive, say N_n at arrival epoch T_n. We assume that $\{N_n\}$ is independent of $\{T_n\}$, and also that $\{N_n\}$ is stationary and $\mathbb{E} N_1 = \ell$. Take $A_n^1 = \ell$ and $A_n^2 = N_n, n \in \mathbb{N}$, then A_n^1 is the admission sequence with fixed (or frozen) batch size. Clearly,

$(A_1^1, \ldots, A_n^1) \leq_{cx} (A_1^2, \ldots, A_n^2)$, $n \in \mathbb{N}$ and part 1) of comparison Lemma 62 and Corollary 7 applies.

If $\ell = 1$ then $(B_1^1, \ldots, B_n^1) = (1, \ldots, 1)$ and in order to show that

$$(B_1^1, \ldots, B_n^1) \leq_{cx} (B_1^2, \ldots, B_n^2)$$

it suffices to verify that for $k = 1, \ldots, n$,

$$\mathbb{E} \, B_k^2 = 1.$$

Since A_n^2 is stationary, it follows that

$$\mathbb{E} \, A_1^2 \cdot \mathbb{E} \, B_1^2 = 1,$$

and $\mathbb{E} \, B_k = \mathbb{E} \, B_1 = 1$. Hence, in this case also, the actual traveling times are smaller in icx-order for the fixed batch sizes.

11.3.3 Application 3: Fluid scaling improves the performance

Consider the following transformations of the time variable t and the state variable x,

$$t \longrightarrow Nt$$
$$x \longrightarrow \frac{x}{N}$$

Fluid limits are obtained by taking limits for $N \to \infty$. Here we take a fixed $N \in \mathbb{N}$. If we have a MAP^1 with finite state space E and transition rates λ_{xy}, $x, y \in E$, and if we divide the time-variable by N, then we get a MAP^2 with transition rates $\frac{1}{N} \lambda_{xy}$. After uniformizing both processes such that the transition times in both processes are a Poisson process with the same parameter λ, say MAP$^1(\lambda)$ and MAP$^2(\lambda)$, we have that a real transition in MAP$^1(\lambda)$ (i.e. a transition in MAP^1) is with probability $\frac{1}{N}$ a real transition in MAP$^2(\lambda)$ (i.e. a transition of MAP^2). Clearly, we can couple the MAP$^1(\lambda)$ and MAP$^2(\lambda)$ such that if T_n are the arrival epochs of MAP$^1(\lambda)$ then the potential arrival epochs of MAP$^2(\lambda)$ are $\{T_n\}$, and the admission sequence is,

$$A_n^2 = \begin{cases} 1 & \text{with probability } \frac{1}{N} \\ 0 & \text{else,} \end{cases}$$

where the A_n^2 are i.i.d. and independent of T_n. If we take $A_n^1 = 1$ then A_n^1 is the admission sequence for MAP$^1(\lambda)$.

As in application 2 we have

$$(A_1^1, \ldots, A_n^1) \leq_{cx} (\overline{A}_1^2, \ldots, \overline{A}_{n'}^2),$$

where $\overline{A}_k^2 = NA_k^2$, $k \geq 1$.

The scaling of the state can be done by considering the original service requirements as a number of packets (possibly of random size), and taking N arrivals instead of one arrival. This gives the $\overline{A}_n^2$ as admission sequence. So, the process corresponding to the A_n^1 can be seen as a fluid scaling of the $\overline{A}_n^2$-induced process. Mathematically, it is the same comparison as in application 2. The comparison lemma 62 and corollary 7 imply that, the performance of the fluid scaled process is better than that of the original process in icx-ordering.

11.4 A second comparison lemma

In comparison lemma 62 we had admission sequences $\{A_n^i\}$, $i = 1, 2$, where A_n^i gives the number of arrivals at T_n. This means that A_n^i is an integer. In the comparison lemma 63, below, the admission sequences are $\{p_n^i\}$, where p_n^i may be any nonnegative real number. For $p^i = (p_1^i, p_2^i, \ldots)$ with $p_n^i \geq 0$, $n = 1, 2, \ldots$ we define an integer admission sequence $\{A_n^i(p^i)\}$ by,

$$A_n^i(p^i) \triangleq \left\lfloor \sum_{j=1}^{n} p_j^i + \theta \right\rfloor - \left\lfloor \sum_{j=1}^{n-1} p_j^i + \theta \right\rfloor,$$

where θ is a random variable, uniformly distributed on $[0,1)$, and where $\lfloor x \rfloor$ denotes the largest integer smaller than or equal to $x \in \mathbb{R}_+$. Note that $A_n^i(p^i)$ is random and integer valued, it gives the number of arrivals at T_n. For the inter-arrival-lengths we proceed similarly given $q^i = (q_1^i, q_2^i, \ldots)$ with $q_n^i \geq 0$, $n = 1, 2, \ldots$, we define

$$B_n^i(q^i) \triangleq \left\lfloor \sum_{j=1}^{n} q_j^i + \theta \right\rfloor - \left\lfloor \sum_{j=1}^{n-1} q_j^i + \theta \right\rfloor,$$

where θ is uniformly distributed on $[0,1)$. Note that if $p_1^i, p_2^i, \ldots, p_n^i$ are all integer valued then $A_k^i(p^i) = p_k^i$ for $1 \leq k \leq n$, and any $\theta \in [0,1)$. Similarly, as in comparison lemma 62, we consider the potential traveling times

$$_aW_n^i \triangleq {}_aW_n(A_1^i(p^i), \ldots, A_n^i(p^i)), \quad i = 1, 2$$

and the actual traveling times

$$_bW_n^i \triangleq {}_bW_n(B_1^i(q^i), \ldots, B_n^i(q^i)), \quad i = 1, 2.$$

Lemma 63. *The following implications hold for any $n = 1, 2, \ldots$:*

1) $(p_1^1, \ldots, p_n^1) \leq_{cx} (p_1^2, \ldots, p_n^2) \Rightarrow ({}_aW_1^1, \ldots, {}_aW_n^1) \leq_{icx} ({}_aW_1^2, \ldots, {}_aW_n^2)$
2) $(q_1^1, \ldots, q_n^1) \leq_{cx} (q_1^2, \ldots, q_n^2) \Rightarrow ({}_bW_1^1, \ldots, {}_bW_n^1) \leq_{icx} ({}_bW_1^2, \ldots, {}_bW_n^2)$.

Proof. We prove part 1), the proof of part 2) is similar. Since

$$(p_1^1, \ldots, p_n^1) \leq_{\mathrm{cx}} (p_1^2, \ldots, p_n^2),$$

we may, by Strassen's representation theorem, assume without loss of generality that

$$\mathbb{E}\left((p_1^2, \ldots, p_n^2) \mid (p_1^1, \ldots, p_n^1)\right) = (p_1^1, \ldots, p_n^1).$$

Theorem 18 together with Theorem 1 imply that for h an increasing convex function,

$$_aW_k(p_1^i, \ldots, p_n^i) \triangleq \mathbb{E}\, h(_aW_k^i(A_1^i(p^i), \ldots, A_n^i(p^i))),$$

is a convex function of $(p_1^i, \ldots, p_n^i)$. The rest of the proof goes similarly as the proof of comparison lemma 62 with $(p_1^i, \ldots, p_n^i)$ substituted for $(A_1^i, \ldots, A_n^i)$.
□

Also in this setting we can consider the Loynes stochastic variables, assuming that p_n^i is defined for all $n \in \mathbb{Z}$,

$$_i\overline{W}_n^j \triangleq {}_iW_n(p_{-n}^j, \ldots, p_{-1}^j)$$

and

$$_iW_\infty^j \triangleq \lim_{n \to \infty} {}_i\overline{W}_n^j.$$

Assumption 67 $A_n^i(p^i)$ *is a stationary sequence in* $n \in \mathbb{Z}$ *for* $i = 1, 2$.

With the same proof as in Corollary 7 we then find

Corollary 8. *Under assumption 67, the following holds*
1) $(p_1^1, \ldots, p_n^1) \leq_{\mathrm{cx}} (p_1^2, \ldots, p_n^2)$ *for all* $n \geq 1 \Rightarrow {}_aW_\infty^1 \leq_{\mathrm{icx}} {}_aW_\infty^2$.
2) $(q_1^1, \ldots, q_n^1) \leq_{\mathrm{cx}} (q_1^2, \ldots, q_n^2)$ *for all* $n \geq 1 \Rightarrow {}_bW_\infty^1 \leq_{\mathrm{icx}} {}_bW_\infty^2$.

11.5 A stochastic lower bound on the traveling times

In section 11.2 we found that the intuitive argument, that queues in a random environment should be bounded below by the corresponding queues where the environment is "frozen" to its mean values, is not generally true. As an application of the second comparison lemma we will derive in this section a lower bound in the icx-ordering. The queueing model is a FSEG with a MAP-source. We will construct a more regular (in fact the most regular) arrival process with the same arrival intensity as the MAP-source. This will provide the lower bound. As we will see this regular arrival process can be approximated by a renewal arrival process with Erlang-distributed inter-arrival times. Without loss of generality we may assume that the MAP has transition times $\{T_n\}$ which form a Poisson(λ) process. Let $\{X_n\}$ be the Markov process with transition probabilities λ_{xy} which governs the transitions of the MAP, i.e. X_n is the state at T_n. We assume that the Markov process is stationary, and we denote by π_x the stationary probability on state $x \in E$. The probability on an arrival at T_n is,

$$p \stackrel{\Delta}{=} \sum_x \sum_y \pi_x \lambda_{xy} r_{xy}. \tag{11.4}$$

The arrival intensity of the stationary MAP is then $\overline{\lambda} \stackrel{\Delta}{=} p\lambda$. The MAP corresponds to the following admission sequence,

$$A_n^2 = \begin{cases} 1 & \text{with probability } r_{X_{n-1}X_n} \\ 0 & \text{else.} \end{cases}$$

Since A_n^2 is 0 or 1, hence integer valued for all n, we can also use the p-representation, i.e. take $p_n^2 = A_n^2$, $n \in \mathbb{N}$ then

$$A_n(p^2) = A_n^2 \quad \text{(for all } \theta\text{)}.$$

Take $p^1 = (p, p, \ldots)$ and $A_n^1 \stackrel{\Delta}{=} A_n^1(p^1)$. Then A_n^1 (for fixed θ) is a bracket sequence with rate p. It follows from Theorem 20 that the lower bound in icx-ordering is obtained if we use the arrival process on T_n with A_n^1 as admission sequence. Let us call this the Regular Arrival Process with parameters (p, λ) (RAP(p, λ)) . The Markov Arrival Process with stationary distribution π_x, $x \in E$ and arrival probabilities r_{xy}, $x, y \in E$, we denote by MAP(π, r). Analogous to the $\cdot/G/1$-notation, let us denote the FSEG (with stationary sequences T_n and S_n) by $\cdot/G/SEG$ and $_iW_\infty(\cdot/G/SEG)$, $i = a, b$ for the potential ($i = a$) or actual traveling time ($i = b$) (to a fixed transition).

Then we have the following theorem, which is an application of comparison Lemma 63.

Theorem 68. *For $i = a, b$,*

$$_iW_\infty(\text{RAP}(p, \lambda)/G/SEG) \leq_{\text{icx}} {_iW_\infty}(\text{MAP}(\pi, r)/G/SEG)$$

Proof. For $i = a$ we apply part 1) of comparison Lemma 63. Therefore we have to show that

$$\mathbb{E}\left((p_1^2, \ldots, p_n^2) \mid (p_1^1, \ldots, p_n^1)\right) = (p_1^1, \ldots, p_n^1).$$

But $p_k^1 = p$ for all k, therefore it suffices to show that $\mathbb{E}\, p_k^2 = p$, for all k. Indeed, this holds since,

$$\mathbb{E}\, p_k^2 = \sum_x \sum_y \pi_x \lambda_{xy} r_{xy} = p.$$

In order to apply Corollary 8 we have to verify assumption 67. Indeed, the admission sequence A_n^2 is stationary since X_n is assumed to be stationary. Since $\int_0^1 \lfloor x + \theta \rfloor d\theta = x$ implies

$$\mathbb{E}\, A_n^1(p) = \int\limits_0^1 (\lfloor np + \theta \rfloor - \lfloor (n-1)p + \theta \rfloor)\mathrm{d}\theta = p,$$

it follows that $A_n^1(p)$ is a stationary sequence, and assumption 67 applies for both sequences. For $i = b$, we consider the B_n-sequence corresponding to the $A_n^1(p)$-sequence. Lemma 21 guarantees that it is a bracket sequence with rate $1/p$. Since $A_n^1(p)$ is stationary, B_n^1 is also stationary. Hence,

$$B_n^1 = B_n(q) \quad \text{with} \quad q = (1/p, 1/p, \dots).$$

Since the A_n^2 sequence is integer valued, also the corresponding B_n^2 sequence is integer valued. It remains to verify that $\mathbb{E}\, B_n^2 = 1/p$, but this is a standard result for stationary MAP-processes. $\square$

It is well-known that a MAP-process with transition times $\{T_n\} \stackrel{d}{=}$ Poisson(λ) can be represented also as one with transition times $\{T_n^1\} \stackrel{d}{=}$ Poisson($N\lambda$) with the same π as stationary distribution. If we want to keep the arrival intensity equal to $\overline{\lambda}$ then we have to divide the p and the r_{xy} by N, hence $1/p$ is multiplied by N. Now suppose $1/p$ is rational, say $\frac{N_1}{N_2}$, then $N_2/p = N_1$ is an integer. The corresponding regular arrival process has inter-arrival lengths of N_1 steps, hence its inter-arrival times are Erlang distributed with N_1 phases of exponential-distributed length with parameter $N_2\lambda$.

Using Theorem 68 we will show the following result:

Theorem 69. *For $i = a, b$, and any real number $0 < c \leq 1$ it holds*

$$_iW_\infty(\mathrm{RAP}(p, \lambda)/\mathrm{G}/\mathrm{SEG}) \leq_{icx} {}_iW_\infty(\mathrm{RAP}(p/c, \lambda c)/\mathrm{G}/\mathrm{SEG}).$$

Proof. RAP($p/c, \lambda c$) can be seen as a MAP with transition times $\{T_n\}$ which form a Poisson(λ) process. With probability c a transition then is a real transition (i.e. a transition of the RAP($p/c, \lambda c$) process). The stationary admission sequence in RAP($p/c, \lambda c$) has rate p/c on an arrival at a (real) transition. Hence the stationary probability on an arrival at T_n is $c \cdot p/c = p$, and Theorem 68 applies. $\square$

As a consequence of Theorem 68 we have that for a MAP(π, r) with p (as in 11.4) rational, say $p = N_2/N_1$, the FSEG with renewal input with N_1 phases of exponential distributed length with parameter $N_2\lambda$ provides a icx-stochastic lower bound on the actual and potential stationary traveling times. By Theorem 69 RAP($pc, \lambda/c$) for any $0 < c \leq 1$ provides also a icx-lower bound. Hence, since RAP($pc, \lambda/c$) is arbitrarily close to a renewal input for c sufficiently small, we get an approximation for irrational p. These facts have been used, without proof, in papers on optimal routing to parallel queues (cf. [43]). Clearly the limit-process of RAP($p/c, \lambda c$) for $c \to \infty$, is the renewal-process with constant inter-arrival-time equal to $1/p\lambda$ (notation $\mathcal{D}(1/p\lambda)$).

Corollary 9. *For $i = a, b$,*

1)$_i W_\infty(\text{RAP}(p/c, \lambda c)/\text{G}/\text{SEG})$ is monotone decreasing in c.

2) For any $c \geq 1$:

$$_i W_\infty(\mathcal{D}(1/p\lambda)/\text{G}/\text{SEG}) \leq_{\text{icx}} {}_i W_\infty(\text{RAP}(p/c, \lambda c)/\text{G}/\text{SEG})$$
$$\leq_{\text{icx}} {}_i W_\infty(\text{MAP}(\pi, r)/\text{G}/\text{SEG}).$$

Proof. Using a continuity argument (see [67]), the assertions 1) and 2) are direct consequences of Theorems 68 and 69.

12 Simplex convexity

12.1 Introduction

So far, the notion of multimodularity is related to a particular base, namely $(\mathcal{F})$. One may wonder what happens if the multimodular base $(\mathcal{F})$ is replaced by another set of vectors.

This question is addressed in this chapter, using results from [8], where the notion of multimodularity is extended to all possible initial bases. The notion of multimodular triangulation was introduced in [27] as a generalization of the original concept of atoms in [59]. Here, following [8], we provide a new sight on multimodular triangulations by using a geometrical point of view which is simpler (we do not use the set indexing techniques or lower envelops) and more general (we do not restrict the triangulations to integer points). From an arbitrary vectorial base of $\mathbb{R}^n$, we show how to build the corresponding multimodular atoms partitioning the space and how to define the associated simplex convexity property which generalizes multimodularity.

From there, we define the cones associated with these atoms and the cone distance compatible with simplex convexity. Finally, we show how to exploit the cone distance to define a partial order relation between arbitrary admission policies in G/G/1 queues.

12.1.1 Organization of the chapter

This chapter is organized as follows. Section 12.2 defines a generalized notion of atoms covering the whole space. Section 12.3 shows that starting with a multimodular triangulation with simplexes formed by a set of base vectors, satisfying simplex convexity is equivalent to the multimodular inequality. Section 12.4 shows how one can restrict multimodularity to some sub-spaces called sub-meshes. Finally, Section 12.5 defines the cone ordering between binary sequences and Section 12.6 provides an application to admission control in queues.

12.2 Multimodular Triangulations

Let us start with a matrix D of size $(n+1) \times n$ and of rank n such that the rows of matrix D define $n+1$ vectors $(s_0, \cdots, s_n)$ satisfying $s_0 + \cdots + s_n = 0$. Such a matrix will be called a multimodular (m.m.) matrix in the following.

Any multimodular matrix D can be constructed starting with a $n \times n$ matrix M with full rank and appending minus the sum of all the rows of M as the last row of D.

Definition 17. *The mesh M_D associated with the m.m. matrix D is the set of all the points* $\{a_0 s_0 + a_1 s_1 + \cdots + a_n s_n, \quad a_i \in \mathbb{Z}, \quad i = 0, \cdots, n\}$.

Lemma 64. *The following properties are true.*
i) A point in $\mathbb{R}^n$ has a unique non-negative decomposition in $(s_0 \cdots s_n)$ (up to the addition of $(a_0 \cdots a_n)$ with $a_0 = \cdots = a_n$).
ii)A point in M_D has a unique non-negative decomposition in $(s_0 \cdots s_n)$ (up to the addition of $(a_0 \cdots a_n)$ with $a_0 = \cdots = a_n$).

Proof. i) Since $s_1, \cdots, s_n$ is a base of $\mathbb{R}^n$, then for any point x in $\mathbb{R}^n$ $x = \alpha_1 s_1 + \cdots + \alpha_n s_n$. for some $\alpha_i \in \mathbb{R}$, $i = 0, ..., n$. Let α_i be the minimal coordinate. If $\alpha_i < 0$, then

$$x = \alpha_1 s_1 + \cdots + \alpha_n s_n - \alpha_i(s_0 + \cdots + s_n)$$
$$= -\alpha_i s_0 + (\alpha_1 - \alpha_i)s_1 + \cdots + (\alpha_n - \alpha_i)s_n, \qquad (12.1)$$

where all the coordinates are non-negative. As for uniqueness, let $x = \alpha_0 s_0 + \cdots + \alpha_n s_n = \beta_0 s_0 + \cdots + \beta_n s_n$ where we may assume that all coordinates are non-negative and $\alpha_i = \beta_j = 0$. If $i = j$, then $\alpha = \beta$ because $(s_0, \cdots, s_n) \backslash s_i$ is a base of $\mathbb{R}^n$. If no coordinates are jointly null, then we can write $x = (\beta_0 - \beta_i)s_0 + \cdots + (\beta_n - \beta_i)s_n$ which means for the jth coordinate in the base $(s_0, \cdots, s_n) \backslash s_i$, is $\alpha_j = -\beta_i$, which is impossible by non-negativity.

ii) A point in M_D has a unique decomposition in $(s_1, \cdots, s_n)$, this decomposition being in $\mathbb{Z}$. By using the same method as in 12.1, we transform this decomposition into a non-negative integer decomposition in $(s_0, s_1, \cdots, s_n)$. $\square$

Definition 18. *A D-atom is a simplex in $\mathbb{R}^n$, made of the $n+1$ points*

$$p_0 = a,$$
$$p_1 = a + s_{\xi(0)},$$
$$p_2 = a + s_{\xi(0)} + s_{\xi(1)} \qquad (12.2)$$
$$\vdots$$
$$p_n = a + s_{\xi(0)} + s_{\xi(1)} + \cdots + s_{\xi(n-1)}$$

where $a \in M_D$ (the root) and ξ is a permutation of $\{0, \cdots, n\}$. This atom will be denoted $\mathcal{S}(p_0, \cdots, p_n)$.

Note that an atom is indeed a simplex since D is of rank n and that $\mathcal{S}(p_0,\cdots,p_n)$ and $\mathcal{S}(p_1,\cdots,p_n,p_0)$ are two notations for the same atom, when starting with p_0 (resp. p_1) as a root.

Definition 19. *A collection of simplexes is a triangulation of E (an arbitrary subset of $\mathbb{R}^n$) if*

- *E is the union of all the simplexes.*
- *The intersection of two simplexes is either empty or a common face.*

Theorem 70. *The set of all the D-atoms forms a triangulation of $\mathbb{R}^n$, called a multimodular triangulation.*

Proof. Let x be a point in $\mathbb{R}^n$. By Lemma 64, $x = \alpha_0 s_0 + \cdots + \alpha_n s_n$, with nonnegative coordinates, one of which is 0. We construct $a = \lfloor \alpha_0 \rfloor s_0 + \cdots + \lfloor \alpha_n \rfloor s_n$ and ξ such that $\alpha_{\xi(i)} - \lfloor \alpha_{\xi(i)} \rfloor \geq \alpha_{\xi(i+1)} - \lfloor \alpha_{\xi(i+1)} \rfloor$. We define $\beta_n = 0$, $\beta_{n-1} = \alpha_{\xi(n-1)} - \lfloor \alpha_{\xi(n-1)} \rfloor$, $\beta_i = \alpha_{\xi(i)} - \lfloor \alpha_{\xi(i)} \rfloor - (\alpha_{\xi(i+1)} - \lfloor \alpha_{\xi(i+1)} \rfloor)$, all of them verify $0 \leq \beta_i \leq 1$ and $\sum_{i=0}^{n-1} \beta_i \leq 1$. We have $x = a + \beta_0 s_{\xi(0)} + \cdots + \beta_{n-1}(s_{\xi(0)} + \cdots + s_{\xi(n-1)})$. Therefore, x belongs to the atom with root a and permutation ξ.

Now, assume that a point x belongs to the interior of two different atoms with respective roots a and b and permutations ξ and τ. Due to shift invariance, we may assume that $b = 0$. Moreover, with loss of generality take τ to be the identity.

$$x = a + \alpha_0 s_{\xi(0)} + \cdots + \alpha_{n-1}(s_{\xi(0)} + \cdots + s_{\xi(n-1)})$$
$$= \beta_0 s_0 + \cdots + \beta_{n-1}(s_0 \cdots + s_{n-1})$$

with $\sum_{i=0}^{n-1} \beta_i \leq 1$ and $\sum_{i=0}^{n-1} \alpha_i \leq 1$. Since x is in the interior of both atoms, we also have $\beta_i > 0$ and $\alpha_i > 0$ for all $i = 0, \cdots, n-1$. Therefore, by uniqueness of the decomposition of x, and writing $a = a_0 s_0 + \cdots + a_n s_n$,

$$a_0 + \sum_{j=\xi^{-1}(0)}^{n-1} \alpha_{\xi(j)} = \beta_0 + \cdots + \beta_{n-1}$$

$$a_1 + \sum_{j=\xi^{-1}(1)}^{n-1} \alpha_{\xi(j)} = \beta_1 + \cdots + \beta_{n-1}$$

$$\vdots$$

$$a_{n-1} + \sum_{j=\xi^{-1}(n-1)}^{n-1} \alpha_{\xi(j)} = \beta_{n-1}$$

$$a_n + \sum_{j=\xi^{-1}(n)}^{n-1} \alpha_{\xi(j)} = 0.$$

Since all the partial sums of the α_i or of the β_i are all smaller than one and since a_i are integer numbers, then, $a_i = 0$ for all $i = 0, \cdots, n$. Both atoms have the same root.

Now, the equality of the partial sums taken one by one imply first that $\sum_{j=\xi^{-1}(n)}^{n-1} \alpha_{\xi(j)} = 0$. Since $\alpha_i > 0$ for all i, then the only possibility is $\xi^{-1}(n) = n$. Considering vectors s_k and s_{k+1}, we have:

$$\sum_{j=\xi^{-1}(k)}^{n-1} \alpha_{\xi(j)} - \sum_{j=\xi^{-1}(k+1)}^{n-1} \alpha_{\xi(j)} = \beta_k$$

$$> 0.$$

This implies that $\xi^{-1}(k) < \xi^{-1}(k+1)$. This means that ξ is the identical permutation. Therefore, both atoms are equal. $\qquad\square$

In the restricted case when the mesh is $\mathbb{Z}^n$ and D the incidence matrix of a graph, this theorem was proved in [27] using an intricate argument of set indexing (see the example presented in Figure 12.1).

Lemma 65. *Combinatorial properties of m.m. triangulations.*

i) A point in M_D belongs to $(n+1)!$ D-atoms.
ii)The unit-cube in M_D is partitioned into $n!$ D-atoms.

Proof. i) This is a straightforward consequence of the definition of atoms.
ii) The unit-cube U in M_D is the set of points of the form $a_1 s_1 + \cdots + a_n s_n$, with $a_i \in \{0, 1\}$ for all $1 \leq i \leq n$. Given a permutation ξ, there exists a point $b \in U$ such that the atom $\mathcal{S}(b, b + s_{\xi(0)}, \cdots, b + s_{\xi(n-1)})$ is included in U. The point $b = b_1 s_1 + \cdots + b_n s_n$ is chosen in the following way:

$$b_i = \begin{cases} 0 & \text{if } \xi^{-1}(i) < \xi^{-1}(0), \\ 1 & \text{otherwise.} \end{cases} \tag{12.3}$$

Each atom in U has $n + 1$ vertices, hence $n + 1$ representations of the form $\mathcal{S}(b, b + s_{\xi(0)}, \cdots, b + s_{\xi(n-1)})$. Finally, combining this with part i), $(n + 1)!/(n + 1)$ atoms are contained in U. Since all the atoms triangulate $\mathbb{R}^n$, those in U triangulate U. $\qquad\square$

Some multimodular triangulations have a special interest. The most used ones are the triangulations with a m.m. matrix D being the incidence matrix of a graph.

An oriented tree is a graph $G = (V, E)$ with $n + 1$ nodes and n arcs. It has an incidence matrix D of size $(n + 1) \times n$ defined by

$$D_{i,j} = \begin{cases} +1 & \text{if vertex } i \text{ is the start point of edge } j \\ -1 & \text{if vertex } i \text{ is the end point of edge } j \\ 0 & \text{otherwise} \end{cases}$$

First note that whenever D has rank n, then the graph has to be a tree. Also, since D is totally unimodular, then its mesh M_D is $\mathbb{Z}^n$ (see for example [54] for a detailed presentation on totally unimodular matrices).

The L-triangulation is the triangulation associated with a linear graph (see Figure 12.1 for an illustration in $\mathbb{R}^2$). The associated m.m. matrix is

$$D = \begin{pmatrix} 1 & 0 & 0 & \cdots & 0 & 0 \\ -1 & 1 & 0 & \cdots & 0 & 0 \\ 0 & -1 & 1 & \cdots & 0 & 0 \\ \vdots & \ddots & \ddots & \ddots & \vdots & \vdots \\ 0 & 0 & 0 & \cdots & -1 & 1 \\ 0 & 0 & 0 & \cdots & 0 & -1 \end{pmatrix} \tag{12.4}$$

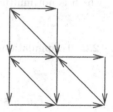

The tree associated with the
L-triangulation in $\mathbb{R}^2$.

The atoms of the
L-triangulation in $\mathbb{R}^2$

Fig. 12.1. The L-triangulation in dimension 2

On the other hand, some triangulations of the space into simplexes are not multimodular triangulations. Such examples in dimension 2 and 3 are given in Figure 12.2. The simplexes in the triangulation displayed in 12.2(a) use 4 different base vectors, s_1, s_2, s_3, s_4. The triangulation in 12.2(b) decomposes the unit cube in 5 simplexes instead of $3!=6$ for any multimodular triangulation.

12.3 Multimodular Functions

Let D be a m.m. matrix with row vectors, $(s_0, \cdots, s_n)$. Sometimes in the following, the reference to D may be omitted. Everything implicitly refers to D such as multimodularity and atoms.

Definition 20. *A function $f : M_D \to \mathbb{R}$ is D-multimodular if and only if for all $a \in M_D$, and for all $0 \leq i < j \leq n$,*

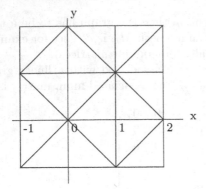 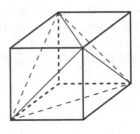

(a) Triangulation of $\mathbb{R}^2$

which is not multimodular

(b) Triangulation of the cube in $\mathbb{R}^3$

into a minimal number of simplices.

It is not multimodular

Fig. 12.2. Triangulations of $\mathbb{R}^2$ and $\mathbb{R}^3$ which are not multimodular triangulations

$$f(a + s_i) + f(a + s_j) \geq f(a) + f(a + s_i + s_j). \tag{12.5}$$

From f, we construct a function $\tilde{f} : \mathbb{R}^n \to \mathbb{R}$ by linear interpolation of f over the atoms defined by D.

Note that $\tilde{f}$ depends on D.

Theorem 71. *f is D-multimodular if and only if $\tilde{f}$ is convex.*

The following proof is completely different from the former proofs and gives a generalization to more general meshes. The hard part ("only if"), in the restricted case of the L-triangulation, was done in [59]. The extension to the more general case of tree triangulations as well as the easy reverse part ("if" part) were presented in [27]. A version of the proof presented here can be found for the restricted case of the L-triangulation in Chapter 1.

A function f which is D-multimodular may also be called *simplex convex*, considering Theorem 71.

Proof. "only if": The function $\tilde{f}$ is continuous by definition. Moreover, along any direction δ, it has only a discrete number of isolated points where it is not differentiable. By using the characterization of convexity given in [96], we will check convexity at a point z by showing that for point z, and any direction δ, the right derivative is greater than or equal to the left derivative. It obviously suffices to check at points that are on the boundary of an atom, since, by definition, $\tilde{f}$ is linear in the interior of atoms.

Hence, we first assume that the point z is on the interior of a face (of dimension $n-1$) which is common between two adjacent atoms. Without loss of generality, assume that the atoms (defined below by their extreme points) are

$$S = \mathcal{S}(x_0, x_1, ..., x_m) \qquad \text{and} \qquad \overline{S} = \mathcal{S}(x_0, x_1^*, ..., x_m).$$

where x_i satisfy (12.2) and

$$x_1^* = x_0 + s_{i_2}, \qquad x_2 = x_1^* + s_{i_1}.$$

Decompose direction δ in its projection δ_2 in the common face between the two atoms and in the component δ_1 along the direction $(x_1^* - x_1)$. In the direction δ_2, the left and right derivatives are equal. In the direction δ_1, the right derivative is a constant c, depending on the length of δ_1, times $\tilde{f}(x_1^*) - \tilde{f}(z)$. The left derivative is $c(\tilde{f}(z) - \tilde{f}(x_1))$. Omitting the constant c, and using point $z = \frac{1}{2}(x_0 + x_2)$ hence $2\tilde{f}(z) = f(x_0) + f(x_2)$, we get for the difference

$$(\tilde{f}(x_1^*) - \tilde{f}(z)) - (\tilde{f}(z) - \tilde{f}(x_1))$$
$$= (f(x_1^*) - f(x_0)) - (f(x_2) - f(x_1)) \qquad (12.6)$$

The fact that (12.6) is nonnegative follows by applying 12.5 with $x = x_0$, and

$$x_1^* = x_0 + s_{i_2}$$
$$x_2 = x_0 + s_{i_2} + s_{i_1}$$
$$x_1 = x_0 + s_{i_1}.$$

It now remains to consider the case where the direction δ in point z crosses from atom S to atom $\overline{S}$, and $S \cap \overline{S}$ is of dimension at most $m-2$. In that case, we consider the cylinder C in direction δ containing the hyper-face F of S, opposite of z in direction δ.

The intersection of C with an arbitrary atom A is of dimension, say k and its projection on F along direction δ is the intersection of the projections of C and A and has dimension at most k. Therefore, F is almost everywhere (in Lebesgue measure) covered with such projections of dimension $m-1$. Therefore, we can find an affine line L with direction δ, included in C, that intersects S and $\overline{S}$, the projection of which is a point in F not belonging to intersections of dimensions smaller than $m-1$. Therefore, we can claim that L only intersects faces of atoms of dimension $m-1$. The convexity in point z and direction δ now follows from the convexity in points z_i corresponding to the intersections of line L with all the intermediate atoms between S and the last atom, $\overline{S}$.

This construction is illustrated for dimension 2 by Figure 12.3.

"if": Consider an arbitrary point x_0 and any two distinct elements s_i, s_j in $\mathcal{F}$. We have to show that

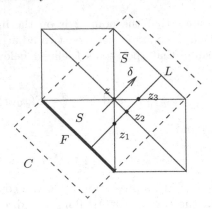

Fig. 12.3. Construction of cylinder C and line L in dimension 2.

$$f(x_0) + f(x_2) - f(x_1) - f(x_1^*) \leq 0, \qquad (12.7)$$

where $x_1 \triangleq x_0 + s_i$, $x_1^* \triangleq x_0 + s_j$, $x_2 \triangleq x_1 + s_j = x_1^* + s_i$.

Define $z \triangleq \frac{1}{2}(x_1 + x_1^*) = \frac{1}{2}(x_0 + x_2)$ and consider the line segment $x_1 \to z \to x_1^*$. The left derivative (l.d.) and right derivative (r.d.) in z are given by

$$l.d. = \widetilde{f}(\frac{1}{2}(x_1 + x_1^*)) - \widetilde{f}(x_1) = \frac{1}{2}f(x_0) + \frac{1}{2}f(x_2) - f(x_1),$$

$$r.d. = \widetilde{f}(x_1^*) - \widetilde{f}(\frac{1}{2}(x_0 + x_2)) = f(x_1^*) - \frac{1}{2}f(x_0) - \frac{1}{2}f(x_2).$$

Since $\widetilde{f}$ is convex, $r.d. - l.d.$ is non-negative, and hence (12.7) holds.   □

Lemma 66. *We consider all the n-periodic sequences $a = (a_i)_{i \in \mathbb{N}}$ with values in $\mathbb{N}$ satisfying*

$$\sum_{i=1}^{n} a_i = k. \qquad (12.8)$$

Let f be a multimodular function defined on a mesh M_D of dimension n. Then, the quantity

$$\sum_{i=1}^{n} \widetilde{f}(a_i s_1 + \cdots + a_{i+n-1} s_n) \qquad (12.9)$$

is minimized at point $r = (\frac{k}{n}s_1 + \cdots + \frac{k}{n}s_n)$.

Proof. Let P be the set of all integer sequences a, which are n-periodic and such that within one period, they add up to k, $\sum_{i=1}^{n} a_i = k$. By periodicity, and using Jensen inequality,

$$\min_{a \in P} \frac{1}{n} \sum_{i=1}^{n} \widetilde{f}(a_i s_1 + \cdots + a_{i+n-1} s_n) = n\widetilde{f}(\frac{k}{n}s_1 + \cdots + \frac{k}{n}s_n).$$

□

12.4 Sub-meshes

Definition 21. *A sub-mesh P of M_D is the intersection of M_D with a convex set of $\mathbb{R}^n$ which is the union of (faces of) D-atoms.*

Since any union of D-atoms which forms a convex set is a sub-mesh by definition, typical sub-meshes are: the positive quadrant: $\{a_0 s_0 + \cdots + a_{n-1} s_{n-1}, a_i \in \mathbb{N}\}$ and the unit cube $\{a_0 s_0 + \cdots + a_{n-1} s_{n-1}, a_i \in \{0, 1\}\}$.

The hyper-plane $\{a_0 s_0 + \cdots + a_{n-1} s_{n-1}, \sum_i a_i = k, a_i \in \mathbb{Z}\}$ is a sub-mesh of dimension $n - 1$ of the L-triangulation. An example of such a sub-mesh in dimension 3, $P = \{x + y + z = 2, x, y, z \in \mathbb{N}\}$, is given in Figure 12.4,

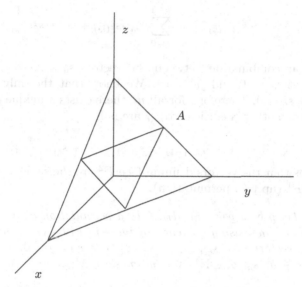

Fig. 12.4. A sub-mesh is a convex union of faces of atoms

Lemma 67. *Any sub-mesh P has the following properties:*
i) The faces defining P form a multimodular triangulation of P.
ii) The vectors of this new multimodular triangulation are disjoint sums of the original vectors.

Proof. i) First, by convexity, P has the same dimension, say k, as any of its faces. It should be obvious that the faces defining P form a triangulation of P. Now, let us show that this is a multimodular triangulation.

Let F^1 (resp. F^2) be a face in P of a simplex S^1 (resp. S^2) with root a (resp. b) and permutation ξ (resp. γ). Without loss of generality, we may assume that $a \in S^1$ and $b \in S^2$ by shifting the starting points in the definitions of atoms S^1 and S^2.

Now, the vertices of F^1 (resp. F^2) are visited in an order depending on ξ (resp. γ), such that

$$
\begin{aligned}
p_0^1 &= a & p_0^2 &= b \\
p_1^1 &= a + s_{\xi(0)} + \cdots + s_{\xi(i_1-1)} & p_1^2 &= b + s_{\gamma(0)} + \cdots + s_{\gamma(j_1-1)} \\
p_2^1 &= p_1^1 + s_{\xi(i_1)} + \cdots + s_{\xi(i_2-1)} & p_2^2 &= p_1^2 + s_{\gamma(j_1)} + \cdots + s_{\gamma(j_2-1)} \\
&\ \vdots & &\ \vdots \\
p_0^1 &= p_k^1 + s_{\xi(i_k)} + \cdots + s_{\xi(n)-1} & p_0^2 &= p_k^2 + s_{\gamma(j_k)} + \cdots + s_{\gamma(n)}.
\end{aligned}
$$

Since both faces are in the same space of dimension k, then for any $m < k$, we have the linear combination

$$
s_{\gamma(j_m)} + \cdots + s_{\gamma(j_{m+1}-1)} = \sum_{\ell=0}^{k-1} \alpha_{m,\ell} s_{\xi(i_\ell)} + \cdots + s_{\xi(i_{\ell+1}-1)}.
$$

This is a linear combination between the vectors $s_0, s_1, \cdots, s_n$ (with the convention that $i_0 = 0$ and $j_0 = 0$). We know that the only relation is $s_0 + s_1 + \cdots + s_n = 0$. Therefore, for all m, there exists a unique q such that $\alpha_{m,q} = 1$. (all the other coefficients $\alpha_{m,\ell}$ are null).

$$
s_{\gamma(j_m)} + \cdots + s_{\gamma(j_{m+1}-1)} = s_{\xi(i_q)} + \cdots + s_{\xi(i_{q+1})-1}.
$$

This means that the vectors defining face F^2 are the same as the vectors defining face F^1 (up to a permutation).   □

Lemma 68. *Let p be a point in M_D, ξ be a permutation of $\{0, \cdots, n\}$ and $\{i_k\}_{k=0,\cdots m}$ be an increasing sequence of integers with $i_0 = 0$ and $i_m = n$. We define the vectors $\bar{s}_j = s_{\xi(i_j)} + \cdots s_{\xi(i_{j+1}-1)}$ for all $j = 0, \cdots m$. The set $p + \{a_0 \bar{s}_0 + \cdots + a_m \bar{s}_m, a_i \in \mathbb{N}\}$ is a sub-mesh of M_D.*

Proof. First note that the set P is a convex subset of M_D of dimension m. Now we consider the triangulation of P defined by the vectors $\bar{s}_0, \cdots, \bar{s}_m$, each atom of this triangulation is a face of a D-atom.   □

Thus the hyper-plane $P = \{a_0 s_0 + \cdots + a_{n-1} s_{n-1}, \sum_i a_i = k\}$ is a sub-mesh of dimension $n - 1$ of the L-triangulation by choosing $\bar{s}_0 = s_1, \cdots, \bar{s}_{n-2} = s_{n-1}$ and $\bar{s}_{n-1} = s_n + s_0$, which yields $P = (k, 0, 0 \cdots, 0) + \{x_0 \bar{s}_0 + \cdots + x_{n-1} \bar{s}_{n-1}, x_i \in \mathbb{N}\}$.

Lemma 69. *A function f which is D-multimodular is multimodular on any sub-mesh P with respect to the induced multimodular matrix on P.*

Proof. Let a be a point in P and let u and v be two arbitrary rows for the multimodular matrix of P. By Lemma 67, $u = s_{i_1} + \cdots + s_{i_k}$ and $v = s_{j_1} + \cdots + s_{j_m}$ where the sets $\{s_{i_1}, \cdots, s_{i_k}\}$ and $\{s_{j_1}, \cdots, s_{j_m}\}$ are pairwise distinct. Therefore, we have

$$f(a) + f(a + u + v) = f(a) + f(a + s_{i_1} + \cdots + s_{i_k} + s_{j_1} + \cdots + s_{j_m})$$
$$\leq f(a + s_{i_1} + \cdots + s_{i_k}) + f(a + s_{j_1} + \cdots + s_{j_m})$$
$$= f(a + u) + f(a + v).$$

□

Corollary 10. *The function f is multimodular in a sub-mesh P if and only if $\tilde{f}$ is convex on $\tilde{P}$, the convex hull of P.*

Proof. It should be clear from the proof of Theorem 71 that the equivalence of the multimodularity of f and the convexity of $\tilde{f}$ still holds if we restrict the function f to P.   □

A second corollary of Theorem 71 concerns the minimization of multi-modular functions. For a function defined on P, we call x a local minimum on P if $f(x) \leq f(x + \varepsilon_i s_i)$ (with $\varepsilon_i \in \{-1, 1\}$) for all i such that $x + \varepsilon_i s_i$ is in P.

Corollary 11. *Let the function f be multimodular in P. Then a local minimum is a global minimum on P.*

Proof. If f is multimodular in P, then $\tilde{f}$ is convex in P, and is linear on the (faces of) atoms forming P. The graph of $\tilde{f}$ (i.e. $\{x : \exists y$ s.t. $x \geq \tilde{f}(y)\}$) is a convex polytope. Therefore, all the local minima are global and are extreme points of atoms.   □

12.5 Cones

Now, the convex space P of dimension n will be divided into $(n + 1)!$ cones, all starting at point h, any point of the mesh of P. Consider one atom $S = \mathcal{S}(p_0, p_1, \cdots, p_n)$ containing $h = p_0$ as a vertex. Let ξ and τ be the permutations on $\{0, \cdots n\}$ such that $\tau(0) = 0$ and

$$p_{\tau(1)} = h + s_{\xi(0)} \tag{12.10}$$
$$p_{\tau(2)} = p_{\tau(1)} + s_{\xi(1)} \tag{12.11}$$
$$\vdots = \vdots \tag{12.12}$$
$$p_{\tau(n)} = p_{\tau(n-1)} + s_{\xi(n-1)} \tag{12.13}$$
$$h = p_{\tau(n)} + s_{\xi(n)}. \tag{12.14}$$

Now, we define for all $1 \leq i \leq n$, $b_i = \sum_{j=1}^{i} s_{\xi(j)}$ and $b = (b_1, \cdots, b_n)$. Therefore, $p_{\tau(i)} = h + b_i$. The vectors $(b_1, \cdots, b_n)$ will be called the *generators* of the cone.

The cone associated with S, denoted $C(S)$ is made of all the points p of P such that $p = h + c^t b$ where c is a non-negative vector in $\mathbb{N}^n$. First, note that, for each p in the cone, vector c^t is uniquely defined since $(b_1, \cdots, b_n)$ are independent vectors. Second, note that when we consider all the atoms containing h as a vertex, then all the $(n+1)!$ associated cones will cover P. If two adjacent atoms share a face, the two corresponding adjacent cones will also share a "face" (of dimension $n-1$).

For any point p in P, p will be in at least one cone and we will have $p = h + c^t(p)b$, where b is uniquely defined on the support of $c(p)$. note that p can be on the boundary of several adjacent cones.

We shall denote $d(h, p) = c_1(p) + \cdots + c_n(p)$ and call it the cone-distance from h to p.

All the previous remarks show that $d(h, p)$ is well defined.

12.5.1 Minimization

In this section, we consider the case where $\tilde{P} = \mathbb{R}^n$, where $\tilde{P}$ is the convex hull of P, and $h = 0$. We also consider a function f multimodular with respect to the row vectors of D, $s_0, \cdots, s_n$.

We focus on one arbitrary cone, C, defined by the permutation γ. This means that the generators of C are the vectors

$$b_0 = s_{\gamma(0)},$$
$$b_1 = s_{\gamma(0)} + s_{\gamma(1)},$$

$$\vdots$$

$$b_{n-1} = s_{\gamma(0)} + s_{\gamma(1)} + \cdots + s_{\gamma(n-1)}.$$

Any point in C has non-negative coordinates $\beta_0, \cdots, \beta_{n-1}$ in the base $b_0, \cdots, b_{n-1}$.

We call Φ the linear transformation which is the passage from the base $b_0, \cdots, b_{n-1}$ to the base $s_{\gamma(0)}, s_{\gamma(1)}, \cdots, s_{\gamma(n-1)}$.

Lemma 70. *Let k be an integer. The set C_k of points in C such that $\beta_0 + \cdots + \beta_{n-1} = k$ is a sub-mesh of M_D.*

Proof. Set $p = ks_{\gamma(0)}$ and $\bar{s}_0 = s_{\gamma(1)}, \cdots, \bar{s}_{n-2} = s_{\gamma(n-1)}, \bar{s}_{n-1} = s_{\gamma(0)} + s_{\gamma(n)}$. Then the sub-mesh $p + \{\sum_i a_i \bar{s}_i, a_i \in \mathbb{N}\} \cap C$ is precisely the set C_k. $\square$

The following lemma is some kind of generalization of Theorem 6 from the L-triangulation and the positive quadrant to any multimodular triangulation and one of its cones.

Lemma 71. *Let f be a m.m. function, then the quantity*

$$\frac{1}{n} \sum_{i=1}^{n} f \circ \Phi(\beta_i, \cdots, \beta_{i+n-1})$$

is minimized over the set C_k at all the points $\beta(\theta)$, $0 \leq \theta \leq 1$ of coordinates

$$\beta_i(\theta) = \lfloor i\frac{k}{n} + \theta \rfloor - \lfloor (i-1)\frac{k}{n} + \theta \rfloor.$$

Proof. Using Corollary 10, the function $\frac{1}{n} \sum_{i=1}^{n} \widetilde{f} \circ \Phi(\beta_i, \cdots, \beta_{i+n-1})$ is convex. Moreover, it is minimized at point $r = (\beta_0 = \frac{k}{n}, \cdots, \beta_{n-1} = \frac{k}{n})$ over C_k (see Lemma 66).

This function is linear on the atoms of the sub-mesh C_k. Therefore, it is also minimum at all the vertices of an atom containing the point r. The vertices of this atom contain the points $\beta(\theta)$ with coordinates

$$\beta_i(\theta) = \lfloor (i+1)\frac{k}{n} + \theta \rfloor - \lfloor i\frac{k}{n} + \theta \rfloor,$$

when θ varies from 0 to 1.

To show this, first note that these points are all in the sub-mesh, since all their coordinates add up to k and since $\Phi(\beta(\theta))$ is integer valued.

Second, let $f_i = 1 - (i+1)\frac{k}{n} + \lfloor (i+1)\frac{k}{n} \rfloor$, ordered in the increasing order.

By construction, when θ varies from 0 to 1, then the points $\beta(\theta)$ have at most $n-1$ values. The point $\beta(\theta)$ changes at all the points of the form f_i for $i = 0, \cdots, n-2$. At $\theta = f_i$ we add $-s_{i+1}$ (if $f_i = f_j$ then when $\theta = f_i$ we add $-s_{i+1} - s_{j+1}$). Therefore, all the points $\beta(\theta)$ form a face of an atom. Noting that

$$r = \frac{1}{n-1}\beta(0) + \frac{1}{n-1}\beta(f_0) + \frac{1}{n-1}\beta(f_1) + \cdots + \frac{1}{n-1}\beta(f_{n-2}),$$

shows that r belongs to that face. $\qquad\square$

12.5.2 Cone ordering and monotonicity

Now, we define a partial order on P by choosing $h = r$. We only consider multimodular functions for which r is a minimal point.

First, this partial order (called cone ordering) is defined in a different manner on each cone.

In a given cone C, with generating vectors $b_1, \cdots, b_n$, we say that $x \leq_C y$ if $c(x) \leq c(y)$ component-wise. Note that a cone with this partial order is a lattice which is isomorphic to $\mathbb{N}^n$ with the classical component-wise order.

Theorem 72. *If f is a multimodular function on P, then $x \leq_C y$ implies that $f(x) \leq f(y)$. In other words, f is monotone with respect to the partial order $\leq_C$.*

Proof. Since x and y are comparable, this means that they are in the same cone. From now on, $b_1, \cdots, b_n$ will be the generators of this cone.

First note that we can assume that $d(x, y) = 1$. If not, then we prove step by step along the path from x to y along the direction of the generators, say $x = x_1 \leq_C \cdots \leq_C x_m = y$ that $f(x) = f(x_1) \leq \cdots \leq f(x_m) = f(y)$.

The proof will now proceed by induction on $d(r, x)$.

First note that the property is true if $d(x, r) = 0$, since r is the argmin of f on S.

Now, let us assume that we have $d(x, r) \geq 1$. Pick a point z such that $x = z + b_i$ in cone C. Equivalently, we have $c(y) + e_i = c(x)$ and $c(y) \leq 0$. Note that $d(z, r) = d(x, r) - 1$ and $z \leq_C x$. By induction, this means $f(z) \leq f(x)$.

Since $d(x, y) = 1$, there exist j such that $y = x + b_j$. Now we have two cases, since we may not be able to choose i such that $i = j$.

- If $i = j$, then by convexity of f,

$$f(z + b_i) - f(z) \leq f(z + b_i + b_i) - f(z + b_i).$$

We also know by induction that $f(z + b_i) - f(z) \geq 0$. This means that $f(x) \leq f(y)$.

- If $i \neq j$, then we choose yet another point, w, such that $w = z + b_j$. We can assume that $i > j$ (the case $j < i$ is similar by inverting the role played by b_i and b_j in the following).

Since b_i is a sum of base vectors, it is also a sum of opposites of base vectors, since all base vectors add up to 0. Note that all these base vectors are distinct from the base vectors involved in b_j.

We have:

$$b_j = s_{\xi(1)} + \cdots + s_{\xi(j)}, \tag{12.15}$$

$$b_i = -s_{\xi(i+1)} - s_{\xi(i+2)} - \cdots - s_{\xi(n+1)}. \tag{12.16}$$

Therefore,

$$
\begin{aligned}
f(w) - f(z) &= f(x + s_{\xi(1)} + \cdots + s_{\xi(j)} + s_{\xi(i+1)} + \cdots + s_{\xi(n+1)}) \\
&\quad - f(x + s_{\xi(i+1)} + \cdots + s_{\xi(n+1)}) \\
&\leq f(x + s_{\xi(1)} + \cdots + s_{\xi(j)}) - f(x), \\
&= f(y) - f(x).
\end{aligned}
\tag{12.17}
$$

where Inequality 12.17 is a direct consequence of the definition of multi-modularity. Since $d(r, w) = d(r, z) + 1$, by induction we have, $f(w) - f(z) \geq 0$, then this implies $f(y) - f(x) \geq 0$.

$\square$

12.6 Application: periodic admission sequences in $G/G/1/\infty$ tandem queues

We consider queues in tandem with general stationary service times. As for the arrival sequence, let $(u_i)_{i \in \mathbb{N}}$ be a stationary process. The integer sequence $\{a_i\}_{i \in \mathbb{N}}$ is the *admission sequence* into the queues. The interarrival times of customers in the queue is a sequence $(\tau_i)_{i \in \mathbb{N}}$ defined by:

$$\tau_i = \sum_{j=a_1+\cdots+a_{i-1}}^{a_1+\cdots+a_i} u_j.$$

The construction of the inter-arrival sequence is illustrated in Figure 12.5.

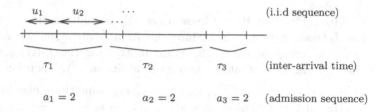

Fig. 12.5. The construction of the inter-arrival sequence from a sequence u and admission $a = (2, 2, 2, \cdots)$.

In the following, the admission sequence will be assumed to be periodic with period n.

As to introduce multimodularity, we choose the L-triangulation of $\mathbb{Z}^n$. The atoms given by the L-triangulation with row vectors $s_i = -e_i + e_{i+1}$, and $s_0 = e_1, s_n = -e_n$, as in 12.4.

The sub-set of $\mathbb{Z}^n$ that we will work with is $P = \{(a_0, \cdots, a_n), a_i \geq 0 \; \forall i, \sum_{i=0}^n a_i = k\}$, where k is a given integer.

The set P corresponds to all admission sequences with k admitted customers among n slots.

Lemma 72. $\tilde{P}$ *is a convex union of hyper-faces of atoms in* $\mathbb{Z}^n$.

Proof. Let us consider the constraints one by one.

- The constraints $a_j \geq 0$ restrict P to $\mathbb{N}^n$ which is made of a convex union of atoms.
- Now, let us look at the constraint $\sum_{i=0}^n a_i = k$. This constraint is a convex union of faces of atoms.

To finish the proof, remark that the intersection of convex union of faces of atoms is a convex union of faces of atoms. $\qquad \square$

The atoms on P are defined by the vectors $s'_1 = s_1, \cdots, s'_{n-1} = s_{n-1}$ as for the L-triangulation of $\mathbb{Z}^n$, and a new vector $s'_0 = s_0 + s_n = e_n - e_1$.

If f is a multimodular function, $f : \mathbb{Z}^n \to \mathbb{R}$, we will consider the restriction of f to P which is also multimodular on P with its own atoms (see Lemma 69). By Corollary 11 f has a global minimum on P. In the following, this minimum will be called r.

Theorem 73. *The average expected waiting time in a stochastic event graph is a multimodular function on P.*

Proof. From the vector $a = (a_1, \cdots, a_n)$ we construct an infinite sequence $\alpha = a^\omega$. Let $g_N(a_1, \cdots, a_n) \overset{\text{def}}{=} \frac{1}{N} \sum_{k=1}^{N} \mathcal{W}_k(\alpha_1, \cdots, \alpha_k)$, where $\mathcal{W}_k$ is the expected total sojourn time of the kth customer, and let $G(a_1, \cdots, a_n) \overset{\text{def}}{=} \lim_{N \to \infty} g_N(a_1, \cdots, a_n)$. We also denote p the largest integer such that $pn \leq N$.

From Theorem 19, we know that $\mathcal{W}_k$ is multimodular with respect to the L-triangulation in $\mathbb{Z}^k$. Since the m.m. matrix for the L-triangulation in $\mathbb{Z}^k$ is a sub-matrix of the m.m. matrix for the L-triangulation in $\mathbb{Z}^N$ then we also know that W_k is multimodular in $\mathbb{Z}^N$. Therefore, the function $H(\alpha_1, \cdots, \alpha_N) \overset{\text{def}}{=} \frac{1}{N} \sum_{k=1}^{N} w_k(\alpha_1, \cdots, \alpha_k)$ is multimodular in $\mathbb{Z}^N$.

For all $0 \leq i \leq n$, $g_N(a + s_i) = H(\alpha + s_i + s_{i+n} + \cdots + s_{i+kn})$, where $k = \lfloor \frac{N-i}{n} \rfloor$. Now, using the general characterization of multimodularity, that is $f(a + S_1 + S_2) - f(a + S_1) \leq f(a + S_2) - f(a)$, for S_1 and S_2 any arbitrary sum of base vectors, with the only restriction that no base vector appears in S_1 and in S_2, then it is immediate to check that $g_N(a + s_i + s_j) - g_N(a + s_i) \leq g_N(a + s_j) - g_N(a)$, for s_i and s_j any arbitrary distinct m.m. row vectors.

Therefore, g_N is multimodular in $\mathbb{Z}^n$. The limit G is also multimodular in $\mathbb{Z}^n$. By using Corollary 69 , G is also multimodular on P. $\qquad\square$

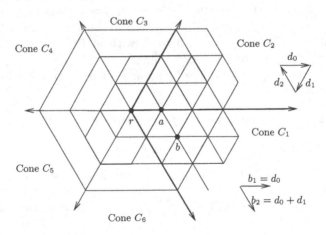

Fig. 12.6. comparison of $b = (1, 1, 4)$ and $a = (1, 2, 3)$

12.6.1 An example

We want to compare the expected waiting time under admission sequence $a = (1, 2, 3)$ and under admission sequence $b = (1, 1, 4)$.

The corresponding space $\tilde{P}$ is the set $\{(x, y, z), x + y + z = 6\}$. The space P is of dimension 2 with induced multimodular vectors $s_1 = (+1, -1, 0)$, $s_2 = (0, +1, -1)$ and $s_0 = (-1, 0, +1)$ (from the L-triangulation of $\mathbb{R}^3$).

The cost function G (expected waiting time) is minimized at point $r = (2, 2, 2)$. This is a direct consequence of a combination of Lemma 71 (used with the L-triangulation and the cone constructed with the n first vectors, which is the positive quadrant) and Theorem 73.

If we consider the cone C_1 generated by $b_1 = s_0$ and $b_2 = s_0 + s_1$, then we have $a = r + b_1$ and $b = r + b_1 + b_2$. Therefore, $a \leq_C b$, which implies $G(a) \leq G(b)$ by Theorem 72. This example is illustrated in Figure 12.6.

13 Orders and bounds for multimodular functions

13.1 Introduction

In Chapter 12 we studied the question which of two deterministic periodic admission sequences (periodic sequences of nonnegative integers) gives the smaller average expected waiting time. A partial order, called the cone order, is introduced there, and it is shown that the average waiting time and more generally any multimodular function is monotone with respect to the cone order. It is natural to define a multimodular order by requiring that any multimodular function is monotone. In contrast to Chapter 11, where we used the convex order for stochastic admission sequences, we consider deterministic admission sequences in this chapter. Note that deterministic admission sequences can only be ordered for all multimodular functions if they are equal. Therefore we consider multimodular functions with a fixed minimal point as we did with the cone order. In Section 13.2 we introduce the multimodular order and we show that the cone order is equivalent to the multimodular order. In Section 13.2.1 the shift invariant counterparts of these orders are studied and it is shown that the regular admission sequence is the minimal point. We introduce a combinatorial notion which we call the unbalance of the admission sequence in Section 13.3. Roughly speaking it measures its 'distance' to the regular admission sequence. We show that the unbalance is a shift invariant and multimodular function. For the definition of the unbalance we introduce in Section 13.3 the graph order, which gives the distance of a given admission sequence to the bracket sequence and thereby defines its unbalance. The relations between the shift invariant orders and the graph order are studied in Section 13.4. We show through counterexamples that they are not equivalent.

For the optimal routing problem to n queues we derived in theorem 26 that a lower bound is obtained by using regular admission sequences (or what is the same bracket sequences) with 'minimizing' routing fractions (densities) as admission sequences to all queues. But generally only in the routing to $n = 2$ queues, the routing fractions will be balanceable (see section 6.2) and only in that case the admission sequences can be glued together and be made to a feasible routing policy. In Section 13.5 we derive an upper bound for the difference between the average expected waiting time for customers which are routed to one queue with a given periodic admission sequence which has a fixed routing density, and the lower bound corresponding to the regular admission sequence with the same density. In Section 13.6 we derive then an upperbound for the average expected waiting time for a routing policy to n queues as a function of its unbalance. For a routing policy the unbalance is the sum of the unbalances of the induced admission sequences to the queues $1, \ldots, n$.

The difference between the upper and lower bound depends on the average interarrival times but it is insensitive for their distributions and the service time distributions. The upper bound is tight for the heavy traffic situation in which the interarrival and service times are deterministic and the traffic load is 1 for each of the queues. This means that the upper bound is correct for the deterministic heavy traffic case. Clearly, it holds for low traffic that the average expected waiting time is close to the lower bound.

All this is mainly based on the unbalance which is a combinatorial notion and the analysis in all this chapter is mainly combinatorial, it turns out that the unbalance is also useful in other models than routing to queues. Indeed, the bounds can be generalized from waiting times to similar bounds for sequences of multimodular functions as in Section 1.3 (see the remark at the end of Section 13.4). This chapter is a modified version of [49]. It uses at some places the results of the technical report [68]. A shortened version of this report appeared as [69] and an extended version will be published in two parts [72] and [71].

13.2 The multimodular order and the cone order

Let D be a multimodular matrix with rowvectors $d_0, d_1, \ldots, d_n \in \mathbb{R}^n$ satisfying $d_0 + \ldots + d_n = 0$ and let $M_D \subset \mathbb{R}^n$ be the corresponding mesh. Let $r \in M_D$ and let

$$\mathcal{F}(D, r) = \{f : M_D \to \mathbb{R} \text{ s. t. } f \text{ is } D - \text{multimodular and } f(r) = \min_{x \in M_D} f(x)\}$$

be the set of D - multimodular functions with global minimum point in r. For given multimodular matrix D and point $r \in M_D$ we will use the multimodular order $\leq_{mm}$ and a cone order $\leq_C$ on M_D. The multimodular order is defined below and the cone order is defined in Section 12.5.

Definition 74 *Let the multimodular matrix D and global minimum point $r \in M_D$ be given. Then for $x, y \in M_D$ we say that $x \leq_{mm} y$ if $f(x) \leq f(y)$ for every $f \in \mathcal{F}(D, r)$.*

It can be shown that both the multimodular order $\leq_{mm}$ and the cone order $\leq_C$ are reflexive, antisymmetric and transitive and thus they are partial orders on M_D. The following theorem says that these partial orders on M_D are equivalent.

Theorem 75 *For given multimodular matrix D and global minimum point $r \in M_D$ we have for $x, y \in M_D$ that $x \leq_{mm} y$ if and only if $x \leq_C y$.*

In Theorem 72 it is shown for $x, y \in M_D$ that $x \leq_C y$ implies that $x \leq_{mm} y$. Thus to prove Theorem 75 we have to show that $x \leq_{mm} y$ implies that $x \leq_C y$. By translation we can assume without loss of generality that that the global minimum point $r = (0, 0, \ldots, 0) \in M_D$. Let $\mathcal{C} = \{C_i\}_{i=1}^{(n+1)!}$ be the set of cones corresponding to D and root $r = (0, 0, \ldots, 0)$ (the origin) as defined in Chapter 12 . Then $\{\cup C_i : C_i \in \mathcal{C}\} = \mathbb{R}^n$ and for $C_i \in \mathcal{C}$ let C_i' be the intersection of C_i and M_D. We have the following proposition (see Lemma 64 and Section 12.5).

Proposition 18. *Let $C_i \in \mathcal{C}$. Then there exist $b_1, b_2, \ldots, b_n \in \mathbb{R}^n$ and a bijection $\sigma : \{1, 2, \ldots, n+1\} \to \{0, 1, \ldots, n\}$ such that*

1. *$b_1, b_2, \ldots, b_n$ are linearly independent.*
2. *$b_j = \sum_{i=1}^{j} d_{\sigma(i)}$ for $j = 1, 2, \ldots, n$.*
3. *$C_i = \{\sum_{i=1}^{n} \lambda_i \cdot b_i : \lambda_i \geq 0$ for $i = 1, 2, \ldots, n\}$.*
4. *$M_D = \{\sum_{i=1}^{n} \lambda_i \cdot b_i : \lambda_i \in \mathbb{Z}$ for $i = 1, 2, \ldots, n\}$.*
5. *$C_i' = \{\sum_{i=1}^{n} \lambda_i \cdot b_i : \lambda_i \in \mathbb{Z}_{\geq 0}$ for $i = 1, 2, \ldots, n\}$.*

Remark 28. For a convex cone $C \subset \mathbb{R}^n$ we say that a set of vectors $v_1, v_2, \ldots, v_k$ are generators of C if $v_1, v_2, \ldots, v_k$ are linearly independent and $C = \{\sum_{i=1}^{k} \lambda_i \cdot v_i : \lambda_i \geq 0$ for $i = 1, 2, \ldots, k\}$. In that case we say that C is a k - dimensional cone. In particular we have for $b_1, b_2, \ldots, b_n$ as in Proposition 18 that they are generators of the n -dimensional convex cone C_i.

Let $b_1, b_2, \ldots, b_n$ be generators of cone $C_i \in \mathcal{C}$ as in Proposition 18. If $x \in M_D$ then we have by Proposition 18 that there exist unique $x_i \in \mathbb{Z}$ such that $x = \sum_{i=1}^{n} x_i \cdot b_i$. For $k = 1, 2, \ldots, n$ we define the function $f_k : M_D \to \mathbb{Z}_{\geq 0}$ by

$$f_k(x) = \max(x_k, 0). \qquad (13.1)$$

For $i = 1, 2, \ldots, (n+1)!$ let $\mathcal{G}(C_i)$ be the set of functions $\{f_1, f_2, \ldots, f_n\}$ for cone C_i and let $\mathcal{G} = \cup_{i=1}^{(n+1)!} \mathcal{G}(C_i)$. For $C_i \in \mathcal{C}$ with generators $b_1, b_2, \ldots, b_n$ as in Proposition 18 we say for $x = \sum_{i=1}^{n} x_i \cdot b_i \in \mathbb{R}^n$ that $\mathrm{supp}_i(x) = \{k \in \{1, 2, \ldots, n\}$ for which $x_k > 0\}$. Thus $k \in \mathrm{supp}_i(x)$ if and only if $f_k(x) > 0$ for $f_k \in \mathcal{G}(C_i)$. Note that we assume for $k \in \mathrm{supp}_i(x)$ not only that $x_k \neq 0$ but that $x_k > 0$ which is stronger than in the standard definition of support.

We have the following proposition.

Proposition 19. *For given multimodular matrix D and global minimum point $r = (0, 0, \ldots, 0) \in \mathbb{R}^n$ we have that $\mathcal{G} \subseteq \mathcal{F}(D, r)$.*

Proof. For $f \in \mathcal{G}$ we have that $f = f_k \in \mathcal{G}(C_i)$ for some $k \in \{1, 2, \ldots, n\}$ and $i \in \{1, 2, \ldots, (n+1)!\}$. Let $b_1, b_2, \ldots, b_n$ be the generators of cone C_i as in

Proposition 18. Then for $x = \sum_{i=1}^{n} x_i \cdot b_i \in M_D$ we have that $f(x) = f_k(x) = \max(x_k, 0)$. Since $f(x) \geq 0$ for every $x \in M_D$ and $f(r) = 0$ it follows that r is a global minimum point of f. It remains to prove that f is a D-multimodular function. So, we have to show that for every $x \in M_D$, $i, j \in \{0, 1, \ldots, n\}$ with $i \neq j$ it holds that

$$f(x + d_i) + f(x + d_j) \geq f(x) + f(x + d_i + d_j). \tag{13.2}$$

According to property 2 of Proposition 18 there exists a bijection σ : $\{1, 2 \ldots, n + 1\} \rightarrow \{0, 1, 2, \ldots, n\}$ such that $b_j = \sum_{i=1}^{j} d_{\sigma(i)}$ for $j = 1, 2, \ldots, n$. From this it follows that $d_{\sigma(1)} = b_1$, $d_{\sigma(j)} = b_j - b_{j-1}$ for $2 \leq j \leq n$ and $d_{\sigma(n+1)} = -\sum_{i=1}^{n} d_{\sigma(i)} = -b_n$. Put $l = \sigma(k)$, $m = \sigma(k+1)$. If $i \notin \{l, m\}$ and $j \notin \{l, m\}$ then we have for every $x \in M_D$ that $f(x + d_i) = f(x + d_j) = f(x) = f(x + d_i + d_j)$. Hence (13.2) holds . Suppose that $i \in \{l, m\}$ and $j \notin \{l, m\}$. Then we have for every $x \in M_D$ that $f(x + d_i) = f(x + d_i + d_j)$ and $f(x + d_j) = f(x)$ and thus (13.2) holds. Suppose that $\{i, j\} = \{l, m\}$ and assume without loss of generality that $i = l$ and $j = m$. If $f(x) > 0$ then it follows that $f(x + d_i) = f(x) + 1$, $f(x + d_j) = f(x) - 1$, $f(x + d_i + d_j) = f(x)$ and thus (13.2) holds. If $f(x) = 0$ then it follows that $f(x + d_i) \geq f(x)$, $f(x + d_j) = f(x + d_i + d_j) = f(x) = 0$ and thus (13.2) holds. So, we can conclude that (13.2) holds in every case and thus $f \in \mathcal{F}(D, r)$.   □

Proposition 19 has the following corollaries.

Corollary 76 *If $x, y \in M_D$, $x \leq_{mm} y$ and there exists some cone $C_i \in \mathcal{C}$ such that $x, y \in C_i$ then $x \leq_C y$.*

Proof. Let $b_1, b_2, \ldots, b_n$ be the generators of cone C_i as in proposition 18 and put $x = \sum_{i=1}^{n} x_i \cdot b_i$ and $y = \sum_{i=1}^{n} y_i \cdot b_i$. Then we have by Proposition 18 that $x_i, y_i \in \mathbb{Z}_{\geq 0}$ for $i = 1, 2, \ldots, n$. So, for $f_k \in \mathcal{G}(C_i)$ we have that $f_k(x) = x_k$ and $f_k(y) = y_k$ for $k = 1, 2, \ldots, n$. Hence by Proposition 19 and $x \leq_{mm} y$ we have that $0 \leq x_k = f_k(x) \leq f_k(y) = y_k$ for $k = 1, 2, \ldots, n$. Thus $x \leq_C y$.   □

Corollary 77 *If $x, y \in M_D$, $x \leq_{mm} y$ then $\mathrm{supp}_i(x) \subset \mathrm{supp}_i(y)$ for $i = 1, 2, \ldots, (n + 1)!$.*

Corollary 77 can be proved in the same way as Corollary 76. The following corollary follows immediately from Corollary 76 and Corollary 77.

Corollary 78 *If $x, y \in M_D$, $x \leq_{mm} y$ and there exists some cone $C_i \in \mathcal{C}$ such that x is an internal point of C_i then $x \leq_C y$.*

To prove Theorem 75 it suffices by Corollary 76 to prove that there exists some cone $C_i \in \mathcal{C}$ such that $x, y \in C_i$ if $x \leq_{mm} y$. If $x = r$, $y = r$ or $y = \lambda \cdot x$ with $\lambda > 0$ then it follows directly that $x, y \in C_i$ for some $C_i \in \mathcal{C}$. Suppose that $y = \lambda \cdot x$ with $\lambda < 0$ and $x \neq r$. Let $C_i \in \mathcal{C}$ be a cone containing x with generators $b_1, b_2, \ldots, b_n$. Then $x = \sum_{i=1}^{n} x_i \cdot b_i$ and $y = \sum_{i=1}^{n} \lambda \cdot x_i \cdot b_i$ with $x_i \geq 0$ for $i = 1, 2, \ldots, n$. Moreover we have that $x_k > 0$ for some

$k \in \{1, 2, \ldots, n\}$ since $x \neq r$. Then for $f_k \in \mathcal{G}(C_i)$ we have that $f_k(x) = x_k > 0$ and $f_k(y) = 0$, which contradicts $x \leq_{mm} y$. So, we can assume that x and y are linearly independent vectors in $\mathbb{R}^n$ and let $H = span(x, y)$ the two dimensional subspace that contains x and y. For $C_i \in \mathcal{C}$ let C_i^* be the intersection of C_i and H. Then we have that the C_i^* are convex cones of dimension smaller or equal than two in the subspace H. To prove Theorem 75 we have the following lemmas.

Lemma 73. *Suppose that $C_i^* = C_i \cap H$ is a two dimensional cone generated by u and v. Then $\text{supp}_i(u)$ is not a subset of $\text{supp}_i(v)$ and $\text{supp}_i(v)$ is not a subset of $\text{supp}_i(u)$.*

Proof. Let $b_1, b_2, \ldots, b_n$ be the generators of C_i as in proposition 18. Then $u = \sum_{k \in supp_i(u)} u_k \cdot b_k$ with $u_k > 0$ and $v = \sum_{k \in supp_i(v)} v_k \cdot b_k$ with $v_k > 0$. For a small positive number ε we have that $w := v - \varepsilon \cdot u \in H$ and if $\text{supp}_i(u) \subseteq \text{supp}_i(v)$ then it is easily seen that $w = \sum_{k \in supp_i(v)} w_k \cdot b_k$ with $w_k > 0$ and thus $w \in C_i^*$. However, $w \in C_i^*$ contradicts the fact that u and v are generators of C_i^*. For $\text{supp}_i(v) \subseteq \text{supp}_i(w)$ we get similarly a contradiction. $\qquad \square$

Lemma 74. *Let $u, v, w \in H$ and let A_1 be a two dimensional cone generated by u and v, A_2 is a two dimensional cone generated by u and w and A_3 is a one dimensional cone generated by u. Let $w = a_1 \cdot u + a_2 \cdot v$. Then $A_1 \cap A_2 = A_3$ if and only if $a_2 < 0$.*

Proof. It is obvious that $A_3 \subseteq A_1 \cap A_2$. Let $x = \nu \cdot u + w$, where $\nu \geq \max(-a_1, 0)$. Then it is easily seen that $x \in A_2 \setminus A_3$. It follows that $x = (\nu + a_1) \cdot u + a_2 \cdot v$ and thus $x \in A_1 \cap A_2$ if $a_2 \geq 0$. So, $A_1 \cap A_2 = A_3$ implies that $a_2 < 0$. Conversely, suppose that

$$x' = \lambda \cdot u + \lambda' \cdot w \in A_1 \cap A_2.$$

Since $x' \in A_2$ we have that $\lambda \geq 0$ and $\lambda' \geq 0$. Then

$$x' = \lambda \cdot u + \lambda' \cdot (a_1 \cdot u + a_2 \cdot v) = (\lambda + \lambda' \cdot a_1) \cdot u + \lambda' \cdot a_2 \cdot v$$

and since $x' \in A_1$ it follows that $\lambda' \cdot a_2 \geq 0$. Thus if $a_2 < 0$ then $\lambda' = 0$ since $\lambda' \geq 0$ and $a_2 \cdot \lambda' \geq 0$. Hence $x' \in A_3$. $\qquad \square$

Proof. (Proof of Theorem 75) We have that $\cup_{i=1}^{(n+1)!} C_i^* = H$ since $\cup_{i=1}^{(n+1)!} C_i = \mathbb{R}^n$. Therefore one of the following two cases holds

1. There exists some two dimensional cone C_i^* such that in the subspace H we have that x is in the internal of cone C_i^*.
2. There exist two dimensional cones C_i^*, C_j^* such that in the subspace H we have that x is on the border of C_i^*, x is on the border of C_j^* and x is in the internal of $C_i^* \cup C_j^*$.

Suppose we have case 1 and let a and b be generators of the cone C_i^*. Then $a, b \in C_i$ and $x = \lambda \cdot a + \mu \cdot b$ with $\lambda, \mu > 0$. Hence $x \in C_i$ and $\operatorname{supp}_i(x) = \operatorname{supp}_i(a) \cup \operatorname{supp}_i(b)$. Suppose that $y \notin C_i^*$. Then $y = y_1 \cdot a + y_2 \cdot b$ with $\min(y_1, y_2) < 0$ and suppose without loss of generality that $y_1 < 0$. According to Lemma 73 there exist $k \in \operatorname{supp}_i(a) \setminus \operatorname{supp}_i(b)$. Then it follows that $k \notin \operatorname{supp}_i(y)$, $k \in \operatorname{supp}_i(x)$. Thus $\operatorname{supp}_i(x)$ is not a subset of $\operatorname{supp}_i(y)$, but this contradicts Corolary 77. Hence $y \in C_i^*$ and thus $x, y \in C_i$. According to Corollary 76 it follows that $x \leq_C y$.

Suppose we have case 2. Then there exist $a, b \in H$ such that x and a are generators of cone C_i^* and x and b are generators of cone C_j^*. Let u_1, u_2, v_1, v_2 be such that $y = u_1 \cdot x + u_2 \cdot a = v_1 \cdot x + v_2 \cdot b$. If $u_1 < 0$ then it follows anagously to the proof in case 1 that $\operatorname{supp}_i(x)$ is not a subset of $\operatorname{supp}_i(y)$, which yields a contradiction again. Thus $u_1 \geq 0$ and analogously it follows that $v_1 \geq 0$. We will prove that $y \in C_i^* \cup C_j^*$. Suppose that $y \notin C_i^*$. Then it follows that $u_2 < 0$. Let a_1, a_2 be such that $a = a_1 \cdot x + a_2 \cdot b$ By Lemma 74 we have that $a_2 < 0$. Then $y = u_1 \cdot x + u_2 \cdot a = u_1 \cdot x + u_2 \cdot (a_1 \cdot x + a_2 \cdot b) = (u_1 + u_2 \cdot a_1) \cdot x + (u_2 \cdot a_2) \cdot b$. Hence $v_2 = u_2 \cdot a_2 > 0$. Thus $y \in C_j^*$ if $y \notin C_i^*$ and thus $y \in C_i^* \cup C_j^*$. Since $x \in C_i^* \cap C_j^*$ it follows that $x, y \in C_i$ or $x, y \in C_j$. Hence $x \leq_C y$ by Corollary 76. □

13.2.1 Shift invariant counterparts

We consider sequences (of nonnegative integers) of a given length $T \in \mathbb{N}$ and given sum $S \in \mathbb{N}$. The set of such sequences is denoted by $P(T, S)$. So, $P(T, S) = \{(x_1, x_2, \ldots, x_T) : x_i \in \mathbb{Z}_{\geq 0} \forall i, \sum_{i=1}^{T} x_i = S\}$ and this is a submesh of dimension $T - 1$ of $\mathbb{Z}^T$. Let D' be a multimodular matrix of size $T \times T$ induced by the submesh $P(T, S)$.

Let $x, x' \in P(T, S)$. Then we say that x and x' are conjugate if they are cyclic permutations of each other. It holds that x and x' are conjugate if there exist finite (possibly empty) sequences v and w such that $x = vw$ and $x' = wv$. It is easily seen that this conjugacy is an equivalence relation on $P(T, S)$ and we write $x \sim x'$ if x and x' are conjugate. We denote by $\tilde{x}$ the conjugacy class of $x \in P(T, S)$, which is the set of all cyclic permutations of x. By $\tilde{P}(T, S)$ we denote the set of conjugacy classes of $P(T, S)$. If $x \sim x'$ then we also say that x and x' are shifts of each other. If $\tilde{x} = \tilde{y}$ then we say that x is a representative of $\tilde{y}$.

Let $f : P(T, S) \to \mathbb{R}$ be a function such that $f(x) = f(x')$ if $x \sim x'$. Then we say that f is a shift invariant function. A shift invariant function $f : P(T, S) \to \mathbb{R}$ induces a function $\tilde{f} : \tilde{P}(T, S) \to \mathbb{R}$ by $\tilde{f}(y) = f(x)$ where $x \in P(T, S)$ is a representative of $y \in \tilde{P}(T, S)$. We denote by $\mathcal{F}^{shift}(D')$ the set of functions mapping $\tilde{P}(T, S)$ to $\mathbb{R}$ which are induced by D' multimodular functions that are shift invariant. We use $\mathcal{F}^{shift}(D')$ to define a partial order

$\leq_{mms}$ on $\widetilde{P}(T,S)$ in the same way as $\mathcal{F}(D,r)$ was used to define the partial order $\leq_{mm}$ on M_D. The partial order $\leq_{mms}$ is called the shift invariant multimodular order.

Definition 79 *Let a multimodular matrix D' induced by the submesh $P(T,S)$ be given. Then for $x, y \in \widetilde{P}(T,S)$ we say that $x \leq_{mms} y$ if $f(x) \leq f(y)$ for every $f \in \mathcal{F}^{shift}(D')$.*

Let $x \in P(T,S)$. Then we say that $x = (x_1, x_2, \ldots, x_T)$ is regular if the induced infinite sequence $x^\infty := (x_1, \ldots, x_T, x_1, \ldots, x_T, \ldots)$ is a bracket sequence and in that case the conjugacy class $\widetilde{x} \in \widetilde{P}(T,S)$ is also called regular. We denote by $R(T,S) \subseteq P(T,S)$ the subset of regular sequences. The following lemma can be proved analogously to the proof for sequences of zeros and ones that is given in [74].

Lemma 75. *Let $T, S \in \mathbb{N}$ be given. Then there exists exactly one element in $\widetilde{P}(T,S)$ which is regular.*

We denote by $\widetilde{\omega} = \widetilde{\omega}(T,S)$ the unique element of $\widetilde{P}(T,S)$ that is regular. Then we have the following theorem.

Theorem 80 *Let a multimodular matrix D' induced by the submesh $P(T,S)$ be given and let $f \in \mathcal{F}^{shift}(D')$. Then a global minimum of f is attained in $\widetilde{\omega}(T,S)$.*

Corollary 81 *We have that $\widetilde{\omega}(T,S)$ is the smallest element for the partial order $\leq_{mms}$ on $\widetilde{P}(T,S)$.*

We have the multimodular order $\leq_{mm}$ and we have the shift invariant multimodular order $\leq_{mms}$ on $\widetilde{P}(T,S)$. We also define a shift invariant cone order $\leq_{Cs}$ on $\widetilde{P}(T,S)$ which is the counterpart of the cone order $\leq_C$.

Definition 82 *Let D' be as in Theorem 80. Then for $u, v \in \widetilde{P}(T,S)$ we say that $u \leq_{Cs} v$ if and only if there exist representatives u' of u and v' of v such that $u' \leq_C v'$, where $\leq_C$ is the cone order for multimodular matrix D' and root r with r being a representative of $\widetilde{\omega}(T,S)$.*

Note. According to Theorem 75 it holds that $u' \leq_C v'$ if and only if $f(u') \leq f(v')$ for every $f \in \mathcal{F}(D',r)$.

Corollary 83 *Let D' be as in Theorem 80. Then for $u, v \in \widetilde{P}(T,S)$ we have that $u \leq_{mms} v$ if $u \leq_{Cs} v$.*

13.3 The graph order and the unbalance

Let $u = (u_1, u_2, \ldots, u_T) \in P(T, S)$ and for $l = 0, 1, \ldots, T - 1$ let $u^{(l)} :=$ $(u_{l+1}, u_{l+2}, \ldots, u_T, u_1, u_2, \ldots, u_l) \in P(T, S)$ be the l-th cyclic permutation of u. For a sequence $u \in P(T, S)$ we define a counting function $\kappa_u : \{0, 1, \ldots, T\} \to \mathbb{Z}$ by $\kappa_u(n) = \sum_{t=1}^{n} u_t$ and we define a discrepancy function $\phi_u : \{0, 1, \ldots, T\} \to \mathbb{Q}$ by $\phi_u(n) = \kappa_u(n) - n \cdot \frac{S}{T}$ for $n = 0, 1, \ldots, T$.

Lemma 76. *For $u \in P(T, S)$ and $l \in \{0, 1, \ldots, T - 1\}$ we have that $\phi_{u^{(l)}}(n) \geq 0$ for $n = 0, 1, \ldots, T$ if and only if $\phi_u(l) = \min_{n=0,1,\ldots,T} \phi_u(n)$ and $\phi_{u^{(l)}}(n) \leq 0$ for $n = 0, 1, \ldots, T$ if and only if $\phi_u(l) = \max_{n=0,1,\ldots,T} \phi_u(n)$.*

Proof. Suppose that $\phi_{u^{(l)}}(n) < 0$ for some $n \in \{0, 1, \ldots, T-1\}$. Then $\phi_u((l + n) \pmod{T}) = \phi_u(l) + \phi_{u^{(l)}}(n) < \phi_u(l)$. Thus $\phi_u(l) = \min_{n=0,1,\ldots,T} \phi_u(n)$ implies that $\phi_{u^{(l)}}(n) \geq 0$ for $n = 0, 1, \ldots, T$. Suppose that $\phi_u(l') < \phi_u(l)$ for some $l \neq l' \in \{0, 1, \ldots, T - 1\}$. Then $\phi_{u^{(l)}}((l' - l) \pmod{T}) = \phi_u(l') - \phi_u(l) < 0$. Thus $\phi_{u^{(l)}}(n) \geq 0$ for $n = 0, 1, \ldots, T$ implies that $\phi_u(l) = \min_{n=0,1,\ldots,T} \phi_u(n)$. Analogously it follows that $\phi_{u^{(l)}}(n) \leq 0$ for $n = 0, 1, \ldots, T$ if and only if $\phi_u(l) = \max_{n=0,1,\ldots,T} \phi_u(n)$. $\qquad \square$

While the multimodular order $\leq_{mm}$ and the cone order $\leq_C$ are partial orders on $\mathbb{Z}^T$ we define next a partial order on $P(T, S)$. This induces partial orders on $\widetilde{P}(T, S)$ from which we derive our bounds for the expected average waiting time.

Definition 84 *For $u, v \in P(T, S)$ we say that $u \preceq v$ if $\kappa_u(n) \leq \kappa_v(n)$ for $n = 1, 2, \ldots, T$.*

We have the following (see [74] where it is proved for sequences of zeros and ones).

Proposition 20. *Let $u \in R(T, S)$. Then $u' \in R(T, S)$ if and only if $u \sim u'$. Moreover the partial order $\preceq$ on $P(T, S)$ induces a total order on $R(T, S)$.*

Since $R(T, S)$ is finite it follows from Proposition 20 that $R(T, S)$ contains a greatest element for this order. We denote this greatest element by $\omega(T, S)$ or just ω if no confusion is possible. For this element ω the partial sums $\kappa_\omega(n)$ for $n = 1, 2, \ldots, T$ are as great as possible under the restriction that ω is regular. For example $\omega(7, 4)$ is the sequence $(1, 1, 0, 1, 0, 1, 0)$. From [74] we have the following lemma which can be used to determine $\omega(T, S)$ quickly.

Lemma 77.

$$\kappa_{\omega(T,S)}(n) = \lceil n \cdot \frac{S}{T} \rceil \text{ for } n = 0, 1, \ldots, T.$$

We have seen in Proposition 20 that the regular sequences $R(T, S)$ form a conjugacy class in $P(T, S)$. It follows that $u \in R(T, S)$ if and only if u is a representative of $\tilde{\omega}(T, S) \in \tilde{P}(T, S)$. Combining Lemma 76, Proposition 20 and Lemma 77 we obtain the following theorem in which the partial order $\preceq$ is used to give a characterising property of the conjugacy class $R(T, k) = \tilde{\omega}(T, S)$ of regular sequences in $\tilde{P}(T, S)$.

Theorem 85 *Every conjugacy class $\tilde{u}$ of $P(T, S)$ contains an upper bound of $R(T, S)$, i.e for every $u \in P(T, S)$ there exists a $v \in P(T, S)$ such that $v \sim u$ and $v \succeq w$ for every $w \in R(T, S)$.*

We have the following preorders $\preceq_{\bar{g}}$, $\preceq_{\underline{g}}$ and $\preceq_g$ on $P(T, S)$ called the primal or upper graph order, the dual or lower graph order and the strong graph order respectively.

Definition 86 *Let $u, v \in P(T, S)$. Then $u \preceq_{\bar{g}} v$ if there exist $u', v' \in P(T, S)$ such that $u' \in \tilde{u}$, $v' \in \tilde{v}$ and $0 \le \phi_{u'}(n) \le \phi_{v'}(n)$ for $n = 0, 1, \ldots, T$ and $u \preceq_{\underline{g}} v$ if there exist $u'', v'' \in P(T, S)$ such that $u'' \in \tilde{u}$, $v'' \in \tilde{v}$ and $0 \ge \phi_{u''}(n) \ge \phi_{v''}(n)$ for $n = 0, 1, \ldots, T$. Further $u \preceq_g v$ if $u \preceq_{\bar{g}} v$ and $u \preceq_{\underline{g}} v$.*

The preorders $\preceq_{\bar{g}}$, $\preceq_{\underline{g}}$ and $\preceq_g$ on $P(T, S)$ are not partial orders. Namely, if $u, v \in P(T, S)$ are cyclic permutations of each other then it is easily seen that $u \preceq_g v$ and $v \preceq_g u$. Thus these orders are not antisymmetric. However, they induce (see Definition 87) partial orders on $\tilde{P}(T, S)$.

Definition 87 *Let $u, v \in \tilde{P}(T, S)$ and $u' \in P(T, S)$ be a representative of u and $v' \in P(T, S)$ be a representative of v. Then we say that $u \preceq_g v$ if and only if $u' \preceq_g v'$ and similar for the orders $\preceq_{\bar{g}}$ and $\preceq_{\underline{g}}$.*

In [74] it is shown that these induced graph orders on $\tilde{P}(T, S)$ are partial orders.

For a sequence $u \in P(T, S)$ we have a primal unbalance $\bar{I}(u)$ and a dual unbalance $\underline{I}(u)$.

Definition 88 *Let $u \in P(T, S)$. Then the primal unbalance of u is*

$$\bar{I}(u) := \frac{1}{T} \cdot \sum_{n=1}^{T} (\kappa_{u'}(n) - \lceil n \cdot \frac{S}{T} \rceil),$$

where $u' \in \tilde{u}$ such that $\phi_{u'}(n) \ge 0$ for $n = 1, 2, \ldots, T$. The dual unbalance of u is

$$\underline{I}(u) := \frac{1}{T} \cdot \sum_{n=1}^{T} (\lfloor n \cdot \frac{S}{T} \rfloor - \kappa_{u'}(n)),$$

where $u' \in \tilde{u}$ such that $\phi_{u'}(n) \le 0$ for $n = 1, 2, \ldots, T$.

Remark 29. For $u \in P(T,S)$ the primal unbalance is well defined. Namely, by Lemma 76 there exist $u' \in \tilde{u}$ such that $\phi_{u'}(n) \geq 0$ for $n = 1, 2, \ldots, T$. Moreover, if $u', u'' \in \tilde{u}$ are such that $\phi_{u'}(n) \geq 0$ and $\phi_{u''}(n) \geq 0$ for $n = 1, 2, \ldots, T$ then $\sum_{n=1}^{T} \kappa_{u'}(n) = \sum_{n=1}^{T} \kappa_{u''}(n)$ (see Theorem 89). Analogously it follows that the dual unbalance is well defined. It is easily seen that for every sequence $u \in P(T,S)$ it holds that $\bar{I}(u) \geq 0$ and $\underline{I}(u) \geq 0$. Namely, for $u' \in P(T,S)$ we have that $\phi_{u'}(n) \geq 0$ if and only if $\kappa_{u'}(n) \geq \lceil n \cdot \frac{S}{T} \rceil$ and $\phi_{u'}(n) \leq 0$ if and only if $\kappa_{u'}(n) \leq \lfloor n \cdot \frac{S}{T} \rfloor$. Moreover, a sequence $u \in P(T,S)$ is regular if and only if $\bar{I}(u) = 0$ if and only if $\underline{I}(u) = 0$ (see [74]).

Theorem 89 *Let $u \in P(T,S)$. Then the following statements are equivalent for $l \in \{0, 1, \ldots, T-1\}$.*

(i) $\bar{I}(u) = \frac{1}{T} \cdot \sum_{n=1}^{T} (\kappa_{u^{(l)}}(n) - \lceil n \cdot \frac{S}{T} \rceil)$.
(ii) $\sum_{n=1}^{T} \kappa_{u^{(l)}}(n) = \max_{i=0,1,\ldots,T-1} \sum_{n=1}^{T} \kappa_{u^{(i)}}(n)$.
(iii) $\min_{n=0,1,\ldots,T-1} \phi_{u^{(l)}}(n) = \phi_{u^{(l)}}(0) = 0$.
(iv) $\phi_u(l) = \min_{i=0,1,\ldots,T-1} \phi_u(i)$.

Proof. By Lemma 76 we have that (iii) implies (iv) and vice versa. We now prove that (ii) implies (iii). Suppose there exists some $t \in \{1, 2, \ldots, T-1\}$ such that $\phi_{u^{(l)}}(t) = -\mu$ with $\mu > 0$. Let x be the prefix of length t of $u^{(l)}$ and let y be the suffix of length $T - t$ of $u^{(l)}$. Then $u^{(l)} = xy$ and let $z := yx = u^{((l+t) \pmod{T})} \in \tilde{u}$. Then $\phi_{u^{(l)}}(n) = \phi_z(n-t) - \mu$ for $n = t, t+1, \ldots, T$ and $\phi_{u^{(l)}}(n) = \phi_z(n+T-t) - \mu$ for $n = 1, 2, \ldots, t-1$. Hence $\sum_{n=1}^{T} \phi_{u^{(l)}}(n) = \sum_{n=1}^{T} \phi_z(n) - T \cdot \mu$ and thus

$$\sum_{n=1}^{T} \kappa_{u^{(l)}}(n) = \sum_{n=1}^{T} \kappa_z(n) - T \cdot \mu < \sum_{n=1}^{T} \kappa_z(n) \leq \max_{i=0,1,\ldots,T-1} \sum_{n=1}^{T} \kappa_{u^{(i)}}(n),$$

which contradicts (ii). From the definition of the primal unbalance it follows directly that (iii) implies (i) and to finish the proof we show that (i) implies (ii). Suppose that $\sum_{n=1}^{T} \kappa_{u^{(l)}}(n) < \max_{i=0,1,\ldots,T-1} \sum_{n=1}^{T} \kappa_{u^{(i)}}(n)$ and let $l \neq l' \in \{0, 1, \ldots, T-1\}$ such that $\sum_{n=1}^{T} \kappa_{u^{(l')}}(n) = \max_{i=0,1,\ldots,T-1} \sum_{n=1}^{T} \kappa_{u^{(i)}}(n)$. Since we have proved that (ii) implies (i) it follows that

$$\bar{I}(u) = \frac{1}{T} \cdot \sum_{n=1}^{T} (\kappa_{u^{(l')}}(n) - \lceil n \cdot \frac{S}{T} \rceil) > \frac{1}{T} \cdot \sum_{n=1}^{T} (\kappa_{u^{(l)}}(n) - \lceil n \cdot \frac{S}{T} \rceil),$$

which contradicts (i). $\qquad \square$

The following result follows immediately from the definitions.

Lemma 78. *Let $u, v \in P(T,S)$. If $u \preceq_{\bar{g}} v$ then $\bar{I}(u) \leq \bar{I}(v)$. If $u \preceq_{\underline{g}} v$ then $\underline{I}(u) \leq \underline{I}(v)$. If $u \preceq_g v$ then $\bar{I}(u) \leq \bar{I}(v)$ and $\underline{I}(u) \leq \underline{I}(v)$.*

We have the following theorem, which shows that the primal and dual unbalance are multimodular with respect to the L - Triangulation (see Section 12.2).

Theorem 90 *The primal unbalance and the dual unbalance are shift invariant and multimodular on $P(T,S)$ with respect to the base $d_1, d_2, \ldots, d_T$ where $d_i = e_i - e_{i+1}$ for $i = 1, 2, \ldots, T-1$, $d_T = e_T - e_1$ and e_i is the i -th unit vector of length T for $i = 1, 2, \ldots, T$.*

Proof. For $u \in P(T,S)$ let $f(u) = \bar{I}(u)$ and $g(u) = \underline{I}(u)$. Thus $f : P(T,S) \to \mathbb{Q}$ is the primal unbalance function and $g : P(T,S) \to \mathbb{Q}$ is the dual unbalance function. From the definition it follows that if $u', u'' \in P(T,S)$ are cyclic permutations of each other then $f(u') = f(u'')$ and $g(u') = g(u'')$. Thus the primal unbalance and dual unbalance are shift invariant. To prove the multimodularity of the primal unbalance function f we have to show that for every $u \in P(T,S)$ and $i, j \in \{1, 2, \ldots, T\}$, $i \neq j$ it holds that

$$f(u + d_i) + f(u + d_j) \geq f(u) + f(u + d_i + d_j). \tag{13.3}$$

We first show that for every $x \in P(T,S)$ and $i \in \{1, 2, \ldots, T\}$ it holds that

$$f(x + d_i) \leq f(x) + \frac{1}{T}. \tag{13.4}$$

Namely, put $y = x + d_i$. Without loss of generality we can assume that $\sum_{n=1}^{T} \kappa_y(n) = \max_{i=0,1,\ldots,T-1} \sum_{n=1}^{T} \kappa_{y^{(i)}}(n)$. Then $f(y) = \frac{1}{T} \cdot \sum_{n=1}^{T} (\kappa_y(n) - \lceil n \cdot \frac{S}{T} \rceil)$ and

$$f(x) = f(y - d_i)$$

$$\geq \frac{1}{T} \cdot \sum_{n=1}^{T} (\kappa_{y-d_i}(n) - \lceil n \cdot \frac{S}{T} \rceil)$$

$$\geq \frac{1}{T} \cdot (\sum_{n=1}^{T} (\kappa_y(n) - \lceil n \cdot \frac{S}{T} \rceil) - 1)$$

$$= f(y) - \frac{1}{T},$$

by Theorem 89. So, (13.4) holds and next we show that (13.3) holds. Without loss of generality we assume that $\min_{n=0,1,\ldots,T-1} \phi_u(n) = \phi_u(0) = 0$. Then by Theorem 89. Suppose that $i \neq T$ and $j \neq T$. Then it is easily seen that $\kappa_{u+d_i}(n) \geq \kappa_u(n)$, $\kappa_{u+d_j}(n) \geq \kappa_u(n)$ and $\kappa_{u+d_i+d_j}(n) \geq \kappa_u(n)$ for $n = 0, 1, \ldots, T$. Hence

$$\min_{n=0,1,\ldots,T-1} \phi_{u+d_i}(n) = \min_{n=0,1,\ldots,T-1} \phi_{u+d_j}(n)$$

$$= \min_{n=0,1,\ldots,T-1} \phi_{u+d_i+d_j}(n)$$

$$= \min_{n=0,1,\ldots,T-1} \phi_u(n)$$

$$= 0.$$

From the definition of the primal unbalance it is easily seen that $f(u + d_i) = f(u + d_j) = f(u) + \frac{1}{T}$ a nd $f(u + d_i + d_j) = f(u) + \frac{2}{T}$ and thus $f(u+d_i) + f(u+d_j) = f(u) + f(u+d_i+d_j)$. It remains to show that (13.3) holds in case $i = T$ or $j = T$ and we can assume that $j = T$ and $i \neq T$. Then from the foregoing we have that $f(u + d_i) = f(u) + \frac{1}{T}$. Moreover by (13.4) we have that $f(u + d_i + d_j) \leq f(u + d_j) + \frac{1}{T}$. Hence

$$f(u + d_i) + f(u + d_j) \geq (f(u) + \frac{1}{T}) + (f(u + d_i + d_j) - \frac{1}{T})$$
$$= f(u) + f(u + d_i + d_j).$$

Thus the primal unbalance function f is multimodular and it follows analogously that the dual unbalance function g is multimodular. □

Note. Since the primal and dual unbalance functions are shift invariant on $P(T, S)$ they induce functions on $\tilde{P}(T, S)$ by $\tilde{I}(v) = \bar{I}(u)$ and $\underline{I}(v) = \underline{I}(u)$ if $u \in P(T, S)$ is a representative of $v \in \tilde{P}(T, S)$. Then by Theorem 90 we have for these induced primal and dual unbalance functions that they are element of $\mathcal{F}^{shift}(D)$, where D is the multimodular matrix having the multimodular base $d_1, d_2, \ldots, d_T$ as row vectors.

Corollary 91 *For this multimodular base* $d_1, d_2, \ldots, d_T$ *we have for* $u, v \in \tilde{P}(T, S)$ *that* $u \leq_{mms} v$ *implies that* $\bar{I}(u) \leq \bar{I}(v)$ *and* $\underline{I}(u) \leq \underline{I}(v)$.

13.4 Relations and counterexamples

13.4.1 The shift invariant cone order does not imply the graph order

Consider the following example. Let $u = (1, 3, 2, 2, 3) \in \mathcal{P}(5, 11)$ and $v = (2, 1, 3, 2, 3) \in \mathcal{P}(5, 11)$. Thus we consider sequences of length 5 and sum 11. It is easily seen that u and v are not ordered for the dual lower graph order (but they are ordered for the primal upper graph order). So, u and v are not graph ordered. However, we now first show that u and v are ordered for shift invariant multimodular functions with respect to the standard base d_0, d_1, d_2, d_3, d_4 where $d_i = e_i - e_{i+1}$ for $i = 1, 2, 3, 4$ and $d_0 = (-1, 0, 0, 0, 1)$. Namely, for a shift invariant multimodular function f we have that $f(u + d_4) = f((1, 3, 2, 3, 2)) = f(v)$ and also $f(u + d_1) = f((2, 2, 2, 2, 3)) = f((2, 2, 2, 3, 2)) = f(u+d_1+d_4)$ by shift invariance. By multimodularity we have that $f(u + d_1) + f(u + d_4) \geq f(u) + f(u + d_1 + d_4)$ and thus $f(u) \leq f(u + d_4) = f(v)$ for every multimodular shift invariant function f.

Conclusion. From this it follows that the shift invariant mutimodular order does not imply the graph order and thus the shift invariant multimodular order does not imply the shift invariant cone order or the shift invariant cone order does not imply the graph order (or both do not hold).

We now investigate whether $u \leq_{Cs} v$. To do this we take the regular sequence $r = (3, 2, 2, 2, 2)$ as root of all the cones we consider and we let u', v' run through all the shifts of u and v respectively. Then we check whether $u' \leq v'$ for some cone. Doing this we find that u and v are indeed ordered for the shift invariant cone order. Namely, for $u' = (3, 1, 3, 2, 2)$ and $v' = (2, 1, 3, 2, 3)$ we have that $u' = r + d_0 + d_1 + d_3 + d_4$ and $v' = r + 2d_0 + d_1 + d_3 + d_4$. Hence, if we consider the cone generated by $b_1 = d_0$, $b_2 = d_0 + d_1$, $b_3 = d_0 + d_1 + d_3$ and $b_4 = d_0 + d_1 + d_3 + d_4$ then $u' = r + b_4$ and $v' = r + b_1 + b_4$. By the way, this u' and v' is the only pair of shifts that are cone ordered for some cone with root in r. Since $u' \leq_C v'$ and thus $u \leq_{Cs} v$ we have the following conclusion.

Conclusion. The shift invariant cone order does not imply the graph order.

Remarks. The explanation for the fact that the ordering of u' and v' in this cone does not imply the lower graph order is the following. For u' you have to start with the second coordinate to get the graph for the lower graph order and for v' you have to start with the first coordinate. This is possible, because for r you have the start at the second coordinate for the lower graph order, while the generating base vectors of this cone "suggest" the lower graph order for starting at the first coordinate. One of the consequences is that this cone does contain other regular sequences than r. Namely, $r + b_1 = (2, 2, 2, 2, 3)$ is for example another regular sequence in this cone. It turns out that the cone order in such cones does not imply the graph order.

If you consider the mirrored sequences of u and v then you get a similar problem with the upper graph order instead of the lower graph order.

Note that u and v are ordered for the unbalance (despite that they are not ordered for the graph order). Namely, for the primal (upper) unbalance $\overline{I}$ we have that $\overline{I}(u) = 1 \leq 2 = \overline{I}(v)$ and for the dual (lower) unbalance $\underline{I}$ we have that $\underline{I}(u) = 1 = \underline{I}(v)$. Of course this order for the unbalance follows immediately from $\tilde{u} \leq_{mms} \tilde{v}$ and Corollary 91.

13.4.2 The shift invariant multimodular order does not imply the shift invariant cone order

A counterexample is given here to show that the shift invariant multimodular order does not imply the shift invariant cone order. In fact it is shown that the shift invariant cone order is not even a partial order, because it is not transitive.

Counterexample. Again we consider sequences in $P(5, 11)$ and we have the same multimodular standard base $\{d_i\}_{i=0,1...,5}$ as before. Let $u = (1, 2, 3, 2, 3)$, $v = (2, 2, 3, 3, 1)$ and $w = (2, 1, 0, 4, 4)$ and as regular sequence

we have again $r = (3, 2, 2, 2, 2)$. Now we have for the shift invariant cone order that $u \leq_{Cs} v$ and $v \leq_{Cs} w$. However, we do not have that $u \leq_{Cs} w$. Namely, $u' = (3, 2, 3, 1, 2) = r + d_3 = r + b_1$ and $v' = (2, 3, 3, 1, 2) = r + d_2 + 2d_3 + d_4 + d_0 = r + b_1 + b_4$, where $b_1 = d_3$, $b_2 = d_3 + d_0$, $b_3 = d_3 + d_0 + d_2$ and $b_4 = d_3 + d_0 + d_2 + d_4$. Hence $u \leq_{Cs} v$.
Further $v'' = (3, 1, 2, 2, 3) = r + d_0 + d_1 = r + b_2$ and $w'' = (2, 1, 0, 4, 4) = r + 4d_0 + 3d_1 + 2d_2 + 2d_4 = r + b_1 + b_2 + 2b_4$, where $b_1 = d_0$, $b_2 = d_0 + d_1$, $b_3 = d_0 + d_1 + d_2$ and $b_4 = d_0 + d_1 + d_2 + d_4$. Hence $v \leq_{Cs} w$. However, there exist no cone with root in r such that some shift of u is smaller in that cone than some shift of w. Hence, u and w are not shift invariant cone ordered, while from $u \leq_{Cs} v$ and $v \leq_{Cs} w$ it follows that u and w are ordered for the shift invariant multimodular order.

Remark 30. So, it turns out that the shift invariant cone order is not a partial order, because it is not transitive. It is natural to consider the smallest preorder which contains the shift invariant cone order and is transitive. It is easily seen that the shift invariant multimodular order implies this order, but we do not know whether the converse is also true in which case this order is equivalent to the shift invariant multimodular order.

13.4.3 The graph order does not imply the shift invariant multimodular order

In this counterexample we show that the graph order does not imply the shift invariant multimodular order.

Consider the sequences
$u = (0, 0, 0, 1, 0, 0, 1, 0, 0, 1, 0, 0, 0, 1, 0, 1, 0, 1, 1, 0, 0, 1, 1, 1, 1, 1, 0, 1, 0, 1, 0, 1) \in P(32, 15)$ and
$v = (0, 0, 0, 0, 0, 0, 0, 0, 0, 0, 0, 1, 1, 1, 0, 1, 1, 0, 1, 1, 0, 1, 1, 0, 1, 1, 0, 1, 1, 1, 0, 1) \in P(32, 15)$. Then we have that $u \leq_g v$. Thus u and v are ordered for the graph order and we have that $\overline{I}(u) < \overline{I}(v)$ and $\underline{I}(u) < \underline{I}(v)$. However, it does not hold that $f(u) \leq f(v)$ for every shift invariant multimodular function $f : P(32, 15) \to \mathbb{R}$. Namely, we consider u and v as period cycles for the splitting sequences corresponding to the admission of arriving customers to some server. Suppose that the interarrival times are deterministic and equal to 1 and the service times are deterministic and equal to $(1 + \epsilon)$, where ϵ is a small positive number. For $x \in P(32, 15)$ let $W(x)$ be the average waiting time of customers admitted according to x. The average waiting time is multimodular with respect to the standard base of the L - Triangulation (see Chapter 12) and since the traffic intensity for the server is smaller than one we have that the average waiting time function W is also shift invariant (see [74]). Thus W is a shift invariant multimodular function. However, for the given interarrival and service times we have that $W(u) = \frac{11}{15} \cdot \epsilon$ and $W(v) = \frac{10}{15} \cdot \epsilon$. Thus $W(v) < W(u)$ for these service times and thus u and v are not ordered for the shift invariant multimodular order.

13.5 Bounding the Difference in Waiting Times for One Queue

In this section we give a bound for the expected average waiting time for customers routed to a single server queue according to a routing sequence of zeros and ones. This bound depends on the unbalance of the routing sequence. To state the results we first have some definitions and notations.

Let u and v be infinite periodic integer sequences. Then we say that u is equivalent to v, $u \sim v$ if there exists a finite sequence w such that w is a period cycle of both u and v.

In our application of routing sequences we have in general that equivalent infinite periodic sequences have the same performance. Let $\mathcal{P}$ be the set of infinite periodic sequences of zeros and ones (modulo the equivalence relation), $\mathcal{R} \subseteq \mathcal{P}$ the subset of regular periodic sequences and for $d \in \mathbb{Q}$, $0 \leq d \leq 1$ let $\mathcal{P}(d) \subseteq \mathcal{P}$ be the subset of sequences with density d. We denote the regular sequence in $\mathcal{P}(d)$ by $\omega(d)$ or just ω. Note that if $u' \in \mathcal{P}(T, k)$ is a period cycle of $u \in \mathcal{P}$ then $u \in \mathcal{P}(d)$ where $d = \frac{k}{T}$. We define the unbalance for infinite periodic sequences of zeros and ones as follows.

Definition 92 *Let $u \in \mathcal{P}$ and let $u' \in \mathcal{P}(T, k)$ be a period cycle of u. Then we define the (primal) unbalance of u as $\overline{I}(u) := \overline{I}(u')$.*

The unbalance is well defined on $\mathcal{P}$. Namely, if u' and u'' are both period cycles of u then $\overline{I}(u') = \overline{I}(u'')$.

Let $\{T_i\}_{i=1,2,...}$ be a sequence of arrival times of customers, with the convention that $T_1 = 0$. Put $\delta_i := T_{i+1} - T_i$ for $i = 1, 2, \ldots$. Then $\{\delta_i\}$ is the sequence of interarrival times. Further these arriving customers are routed to a server according to some routing sequence $u = (u_1, u_2, \ldots)$ of zeros and ones. For such routing sequences we have the counting function $\kappa_u(n) = \sum_{t=1}^{n} u_t$ and we define a partial order $\preceq$ that is similar as the order on $\mathcal{P}(T, S)$. Namely, we say that $u \preceq v$ if $\kappa_u(n) \leq \kappa_v(n)$ for $n = 0, 1, \ldots$. Further we define the the following related function $\nu_u(i) : \mathbb{Z}_{\geq 0} \to \mathbb{Z}$ with $\nu_u(j) = \min\{n \in \mathbb{Z}_{\geq 0} : \kappa_u(n) = j\}$ and we put $\tau_u(j) = \sum_{i=\max(\nu_u(j-1),1)}^{\nu_u(j)-1} \delta_i$ for $j = 1, 2, \ldots$. Then $\tau_u(j)$ is the time elapsed between the routing of the $(j-1)$-th and j-th customer to the server according to routing sequence u. If we put $\lambda_u(j) = \sum_{k=1}^{j} \tau_u(k)$ for $j = 1, 2, \ldots$ then $\lambda_u(j)$ is the time at which the $j - th$ customer is routed to the server according to routing sequence u. We have a sequence of service times $\{\sigma_j\}_{j=1,2,...}$, where σ_j is the service time of the j-th customer that is routed to the server according to the routing sequence. Further we define $W_u(j)$ to be the workload for the server at the moment the j-th customer is routed to the server according to routing sequence u. In other words $W_u(j)$ is the waiting time for the j-th customer that is routed to the server. We assume that the server starts empty at $T_1 = 0$ and thus $W_u(1) = 0$. If the interarrival times $\{\delta_i\}$ and service times $\{\sigma_j\}$ are random variables then we say that $\overline{W}(u)$ is the almost sure long-run average waiting time of customers routed to the server according to routing sequence u if $\lim_{m \to \infty} \frac{1}{m} \cdot \sum_{j=1}^{m} W_u(j) = \overline{W}(u)$ with probability one. From ergodic theory we have the following theorem. See [74] for a proof of this theorem.

Theorem 93 *Suppose that the interarrival times $\{\delta_i\}$ of customers arriving at the system are independent and identically distributed (i.i.d.) random variables with mean δ and the service times $\{\sigma_j\}$ of the considered server are i.i.d. random variables with mean σ and independent of the interarrival times. Further let u' and u'' be routing sequences of zeros and ones that are both representatives of some $u \in \mathcal{P}(d)$ where $d \in \mathbb{Q}$ (thus they have a common period cycle) and $\frac{\sigma}{\delta} \cdot d < 1$. Then $\overline{W}(u')$ and $\overline{W}(u'')$ exist and are finite. Moreover $\overline{W}(u') = \overline{W}(u'')$.*

Let δ, σ and d be as in Theorem 93. Then we say that $\rho := \frac{\sigma}{\delta} \cdot d$ is the traffic intensity for the server. By Theorem 93 all the routing sequences which are representative of some $u \in \mathcal{P}(d)$ have the same long-run average waiting time if the stability condition $\rho < 1$ is fulfilled. Therefore in this case we denote this long-run average waiting time simply by $\overline{W}(u)$. The following theorem which is obtained by a sample path argument is the main result of this section.

Theorem 94 *Let the interarrival times $\{\delta_i\}$ and the service times $\{\sigma_j\}$ be as in Theorem 93. Further let $u \in \mathcal{P}(d)$ for some $d \in \mathbb{Q}$ such that $\rho < 1$ and let $\omega = \omega(d) \in \mathcal{P}(d)$ be the regular sequence of density d. Then*

$$\overline{W}(u) \leq \overline{W}(\omega) + \frac{\delta}{d} \cdot \overline{I}(u). \tag{13.5}$$

Remark 31. From the results of Chapters 2 and 4 we have that $\overline{W}(u) \geq \overline{W}(\omega)$.

We give an outline of the proof of Theorem 94. For a complete proof for this and following results in Sections 13.4 and 13.5, see [74]. We first assume that the interarrival times $\{\delta_i\}_{i=1,2,\ldots}$ and service times $\{\sigma_j\}_{j=1,2,\ldots}$ are fixed sequences of non-negative real numbers. Then we have the following lemma.

Lemma 79. *Let $u = (u_1, u_2, \ldots)$ and $v = (v_1, v_2, \ldots)$ be routing sequences of zeros and ones and suppose that $u \preceq v$. Then*

$$W_v(j) + \lambda_v(j) \leq W_u(j) + \lambda_u(j) \text{ for every } j \in \mathbb{N}. \tag{13.6}$$

Note that $W_u(j) + \lambda_u(j)$ is the moment that the server starts serving the j-th customer that is routed to the server according to u. So, Lemma 79 states that if $u \preceq v$ then the moment when the service of the j-th customer that is routed to the server starts, is for v not later than for u. This can be proved by induction to j.

For every $j \in \mathbb{N}$ we have that $\lambda_u(j) - \lambda_v(j) = \sum_{i=\nu_v(j)}^{\nu_u(j)-1} \delta_i$. This is a sum of interarrival times δ_i, where the number of terms in the sum is $\nu_u(j) - \nu_v(j)$. Therefore, if we put $N_{uv}(m) = \sum_{j=1}^{m}(\nu_u(j) - \nu_v(j))$ for $m = 1, 2, \ldots$ then we have the following Corollary by Lemma 79.

Corollary 95 *Let u and v be routing sequences with $u \preceq v$. For every $m \in \mathbb{N}$ we have that*

$$\sum_{j=1}^{m} W_v(j) \le \sum_{j=1}^{m} W_u(j) + \sum_{j=1}^{m} \sum_{i=\nu_v(j)}^{\nu_u(j)-1} \delta_i. \tag{13.7}$$

In the double sum the number of terms is $N_{uv}(m)$.

It can be shown that if $u = (u')^\infty$ and $v = (v')^\infty$ for some $u', v' \in \mathcal{P}(T, k)$ and $\omega(T, k) \preceq u' \preceq v'$ then

$$\lim_{m \to \infty} \frac{N_{uv}(m)}{m} = \frac{T}{k} \cdot (\overline{I}(v) - \overline{I}(u)). \tag{13.8}$$

Let $u \in \mathcal{P}(d)$ be as in Theorem 94. Then by Theorem 85 there exist $T, k \in \mathbb{N}$ with $d = \frac{k}{T}$ and $u'' \in \mathcal{P}(T, k)$ such that $u' := (u'')^\infty$ is a representative of u and $\omega' \preceq u'$ where $\omega' = \omega(T, k)^\infty$ is a representative of $\omega(d)$. By Theorem 93 we have that both $\overline{W}(u) = \overline{W}(u')$ and $\overline{W}(\omega) = \overline{W}(\omega')$ exist. We make a coupling between the interarrival times for u' and ω' and we also do this for the service times. After the coupling we can apply Corollary 95 and (13.8) and it can be shown that

$$\overline{W}(u') - \overline{W}(\omega') \le \frac{\delta \cdot T}{k} \cdot (\overline{I}(u) - \overline{I}(\omega)) = \frac{\delta}{d} \cdot I(u),$$

·which proves Theorem 94.

Remark 32. From this proof it follows that it is possible to refine Theorem 94 as follows: Let $u, v \in \mathcal{P}(d)$ be as in Theorem 94 and let ω' be a representative of $\omega(d)$ as above. Then, if there exist representatives u' and v' of u and v respectively such that $\omega' \preceq u' \preceq v'$ (The reader can check that this is equivalent to the following: there exist $T, k \in \mathbb{N}$ and period cycles u'' and v'' of u and v respectively such that $d = \frac{k}{T}$, $u'', v'' \in \mathcal{P}(T, k)$ and $u'' \preceq_{\overline{g}} v''$) it holds that

$$\overline{W}(v) - \overline{W}(u) \le \frac{\delta}{d}(\overline{I}(v) - \overline{I}(u)). \tag{13.9}$$

The condition $\rho < 1$ in Theorem 94 is just necessary to apply Theorem 93. However, if both the interarrival times and the service times are deterministic then Theorem 93 also holds if $\rho = 1$ (see [112]). Thus in that case Theorem 94 also holds if $\rho = 1$ and in fact we have for that case the following stronger result from which it follows that the bound of Theorem 94 is tight.

Theorem 96 *Suppose that the interarrival times and service times are deterministic and equal to δ respectively σ. Let $u \in \mathcal{P}(d)$ for some $d \in \mathbb{Q}$ such that $\delta = d\sigma$, hence $\rho = 1$, and let $\omega = \omega(d)$ be the regular sequence of density d. Then*

$$\overline{W}(u) = \overline{W}(\omega) + \frac{\delta}{d} \cdot \overline{I}(u). \tag{13.10}$$

Outline of the proof. Let ω' be a representative of $\omega(d)$ as above and let u' be a representative of u such that $\omega' \preceq u'$. Then it can be shown that $W_{u'}(j) + \lambda_{u'}(j) = W_{\omega'}(j) + \lambda_{\omega'}(j) = (j-1)\sigma$ for every $j \in \mathbb{N}$. Hence we have for every $m \in \mathbb{N}$ that (13.7) holds with equality for u' and ω' with $\delta_i = \delta$ for $i = 1, 2, \ldots$. From this it follows that (13.5) also holds with equality.

Remark 33. The upper bound can be generalized to a sequence of multimodular functions satisfying the relations of Section 1.3. Indeed, it follows from the results in the Section 1.3 that the regular or bracket sequence is the minimal admission sequence. If the density is rational then the corresponding regular admission sequence is periodic. In order to obtain upper bounds as in Theorem 94 and Theorem 99 we need a generalization of Lemma 79, especially the relation (13.6). It is easily seen that for multimodular functions f_k corresponding to $W_u(j)$ it suffices that

$$f_k(a_1, \ldots, a_l + 1, a_{l+1} - 1, \ldots, a_k) \leq f_k(a_1, \ldots, a_l, a_{l+1}, \ldots, a_k) + \delta,$$

for $1 \leq l \leq k - 1$ and some constant δ. Clearly, also other multimodular models treated in this monograph do satisfy this relation e.g. the models of Chapter 9 .

13.6 Routing to Parallel Queues

In this section we derive upper bounds for the expected average waiting time for routing arriving customers to $N \geq 2$ parallel queues according to some periodic routing policy $U = (U_1, U_2, \ldots)$. The routing sequence $U = (U_1, U_2, \ldots)$ can be seen as an $\mathbb{N}$ - word on the alphabet $\{1, 2, \ldots, N\}$. Further to every letter $i \in \{1, 2, \ldots, N\}$ corresponds a sequence of zeros and ones $u^i = (u^i_1, u^i_2, \ldots)$ via the support of the letter, i.e. $u^i_n = 1$ if and only if $U_n = i$. We call u^i the routing or splitting sequence for queue i (see Chapter 6). Extending the notion of unbalance for periodic sequences of zeros and ones we define a (total) unbalance $O(U)$ for periodic integer sequences and corresponding words on some finite alphabet. Then we extend the result of Theorem 94 for one queue to multiple queues by using the unbalance $O(U)$ of U.

Let $d_1, \ldots, d_N \in \mathbb{Q}_{>0}$ with $\sum_{i=1}^{N} d_i = 1$. We denote by $\mathcal{Q}(d_1, d_2, \ldots, d_N)$ all the infinite periodic words U on the alphabet $\{1, 2, \ldots, N\}$ for which $u^i \in$

$\mathcal{P}(d_i)$ for $i = 1, 2, \ldots, N$. Further for $T, k_1, k_2, \ldots, k_N \in \mathbb{N}$ with $\sum_{i=1}^{N} k_i = T$ we denote by $\mathcal{Q}(\{T\}, k_1, k_2, \ldots, k_N)$ all the N-words on the alphabet $\{1, 2, \ldots, N\}$ for which every subword of length T contains exactly k_i letters i for $i = 1, 2, \ldots, N$. Note that $\mathcal{Q}(\{T\}, k_1, k_2, \ldots, k_N) \subseteq \mathcal{Q}(\frac{k_1}{T}, \frac{k_2}{T}, \ldots, \frac{k_N}{T})$. For example if $U = (1, 2, 1, 2, 1, 3, 1, 2, 1, 1, 2, 3)^{\infty}$ then $U \in \mathcal{Q}(\{12\}, 6, 4, 2) \subseteq \mathcal{Q}(\frac{1}{2}, \frac{1}{3}, \frac{1}{6})$. Further $u^1 = (1, 0, 1, 0, 1, 0, 1, 0, 1, 1, 0, 0)^{\infty} \in \mathcal{P}(\frac{1}{2})$, $u^2 = (0, 1, 0, 1, 0, 0, 0, 1, 0, 0, 1, 0)^{\infty} \in \mathcal{P}(\frac{1}{3})$ and $u^3 = (0, 0, 0, 0, 0, 1)^{\infty} \in \mathcal{P}(\frac{1}{6})$.

Let ψ be a routing policy and U the corresponding routing sequence. Then for $t \in \mathbb{N}$ we define $W(t) = W_{\psi}(t)$ as the waiting time of the t-th arriving customer if policy ψ is applied, which is the remaining workload for server U_t at the moment that the t-th customer arrives. We assume that all the servers $i \in \{1, 2, \ldots, N\}$ are empty at $T_1 = 0$ and thus $W(t) = 0$ if $t = \nu_{u^i}(1)$ for some $i \in \{1, 2, \ldots, N\}$. If the interarrival times and service times of the various servers are random variables then we say that $\overline{W}(\psi) = \overline{W}(U)$ is the almost sure long-run average waiting time of the arriving customers routed according to policy ψ if $\lim_{\tau \to \infty} \frac{1}{\tau} \cdot \sum_{t=1}^{\tau} W_{\psi}(t) = \overline{W}(\psi)$ with probability one. Let $\{\delta_i\}$ be the sequence of interarrival times and let $\{\sigma_j^i\}$ be the sequence of service times of server i for $i = 1, 2, \ldots, N$, i.e σ_j^i is the service time of the j-th customer that is routed to server i. We define $\overline{W}^i(\psi) = \overline{W}^i(u^i)$ as the almost sure long-run average waiting time of customers routed to server i if policy ψ is applied in the same way as we did in the previous section. The only differences are that routing sequence u is replaced by routing sequence u^i and the sequence of service times $\{\sigma_j\}$ is replaced with the sequence of service times $\{\sigma_j^i\}$. From Theorem 93 we obtain the following theorem.

Theorem 97 *Suppose that the interarrival times $\{\delta_i\}$ of customers arriving at the system are i.i.d. random variables with mean δ and for every $i \in \{1, 2, \ldots, N\}$ the service times $\{\sigma_j^i\}$ are i.i.d random variables with mean σ_i, and independent of the interarrival times. Let ψ be a routing policy that corresponds to some word $U \in \mathcal{Q}(d_1, d_2, \ldots, d_N)$ such that $\rho_i := \frac{\sigma_i}{\delta} \cdot d_i < 1$ for $i = 1, 2, \ldots, N$. Then $\overline{W}(\psi)$ exists and is finite. Moreover $\overline{W}(\psi) = \sum_{i=1}^{n} d_i \cdot \overline{W}^i(\psi)$.*

Definition 98 *Let $U \in \mathcal{Q}(d_1, d_2, \ldots, d_N)$ for some $d_i \in \mathbb{Q}_{>0}$ with $\sum_{i=1}^{N} d_i = 1$. Then we define the (total) unbalance of U as*

$$O(U) := \sum_{i=1}^{N} \overline{I}(u^i). \tag{13.11}$$

If it is clear from the context what is meant we just say unbalance and not total unbalance. Let the interarrival times $\{\delta_i\}$ and the service times $\{\sigma_j^i\}$ for $i = 1, 2, \ldots, N$ be as in Theorem 97. Further let $d_i \in \mathbb{Q}_{>0}$ for $i = 1, 2, \ldots, N$ with $\sum_{i=1}^{N} d_i = 1$. Recall that $\omega(d_i)$ is the regular sequence of density d_i. If $\overline{W}^i(\omega(d_i))$ exists and is finite for $i = 1, 2, \ldots, N$ then we put

$$\widetilde{R} = \widetilde{R}(d_1, d_2, \ldots, d_N) \overset{\text{def}}{=} \sum_{i=1}^{N} d_i \cdot \overline{W}^i(\omega(d_i)). \tag{13.12}$$

Remark 34. If the interarrival times and service times are random variables then for the existence of $\widetilde{R}$ it suffices that $\rho_i < 1$ for $i = 1, 2, \ldots, N$, while if the interarrival times and service times are deterministic then it suffices that $\rho_i \leq 1$ for $i = 1, 2, \ldots, N$. It is possible to extend the definition of $\widetilde{R}$ and some of the results to the case of irrational d_i. In general $\widetilde{R}$ depends on the distribution of the interarrival times and service times and in some cases it is possible to compute $\widetilde{R}$ explicitly. See Chapter 8 and [112] for an exact computation in case of deterministic interarrival and service times and see [101] for computations and bounds in general.

Combining Theorem 94 and Theorem 97 we obtain the following theorem.

Theorem 99 *Let the interarrival times $\{\delta_i\}$ and the service times $\{\sigma_j^i\}$ for $i = 1, 2, \ldots, N$ be as in Theorem 97. Let $d_i \in \mathbb{Q}_{>0}$ for $i = 1, 2, \ldots, N$ with $\sum_{i=1}^{N} d_i = 1$. Suppose that a routing policy ψ is applied with corresponding word $U \in \mathcal{Q}(d_1, d_2, \ldots, d_N)$ and $\rho_i < 1$ for $i = 1, 2, \ldots, N$. Then*

$$\overline{W}(\psi) - \widetilde{R} \leq \delta \cdot O(U).$$

From Theorem 26 we have that $\overline{W}(\psi) \geq \widetilde{R}$, which is the lower bound obtained by replacing the routing sequence to queue i by the regular routing sequence with the same density for any queue i. Hence we have the following bounds.

$$\widetilde{R} \leq \overline{W}(\psi) \leq \widetilde{R} + \delta \cdot O(U). \tag{13.13}$$

Remark 35. Suppose that we have a queueing system where the arrivals are according to a Poisson process with parameter λ. Suppose that the service times of server i are exponentially distributed with parameter μ_i. If d_i, the fraction of jobs that is routed to server i, equals $\frac{1}{q_i}$ for some $q_i \in \mathbb{N}$, then $\overline{W}^i(\omega(d_i))$ can be calculated in the following way. For the routing sequence $\omega(d_i)$ we have that among every q_i arriving jobs exactly one job is routed to server i. So, the interarrival times at server i consist of q_i Poisson arrivals with parameter λ. Hence the interarrival times for the queue of server i are Erlang distributed, namely according to an $E_\lambda^{q_i}$ distribution. Thus $\overline{W}^i(\omega(d_i))$ is the same as the average waiting time for a $E_\lambda^{q_i}/M/1$ queue, where the parameter of the service times is μ_i. So (see [56]) if $x_i \in (0, 1)$ is a solution of the equation

$$x = (\frac{\lambda}{\lambda + \mu_i - \mu_i \cdot x})^{q_i}, \tag{13.14}$$

then

$$\overline{W}^i(\omega(d_i)) = \frac{x_i}{\mu_i \cdot (1 - x_i)}. \tag{13.15}$$

In the following example we have calculated $\overline{W}^i(\omega(d_i))$ for $i = 1, 2, 3, 4$ by applying (13.15). Further we have explicitly calculated the lower bound $\widetilde{R}$ and upper bound $\widetilde{R} + \delta \cdot O(U)$ of (13.13) for $\overline{W}(\psi)$, where U is the word corresponding to the applied routing policy ψ.

Example 10. We consider a queueing system with 4 parallel servers where the arrivals are according to a Poisson process with parameter $\lambda = 11$. Hence for the mean interarrival time δ we have that $\delta = \frac{1}{11}$. The arriving jobs are routed to the servers according to the policy ψ that corresponds to the word

$$U = (1, 2, 3, 1, 4, 2, 1, 3, 1, 2, 4, 3)^{\infty} \in \mathcal{Q}(\{12\}, 4, 3, 3, 2) \subseteq \mathcal{Q}(\tfrac{1}{3}, \tfrac{1}{4}, \tfrac{1}{4}, \tfrac{1}{6}).$$

Then we have that $\overline{I}(u^1) = \frac{1}{12}$, $\overline{I}(u^2) = 0$, $\overline{I}(u^3) = \frac{1}{12}$, $\overline{I}(u^4) = 0$ and thus $O(U) = \frac{1}{6}$. For every $i \in \{1, 2, 3, 4\}$ the service times are exponentially distributed with parameter μ_i and $\mu_1 = 4$, $\mu_2 = \mu_3 = 3$ and $\mu_4 = 2$. Then we find by (13.15) that $\overline{W}^1(\omega(\tfrac{1}{3})) = 1.7792$ (rounded to 4 decimals) and thus by Corollary 94 we have that $\overline{W}^1(u^1) \leq 1.7792 + 3 \cdot \frac{1}{11} \cdot \frac{1}{12} = 1.8019$. Similarly we have that $\overline{W}^2(\omega(\tfrac{1}{4})) = \overline{W}^2(u^2) = 2.2105$, $\overline{W}^3(\omega(\tfrac{1}{4})) = 2.2105$, $\overline{W}^3(u^3) \leq 2.2408$ and $\overline{W}^4(\omega(\tfrac{1}{6})) = \overline{W}^4(u^4) = 3.0732$. Hence by (13.12) we have that

$$\widetilde{R} = \frac{1}{3} \cdot \overline{W}^1(\omega(\tfrac{1}{3})) + \frac{1}{4} \cdot \overline{W}^2(\omega(\tfrac{1}{4})) + \frac{1}{4} \cdot \overline{W}^3(\omega(\tfrac{1}{4})) + \frac{1}{6} \cdot \overline{W}^4(\omega(\tfrac{1}{6})) = 2.2105$$

and $\widetilde{R} + \delta \cdot O(U) = 2.2105 + \frac{1}{11} \cdot \frac{1}{6} = 2.2257$. So, by (13.13) we have that $2.2105 \leq \overline{W}(\psi) \leq 2.2257$.

The following theorem provides conditions under which the right-hand side inequality of (13.13) actually holds with equality.

Theorem 100 *Let the interarrival times $\{\delta_i\}$ be deterministic equal to δ and the service times $\{\sigma_j^i\}$ be deterministic equal to σ_i for $i = 1, 2, \ldots, N$. Let $d_i \in \mathbb{Q}_{>0}$ for $i = 1, 2, \ldots, N$ with $\sum_{i=1}^{N} d_i = 1$. Let $p_i, q_i \in \mathbb{N}$ be such that $d_i = \frac{p_i}{q_i}$ with $\gcd(p_i, q_i) = 1$ for $i = 1, 2, \ldots, N$. Suppose that a routing policy ψ is applied with corresponding word $U \in \mathcal{Q}(d_1, d_2, \ldots, d_N)$ and $\rho_i = 1$ for $i = 1, 2, \ldots, N$. Then*

$$\overline{W}(\psi) = \widetilde{R} + \delta \cdot O(U) = \delta \cdot \left(\frac{1}{2} - \sum_{i=1}^{N} \frac{1}{2q_i} + O(U) \right).$$

By combining Theorem 96 and Theorem 97 we get the first equality of Theorem 100. See [112] for the second equality which follows from the computation of $\widetilde{R}$ for this case.

Example 11. We consider a queueing system with 3 parallel servers where the interarrival times are deterministic and equal to $\delta = 3$. The arriving jobs are routed to the servers according to the policy ψ that corresponds to the word

$$U = (1, 2, 1, 2, 1, 3, 1, 2, 1, 3)^\infty \in \mathcal{Q}(\{10\}, 5, 3, 2) \subseteq \mathcal{Q}(\frac{1}{2}, \frac{3}{10}, \frac{1}{5}).$$

All the service times are deterministic and for server 1 they are equal to $\sigma_1 = 6$, for server 2 equal to $\sigma_2 = 10$ and for server 3 equal to $\sigma_3 = 15$. Hence $\rho_i = 1$ for $i = 1, 2, 3$. Further $\overline{I}(u^1) = 0$, $\overline{I}(u^2) = \frac{1}{10}$, $\overline{I}(u^3) = \frac{1}{10}$ and thus $O(U) = \frac{1}{5}$. So, according to Theorem 100 we have that

$$\overline{W}(\psi) = 3 \cdot \left(\frac{1}{2} - (\frac{1}{4} + \frac{1}{20} + \frac{1}{10}) + \frac{1}{5} \right) = \frac{9}{10}$$

which can be checked by direct calculation.

14 Regular Ordering

14.1 Introduction

As mentioned before, the control of a stream of incoming customers into several queues in parallel is a difficult problem. When the number of queues is not larger than two, then the problem has been solved in several cases. In the case with full information on the state, the optimal policy has switching curves and can be computed using dynamic programming (see [58]). When the system is controled when no information is available, the optimal policy is a *Sturmian sequence* as shown in Chapter 6 and the exact computation of the optimal policy has been done in the deterministic case in Chapter 7. However, when the number of queues in parallel is larger than two, then the problem becomes more difficult. One of the reasons for this "three queue gap" is that Sturmian sequences in dimension three or more do not exist in most cases (see Chapter 2). The only case where this problem has been solved so far is when the system is fully loaded and deterministic (see [112]).

Here, we will deal with the problem of routing customers in several queues in parallel using a softer approach. As already presented in [9], we will not try to compute the optimal policy, which seems to be a very hard task. Instead, we will introduce a partial order on the routing sequences called the *regular ordering*. The main result of the paper is the following:

If the routing sequence s is more regular than the routing sequence u', then the maximal waiting times under s are smaller than the maximal waiting times under u', for the stochastic order.

This statement deserves several comments.

First, the notion of regularity is close the notion of *balance* (see Chapter 2) in the sense that a balanced sequence is the most regular sequence possible (see Section 14.2.2). An integer sequence is balanced if its partial sums over two arbitrary windows of the same length differs by at most one. They have been extensively investigated in the past [91, 55] and lead to fruitful results in several fields such as theoretical physics [99], combinatorics [86, 29] as well as control theory [59, 47] and Part II of this monograph using their close relation with bracket sequences.

In the previous chapters, it is shown under rather general assumptions that balanced sequences minimize several cost functions in discrete event systems. This result relies on the following theorem.

If $f_n : \mathbb{Z}^n \to \mathbb{R}$ are multimodular functions, then the Cezaro limit $\lim_{N \to \infty} \frac{1}{N} \sum_{n=0}^{N} f_n(u_1, \cdots, u_n)$ *is minimized by a balanced sequence.*

Second, majorization and Schur convex functions seem to be the right notions to use when comparing the dispersion of finite sequences [88]. However, majorization does not take into account the order of elements of the sequences to be compared since it is left invariant by the group of all permutations.

It is often desirable to take the exact positions into account. For many systems, the input sequence $(1,1,1,0,0,0)$ certainly induces a different behavior than the input sequence $(1,0,1,0,1,0)$, which looks more balanced. However, these two sequences cannot be compared by using the classical majorization technique because the former is a permutation of the latter. To overcome this difficulty, we introduce the notion of gap sequences (similar to partial sums, already used in [36]) that takes into account the exact positions in the sequence. This will narrow the gap between the notion of Schur convexity and multimodularity as shown in Appendix 14.6.

Finally, the usual performance measure in queuing networks is the average waiting time. However, in communication models where real time constraints are important (such as voice and video traffic) the maximal waiting time (or sojourn time) are more important than the average. In Section 14.3, we show that for FIFO stochastic event graphs, the maximal waiting time (or sojourn time) is regular preserving. As for computational issues, finding good allocation patterns (*i.e.* very regular sequences), is possible via a mathematical programming problem with convex objective functions as in Corollary 13. A similar procedure for minimizing the average waiting time has been used by Combé and Boxma [43] and in [36] for different performance measures.

The rest of the chapter is organized as follows. In Section 14.2, we introduce the notion of gap sequences and balanced sequences and the regular ordering.

In Section 14.3, we introduce a model of a controlled queue with periodic inter-arrival sequences and stationary service times. We show that the maximum waiting time is regular preserving with respect to the inter-arrival time as well as with respect to the service times. Section 14.3.6 extends the result to a routing problem between several queues. Finally we show in Section 14.4, that the transmission rate on a link with redundancy is also regular preserving.

14.2 Preliminaries

This section is devoted to the definition of gap sequences, balanced sequences and to the introduction of the notion of regularity.

14.2.1 Gap Sequences

We again use the notation $\mathbb{N} = \{0, 1, 2, \cdots\}$ for the set of all non-negative integers and we consider the set $P(T, n)$ 2]P@$P(T, n)$ of all non-negative integer sequences $u = (u_0, u_1, \cdots, u_{T-1})$ of size T which sum exactly to n. Let $u \in P(T, n)$. We call $(p_u(0), p_u(1), \cdots, p_u(n))$ the positions in u of all its partial sums.

$$p_u(0) \stackrel{\text{def}}{=} 0 \text{ and } p_u(i) \stackrel{\text{def}}{=} \inf\{j | \sum_{k=0}^{j} u_k \geq i\}, \quad \forall i \geq 1.$$

We define the vector $d^i(u)$ of the gaps of order $i \geq 1$ in the sequence u, by

$$\forall 1 \leq j \leq n, \quad d_j^i(u) \stackrel{\text{def}}{=} p_u((j+i) \bmod n) - p_u(j). \tag{14.1}$$

When the sequence u is in $\{0, 1\}^T$, then the gaps are never null and measure the size of the number of elements in the sequence u between the ones. This is illustrated in Figure 14.1.

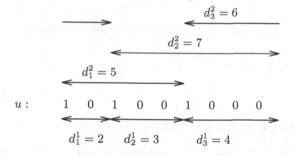

Fig. 14.1. Gaps of order one and two for a sequence u in $P(9, 3)$.

14.2.2 Balanced Sequences

We repeat here some of the material presented in Chapter 2.

Definition 22. *A sequence u is balanced if for all j, k and n,*

$$\left| \sum_{i=j}^{j+n} u_i - \sum_{i=k}^{k+n} u_i \right| \leq 1.$$

A finite sequence u is balanced if the infinite sequence u^∞ is balanced.

Theorem 101 ([91]). *If a sequence u in $P(T, n)$ is balanced, then there exists $\theta \in [0, 1]$ such that for all $j \geq 0$,*

$$u_j = \lfloor (j+1)\frac{n}{T} + \theta \rfloor - \lfloor j\frac{n}{T} + \theta \rfloor.$$

Proof. The proof follows from Chapter 2 adapted to the finite case.   □

If u is a balanced sequence then the associated gap sequences have the following property.

Lemma 80. *If u is a balanced sequence in $P(T, n)$, then for all j, i, $d_j^i(u) \in \{\lfloor \frac{iT}{n} \rfloor, \lceil \frac{iT}{n} \rceil\}$.*

Proof. By definition (everything is implicitely done modulo n) $d_j^i = p(i + j) - p(j)$. By definition of $p(j)$, we have: $p(j) = \inf\{k | \lfloor (k+1)\frac{n}{T} + \theta \rfloor \geq j\}$. Therefore, $\lfloor (p(j) + 1)\frac{n}{T} + \theta \rfloor \geq j$ and $\lfloor p(j)\frac{n}{T} + \theta \rfloor < j$. This implies that

$$j - \frac{n}{T} \leq p(j)\frac{n}{T} + \theta < j.$$

Similarly, we have:

$$j + i - \frac{n}{T} \leq p(j+i)\frac{n}{T} + \theta < j + i.$$

As for d_j^i, we get by subtracting the two previous equations

$$\frac{iT}{n} - 1 < d_j^i < \frac{iT}{n} + 1.$$

□

The following lemma is the converse of Lemma 80.

Lemma 81. *Let u be a sequence such that for all i, all the gaps of order i, d_j^i, differ by at most one. Then, u is balanced.*

Proof. Assume that u is not balanced. There exist two intervals of the same length (say l) w_1 and w_2 such that the sum in w_1, denoted n_1 is larger than the sum in w_2, denoted n_2, plus two: $n_1 \geq n_2 + 2$. This means that there exists j_1 and j_2 such that $d_{j_1}^{n_1} \leq l$ and $d_{j_2}^{n_1} \geq l + 2$.   □

Lemma 82. *A sequence $u \in P(T, n)$ is balanced if and only if $d^1(u)$ is balanced.*

Proof. The "only if" assertion: we consider the sequence $d^1(u)$. Let $w_1 = \{i, \cdots, i+k\}$ and $w_2 = \{j, \cdots, j+k\}$ be two windows of length k. Then the sums over w_1 of the sequence $d^1(u)$ are of the form $d_i^1(u) + \cdots + d_{i+k}^1(u) = d_i^{k-1}(u)$ and the sums over w_2 are $d_j^1(u) + \cdots + d_{j+k}^1(u) = d_j^{k-1}(u)$. By Lemma 80, then these two quantities differ by at most one.

The "if" assertion: $d^1(u)$ is balanced means that there exists θ such that for all j, $d_j^1(u) = \lfloor (j+1)\frac{T}{n} + \theta \rfloor - \lfloor j\frac{T}{n} + \theta \rfloor$. Therefore, $d_j^i = \lfloor (j+i)\frac{T}{n} + \theta \rfloor - \lfloor j\frac{T}{n} + \theta \rfloor$. This means that

$$\frac{iT}{n} - 1 < d_j^i < \frac{iT}{n} + 1.$$

Therefore, for any order i, the gaps of order i differ by at most one. By applying Lemma 81, u is balanced.   □

Remark 36. Note however, that the gaps of order higher than one in a balanced sequence may not be balanced. For example, if $d^1(u) = (2, 3, 3, 2, 3, 3, 3)$ (which is balanced), then $d^2(u) = (5, 6, 5, 5, 6, 6, 5)$ which is not balanced because it contains two consecutive 5 (which sum up to 10) and also two consecutive 6 (which sum up to 12).

14.2.3 Regularity and Schur Convexity

If $x \in \mathbb{R}^n$, then we denote by $(x_{[1]}, x_{[2]}, \cdots, x_{[n]})$ the components of x arranged in decreasing order. For $x, y \in \mathbb{R}^n$, we say that x (resp. strictly) majorizes y (denoted $x \succeq y$) (resp. $x \succ y$) if for all $1 \leq m \leq n - 1$, then

$$\sum_{i=1}^{m} x_{[i]} \geq \sum_{i=1}^{m} y_{[i]} \quad \text{and} \quad \sum_{i=1}^{n} x_{[i]} = \sum_{i=1}^{n} y_{[i]},$$

(resp. with strict inequality for at least one m).

A function $f : \mathbb{R}^n \to \mathbb{R}$ is (resp. strictly) *Schur convex* if $x \succeq y \Rightarrow f(x) \geq f(y)$ (resp. $x \succ y \Rightarrow f(x) > f(y)$). For more details on the theory of majorization, see [88].

Definition 23. *If $u, u' \in P(T, n)$, then u' is more regular than u (denoted $u' \trianglelefteq s$) if for all order i, $d^i(u') \preceq d^i(u)$.*

Lemma 83. *The balanced sequences are the minimal sequences in $P(T, n)$ for the regularity order. Moreover, if u is balanced, then for all $u' \in P(T, n)$, $s \trianglelefteq u'$.*

Proof. This result is a consequence of Lemma 81. Indeed, if n divides iT, then for a balanced sequence σ, $d_j^i(\sigma) = \frac{iT}{n}$, which is minimal. If n does not divide iT, then for the balanced sequence, $d_j^i(\sigma) \in \{\lfloor \frac{iT}{n} \rfloor, \lceil \frac{iT}{n} \rceil\}$ and is also minimal.   □

Balanced sequences in $P(T, n)$ will also be called the "most regular" sequences in $P(T, n)$ in the following.

Example 12. Let us compare three sequences in $P(6, 3)$: $w_1 = (1, 1, 1, 0, 0, 0)$, $w_2 = (1, 1, 0, 1, 0, 0)$ and $w_3 = (1, 0, 1, 0, 1, 0)$, using the regular ordering. The respective gap sequences are:

$$d^1(w_1) = (1, 1, 4) \; d^1(w_2) = (1, 2, 3) \; d^1(w_3) = (2, 2, 2),$$
$$d^2(w_1) = (2, 5, 5) \; d^2(w_2) = (3, 5, 4) \; d^2(w_3) = (4, 4, 4),$$
$$d^3(w_1) = (6, 6, 6) \; d^3(w_2) = (6, 6, 6) \; d^3(w_3) = (6, 6, 6).$$

Using majorization, it is not difficult to check that

$$d^1(w_1) \succ d^1(w_2) \succ d^1(w_3),$$
$$d^2(w_1) \succ d^2(w_2) \succ d^1(w_3),$$
$$d^3(w_1) = d^3(w_2) = d^1(w_3).$$

According to the definition of the regular ordering, this means that $w_3 \trianglelefteq w_2 \trianglelefteq w_1$. Actually, w_3 is balanced and is the most regular sequence in $P(6, 3)$.

Definition 24 (regular-preserving functions). *A function $f : \mathbb{N}^T \to \mathbb{R}$ is regular-preserving (r.p.) if $x \trianglelefteq y \Rightarrow f(x) \leq f(y)$.*

Lemma 84. *A function $f : \mathbb{N}^T \to \mathbb{R}$ is regular-preserving if and only if it can be written under the form:*

$$f(u) = F(H_1(d^1(u)), H_2(d^2(u)), \cdots, H_n(d^n(u))),$$

where $F : \mathbb{R}^n \to \mathbb{R}$ is increasing in all coordinates and $H_i : \mathbb{N}^n \to \mathbb{R}, i = 1 \cdots, n$ are Schur convex.

Proof. The only if assertion is a direct consequence of Lemma 83. As for the if assertion, then f is necessarily Schur convex in all d^i. Now, the combination of all these arguments for all i shows that f is of the form given in the lemma. □

Note that if f is regular preserving on the set $P(T, n)$ then it reaches its minimal value for the balanced sequences.

Remark 37. (Regularity without using the gap sequences)
The definition of regularity, as well as regular preserving functions, can be done in a similar fashion by using the partial sum sequences of the original sequence u instead of the partial sums of its order one gap sequence.

All the results given in this chapter are still true in this new framework.

The passage from one framework to the other is simply done through Equation (14.1) and Lemma 91.

14.3 Application 1: Maximal Waiting Times in Networks

In this section, we show that the maximum waiting time in a queuing system with general stationary inter-arrival and service times is a regular preserving function.

14.3.1 The D/D/1 Model

First, we consider a slotted D/D/1 queue where the arrival sequence u is periodic with n customers arriving every T time slots (each of unit length). Therefore, $u \in P(T, n)^{\infty}$. By definition, the number of customers arriving at time i is w_i. The i-th customer arrives at time $p_i(u)$ and brings a workload of σ in the system.

We will also denote by $C \overset{\text{def}}{=} n\sigma$, the total load brought by n consecutive customers.

We finally require that the queue is initially empty and that the customers are served in a FIFO order (this last assumption can be relaxed to a non-idling server assumption in some cases).

Now, let us consider the waiting time W_i of the ith customer that enters the system. We have

$$W_i(u, \sigma) = \max_{j=1}^{i} \{ \sum_{k=j}^{i-1} (\sigma - d_k^1(u)) + \sigma \}.$$

We define $M(u, \sigma) \overset{\text{def}}{=} \max_i W_i(u, \sigma)$.

Theorem 102. *The function $M(u)$ is regular preserving.*

Proof. First, let us remark that if $C > T$, then $W_i \to \infty$ when $i \to \infty$. Therefore $M(u, \sigma) = \infty$. In the rest of the proof, we will only consider the case where $C \leq T$. In that case, note that after a transient period of length n, the waiting times are periodic: $W_{i+n} = W_n$, if $i \geq n$. Also note that since the system is initially empty, the maximum waiting time is not reached during the initial transient period. So we have

$$M(u, \sigma) = \max_{i=n+1}^{2n} \max_{j=i-n}^{i} \{\sum_{k=j}^{i-1}(\sigma - d_k^1(u)) + \sigma)\}, \tag{14.2}$$

$$= \max_{i=n+1}^{2n} \max_{j=i-n}^{i} \{(\sum_{k=j}^{i}\sigma) - d_j^{i-j}(u)\},$$

$$= \max_{i=n+1}^{2n} \max_{j=i-n}^{i} \{(\sum_{k=j}^{i}\sigma) + d_i^{n-i+j}(u) - T\}, \tag{14.3}$$

$$= \max_{i=1}^{n} \max_{j=1}^{n} \{(\sum_{k=j+i}^{n+i}\sigma) + d_i^j(u) - T\},$$

$$= \max_{j=1}^{n} \max_{i=1}^{n} \{(n - j + 1)\sigma + d_i^j(u) - T\}. \tag{14.4}$$

Equation (14.2) follows from the fact that $C \leq T$. Equation (14.3) follows from Equation (14.13). Changing the order of the two max operators yields Equation (14.4).

Now, the function $d^j \mapsto \max_{i=1}^{n}\{(n - j + 1)\sigma + d_i^j - T\}$ is Schur convex for each j and the function M is increasing in d^j for all j, and thus satisfies the conditions of Lemma 84. This means that M is regular preserving. $\square$

Corollary 12. *Among all inter-arrival sequences in $P(T, n)$, the balanced sequence minimizes the maximum waiting time in the system. Moreover, for two sequences $u, u' \in P(T, n)$, $u \trianglerighteq u' \Rightarrow M(u, \sigma) \geq M(u', \sigma)$.*

Remark 38. The function $M(u)$ is multimodular (see Appendix 14.6). This is a direct consequence of Corollary 13 there.

14.3.2 Characterization using regular preserving functions

The function M when $C = T$ plays an important role. Indeed, we have the following result giving yet another characterization of balanced sequences.

Theorem 103. *If $C = T$, and given $u \in P(T, n)$, if $M(u) \leq M(u')$ for all $u' \in P(T, n)$, then u is balanced.*

Proof. When $C = T$, then $\sigma = T/n$. Equation (14.4) becomes

$$M(u) = \max_{i=1}^{n} \max_{j=1}^{n}\{-jT/n + d_i^j(u)\} + T/n.$$

Note that jT/n is the average value of $d_i^j(u)$. Using Lemmas 80 and 81, u is balanced is equivalent to the fact that $M(u) < 1 + T/n$. $\square$

14.3.3 The average waiting time is not regular preserving

In this section, we will show that unlike for the maximal waiting time, the average waiting time in a stable D/D/1 queue is not regular preserving through an example.

Consider in $P(16, 5)$ the sequences u and u' with the respective one-order gaps,

$$d^1(u) = (4, 4, 3, 1, 4),$$
$$d^1(u') = (4, 4, 2, 2, 4).$$

It is immediate to check that $u \trianglerighteq u'$. However, if one computes the average waiting time in both cases, when the load brought by each customer is equal to 3, one gets:

$$\overline{W}(u) = \frac{18}{5} < \frac{19}{5} = \overline{W}(u').$$

Remark 39. The function $\overline{W}$ is multimodular (see theorem 73). Therefore, it is minimized by a balanced sequence in $P(16, 5)$: $\overline{W}(u) = 3$ if u is a sequence such that $d^1(u) = (3, 3, 3, 3, 4)$.

14.3.4 The G/G/1 Model

Here, we generalize the previous model. The deterministic arrival process is replaced by a stochastic process with stationary interarrival times: $\{\tau_i\}_{i \in \mathbb{N}}$. The load carried by the k-th customer is denoted by σ_k. The sequence $\{\sigma_k\}_{k \in \mathbb{N}}$, is stationary and independent of the arrival times. We denote by $\delta_i(u)$ the time elapsed between the arrival of customer i and customer $i + 1$ for sequence u,

$$\delta_i(u) = \sum_{j=p_u(i)+1}^{p_u(i+1)} \tau_j, \forall i = 0, \cdots, m - 1.$$

Using Lindley's equation, the waiting time of the i-th customer satisfies

$$W_i(u) = \max_{j=0}^{i} \left\{ \sum_{k=j}^{i-1} (\sigma_k - \delta_k(u)) + \sigma_i \right\}.$$

If we consider the maximal waiting time within on period, starting with customer $m + 1$, we define $\Phi_m^n(u) \stackrel{\text{def}}{=} \max_{i=m+1}^{n+m} W_i(u)$.

In a similar fashion as with the D/D/1 case, we define

$$\Phi_{r,m}^n(u) \stackrel{\text{def}}{=} \max_{i=m+1}^{n+m} \max_{j=i-r+1}^{i} \left\{ \sum_{k=j}^{i-1} (\sigma_k - \delta_k(u)) + \sigma_i \right\}.$$

Intuitively, $\Phi^n_{r,m}$ is constructed by going back in time for r customers only, while Φ^n_m goes back to time 0.

Note that if the queue has emptied during the last r arrivals, then both quantities coincide.

We assume that the queuing system is stable, that is $TE(\tau_1) > nE(\sigma_1)$ (this generalizes the previous assumption, $C \leq T$ to the stochastic case). Under this stability condition, the queue empties almost surely in finite time. Therefore, taking r large enough will ensure that $\Phi^n_{r,m}$ and Φ^n_m will coincide. More precisely,

$$\left| \Phi^n_{an,an}(u) - \Phi^n_{an}(u) \right| \longrightarrow_{a\to\infty} 0, \quad a.s. \qquad (14.5)$$

Now, we have,

$$\Phi^n_{an,an}(u) = \max_{i=an+1}^{n+an} \max_{j=i-an+1}^{i} \left\{ \sum_{k=j}^{i-1} (\sigma_k - \delta_k(u)) + \sigma_i \right\},$$

$$= \max_{i=an+1}^{n+an} \max_{j=i-an+1}^{i} \left\{ \sum_{k=j}^{i} \sigma_k + \sum_{k=i-an}^{j-1} \delta_k(u) - \sum_{k=i-an}^{i-1} \delta_k(u) \right\},$$

$$= \max_{i=1}^{n} \max_{j=1}^{an} \left\{ \sum_{k=j+i}^{i+an} \sigma_k + \sum_{k=i}^{j+i-1} \delta_k(u) - \sum_{k=i}^{i+an-1} \delta_k(u) \right\}. \qquad (14.6)$$

Lemma 85. *Let us assume that u and u' are two sequences in $P(T,n)$ such that $u \preceq u'$. Then, $\Phi^n_{an,an}(u) \leq_{st} \Phi^n_{an,an}(u')$, where $\leq_{st}$ is the stochastic order between random variables.*

Proof. For a given pair (i,j) in $\{1, \cdots, n\} \times \{1, \cdots an\}$, we know by definition of the regular ordering that there exits $i' \in \{1, \cdots, n\}$ such that $d_i^j(u) \leq d_{i'}^j(u')$. Indeed, $d_i^j(u) = bT + d_i^r(u)$ where $bn + r = j$, $r < n$.

Now, we define a coupling of the service sequence as well as of the arrival sequence such that $\sigma'_{k+i'} = \sigma_{k+i}$ and $\tau'_{k+p_{u'}(i')} = \tau_{k+p_u(i)}$ for all $k \in \mathbb{N}$. Note that the coupled sequences have the same probability distributions as the original ones because of stationarity and independence between them and because the sequences σ and τ are independent of u and u'.

Using this coupling, and considering Equation (14.6), we have directly

$$\sum_{p=j+i'}^{i'+an} \sigma'_p = \sum_{p=j+i}^{i+an} \sigma_p.$$

Since during a periods, the total number of arrivals under u and u' is the same and using the coupling, we have

$$\sum_{k=i}^{i+an-1} \delta_k(u) = \sum_{p=p_u(i)+1}^{p_u(i+an)} \tau_p$$

$$= \sum_{p=p_u(i)+1}^{p_u(i)+Ta} \tau_p$$

$$= \sum_{p=p_{u'}(i')+1}^{p_{u'}(i')+Ta} \tau'_p$$

$$= \sum_{k=i'}^{i'+an-1} \delta'_k(u).$$

Finally, using the fact that $d_i^j(u) \le d_{i'}^j(u')$, we also get

$$\sum_{k=i}^{j+i-1} \delta_k(u) = \sum_{p=p_u(i)+1}^{p_u(i+j)} \tau_p$$

$$= \sum_{p=p_u(i)+1}^{p_u(i)+d_i^j(u)} \tau_p$$

$$\le \sum_{p=p_{u'}(i')+1}^{p_{u'}(i')+d_{i'}^j(u')} \tau'_k$$

$$= \sum_{k=i'}^{j+i'-1} \delta'_k(u').$$

Using these three relations,

$$\max_{j=1}^{an} \left\{ \sum_{k=j+i}^{i+an} \sigma_k + \sum_{k=i}^{j+i-1} \delta_k(u) - \sum_{k=i}^{i+an-1} \delta_k(u) \right\}$$

$$\le \max_{j=1}^{an} \left\{ \sum_{k=j+i'}^{i'+an} \sigma'_k + \sum_{k=i'}^{j+i'-1} \delta'_k(u') - \sum_{k=i'}^{i'+an-1} \delta'_k(u') \right\}.$$

This is true for all i, and therefore also for the maximum over all possible i from 1 to an. Using the fact that under the coupling, the sequences σ' and τ' are in a "typical" situation, we get $\Phi_{an,an}^n(u) \le_{st} \Phi_{an,an}^n(u')$.   □

Now, the stationary version of the queue is considered. Using a backward coupling argument and Lemma 85, we obtain the following theorem.

Theorem 104. *Let* $\mathbf{W}_i$ *be the stationary waiting time in a stable G/G/1 queue. Then for two sequences* $u, u' \in P(T, n)^\omega$ *with* $u \trianglelefteq u'$, *it holds that*

$$\max_{i=1}^{n} \mathbf{W}_i(u) \leq_{st} \max_{i=1}^{n} \mathbf{W}_i(u').$$

14.3.5 The Event Graph Model

We generalize the previous model to event graph models introduced in Chapter 3. We recall briefly how this generalizes the G/G/1 queue.

- *Networks assumptions.* We replace the single queue by a network made of Q servers (transitions) and P buffers (places) forming a FIFO stochastic event graph with a single entry.
- *Stochastic assumptions.* The service time at node j for the k-th customer is denoted by σ_k^j. The sequence $\{\sigma_k^j\}_{k \in \mathbb{N}, j \in Q}$ is stationary and independent of the arrival sequence. However no independence assumption among the service times is required.
- *Initial state assumptions.* We assume that initially, the event graph is input-deadlocked (this is a generalization of the empty queue assumption).

In Chapter 3 we showed that to any stochastic event graph, we can associate a family of matrices, $A(\ell)$, of size $Q \times Q$. The entry (i, j) in matrix $A(\ell)$ is the maximum over all the sums of the service times of the ℓ-th customer, σ_ℓ^k, in all the nodes k on any path from node j to node i with no initial tokens, except at node j, and is $-\infty$ otherwise.

let $X_q(n)$ be the epoch when the n-th customer completes its service in server q. Using the $(max, +)$ notation, where $\oplus$ stands for max and $\otimes$ for + (see Chapter 3 again), the vector $X(n) = (X_1(n), \cdots, X_Q(n))$ satisfies the following recurrence equation:

$$X(n) = A(n) \otimes X(n-1).$$

Let W_n be a vector , with its q-th component equal to: $(W_n)_q \stackrel{\text{def}}{=} X_q(n) - X_0(n)$. $(W_n)_q$ can be seen as the *traveling time* for customer n between its entrance in the system and its passage in server q.

W_n satisfies the following equation:

$$W_{n+1} = A(n) \otimes D(-\delta_n) \otimes W_n \oplus B(n+1), \tag{14.7}$$

where $D(h)$ is the diagonal matrix with h on the diagonal and $-\infty$ everywhere else, and $B(n)$ is a vector which describes the input connection.

Equation (14.7) can be seen as a vectorial form of the Lindley's equation. Its solution is similar to the solution of the scalar case.

$$W_n(u) = B(n) \oplus \bigoplus_{j=0}^{n-1} C_j(u),$$

with

$$C_j(u) = \bigotimes_{k=j}^{n-1} (A(k) \otimes D(-\delta_k(u))) \otimes B(n-j-1).$$

If we define as previously, $\Phi_m^n = \bigoplus_{i=m+1}^{m+n} \bigoplus_{j=0}^{i} C_j(u)$ and

$$\Phi_{r,m}^n(u) = \bigoplus_{i=m+1}^{m+n} \bigoplus_{j=i-r+1}^{i} \left(\bigotimes_{k=j}^{i-1} (A(k) \otimes D(-\delta_k(u))) \otimes B(i-j-1) \right),$$

we get by a similar technique as in the G/G/1 case (with a vectorial notation),

$$\left| \Phi_{an,an}^n(u) - \Phi_{an}^n(u) \right| \longrightarrow_{a \to \infty} 0, \quad a.s. \tag{14.8}$$

provided the event graph is stable (see [20]).

The quantity $\Phi_{an,an}^n(u)$ can be transformed using the fact that diagonal matrices commute with everything.

$$
\begin{aligned}
\Phi_{an,an}^n(u) &= \bigoplus_{i=an+1}^{an+n} \bigoplus_{j=i-an+1}^{i} \left(\bigotimes_{k=j}^{i-1} (A(k) \otimes B(i-j-1) \otimes D(-\delta_k(u))) \right) \\
&= \bigoplus_{i=an+1}^{an+n} \bigoplus_{j=i-an+1}^{i} \left(\bigotimes_{k=j}^{i-1} (A(k) \otimes B(i-j-1)) \otimes \bigotimes_{k=j}^{i-1} D(-\delta_k(u)) \right) \\
&= \bigoplus_{i=1}^{n} \bigoplus_{j=1}^{an} \left(\bigotimes_{k=j+i}^{i+an-1} (A(k) \otimes B(an-j-1)) \otimes \bigotimes_{k=j+i}^{i+an-1} D(-\delta_k(u)) \right) \\
&= \bigoplus_{i=1}^{n} \bigoplus_{j=1}^{an} \left(\bigotimes_{k=j+i}^{i+an-1} (A(k) \otimes B(an-j-1)) \otimes \bigotimes_{k=i}^{i+an-1} D(-\delta_k(u)) \otimes \right. \\
&\qquad\qquad \left. \bigotimes_{k=i}^{i+j-1} D(\delta_k(u)) \right).
\end{aligned}
\tag{14.9}
$$

Lemma 86. *Let us assume that u and u' are two sequences in $P(T,n)^\omega$ such that $u \trianglelefteq u'$. Then, $\Phi_{an,an}^n(u) \leq_{st} \Phi_{an,an}^n(u')$.*

Proof. The proof is similar to the single queue case, using a coupling argument and the definition of the regular ordering.

For a given pair (i,j) in $\{1, \cdots, n\} \times \{1, \cdots an\}$, we know by definition of the regular ordering that there exits $i' \in \{1, \cdots, n\}$ such that $d_i^j(u) \leq d_{i'}^j(u')$.

Now, we define a coupling of the service sequence such that for each queue q, $\sigma_{k+i'}^{'q} = \sigma_{k+i}^q$. As for the interarrival sequence, we set $\tau_{k+p_{u'}(i')}^{'} = \tau_{k+p_u(i)}$ for all $k \in \mathbb{N}$.

Note that the coupled sequences have the same probability distributions as the original ones because of stationarity and independence between them and because the sequences σ and τ are independent of u and u'.

Using this coupling, we have by definition of $A(k)$,

$$\bigotimes_{k=j+i}^{i+an-1} (A(k) \otimes B(an - j - 1)) = \bigotimes_{k=j+i'}^{i'+an-1} (A(k) \otimes B(an - j - 1)).$$

During a consecutive periods, the total number of arrivals with u and u' is the same. Therefore, we have using the fact that $D(-\delta_k(u))$ are diagonal matrices,

$$\bigotimes_{k=i}^{i+an-1} D(-\delta_k(u)) = \bigotimes_{p=p_u(i)+1}^{p_u(i+an)} D(-\tau_p)$$

$$= \bigotimes_{p=p_u(i)+1}^{p_u(i)+Ta} D(-\tau_p)$$

$$= \bigotimes_{p=p_{u'}(i')+1}^{p_{u'}(i')+Ta} D(-\tau'_p)$$

$$= \bigotimes_{k=i'}^{i'+an-1} D(-\delta'_k(u)).$$

As for the last term,

$$\bigotimes_{k=i}^{j+i-1} D(\delta_k(u)) = \bigotimes_{p=p_u(i)+1}^{p_u(i+j)} D(\tau_p)$$

$$= \bigotimes_{p=p_u(i)+1}^{p_u(i)+d_i^j(u)} D(\tau_p)$$

$$\leq \bigotimes_{p=p_{u'}(i')+1}^{p_{u'}(i')+d_{i'}^j(u')} D(\tau'_k)$$

$$= \bigotimes_{k=i'}^{j+i'-1} D(\delta'_k(u')).$$

These three relations are true for all i and j, therefore, considering Equation (14.9), $\Phi^n_{an,an}(u) \leq_{st} \Phi^n_{an,an}(u')$.   □

Finally, we obtain the following extension of Theorem 104.

Theorem 105. *In a stochastic event graph, the maximum traveling time $M(u)$ over one period of the admission sequence is a regular preserving function (in the stochastic sense).*

14.3.6 Routing Problem

Now, we consider a network of K identical queues in parallel. Customers enter a system composed by K queues in parallel. The routing of the customers is controlled by a sequence of vectors $\{W_i\}$, with $W_i \in \{0,1\}^K$ and $1 \leq i \leq T$: $W_i^j = 1$ means that the i-th customer is routed to queue j. Note that u is a feasible admission sequence as long as $\sum_j U_i^j = 1$ for all i.

Customers arrive in the system at each time unit and each customer brings a constant load of σ.

Figure 14.2 displays the model we are considering.

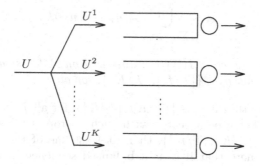

Fig. 14.2. Routing of customers in a multiple queue system

The Constrained Case We consider periodic routing policies U satisfying the following constraint:

$$\forall 1 \leq j \leq K, \quad \sum_i U_i^j = n_j, \tag{14.10}$$

where n_j are fixed with $\sum_j n_j = T$. In other words, $u^j \in P(T, n_j)^\infty$. We denote this set of policies by $\mathcal{A}(n_1, \cdots, n_K)$.

Now, we define $I(u)$ to be the maximum waiting time of any customer in the system, for $U \in \mathcal{A}(n_1, \cdots, n_K)$. $I(U) \overset{\text{def}}{=} \max_{j=1}^K M_j(U^j)$, where M_j is the maximum waiting time in queue j.

We say that the vector $(n_1, \cdots, n_K)$ is *balanceable* if there exists $U \in \mathcal{A}(n_1, \cdots, n_K)$ such that for all j, U^j is balanced. See Chapter 2 for a more detailed discussion on balanceable vectors. From Lemma 83 and 102, we get,

Lemma 87. *If $(n_1, \cdots, n_K)$ is balanceable then $I(U)$ is minimized by a routing policy U such that each U^j is balanced.*

Proof. This result is a direct consequence of Lemmas 83 and 102. □

This result is similar to some extend to the results proved in Chapter 6 where multimodularity of the waiting times in each queue is used. However, we get a new result here which follows directly from the previous lemma and which holds for all $(n_1, \cdots, n_K)$ (*i.e.* not necessarily balanceable).

Theorem 106. *If $V, U \in \mathcal{A}(n_1, \cdots, n_K)$, such that for all j, $V^j \trianglerighteq U^j$, then $I(V) \geq I(U)$.*

The Unconstrained Case Now, we do not fix the number of customers sent to each queue.

We want to consider all admission sequences U in the set

$$\mathcal{A} = \bigcup_{\sum_j n_j = T} \mathcal{A}(n_1, \cdots, n_K).$$

Lemma 88. *The function $I(U)$ is minimized on $\mathcal{A}$ for some balanced $U \in \mathcal{A}(n_1, \cdots, n_K)$, with $n_j \in \{\lfloor T/K \rfloor, \lceil T/K \rceil\}$, for all j.*

Proof. First, note that if $n_j \in \{\lfloor T/K \rfloor, \lceil T/K \rceil\}$, for all j, then $(n_1, \cdots, n_K)$ is balanceable. Let b be a vector with such a property. Second, note that if not all $n_j \in \{\lfloor T/K \rfloor, \lceil T/K \rceil\}$, then at least one of them is larger than $\lceil T/K \rceil$. Finally, note that if U is a balanced sequence in $P(T, n)^\omega$ and if $U' \in P(T, n')^\omega$, with $n' \geq n$, then $M(U', \sigma) \geq M(U, \sigma)$.

Now, the proof goes like this. By Theorem 106, I is minimized on $\mathcal{A}(b)$ by a balanced sequence $U(b)$. Moreover, there is some j such that $I(U(b)) = M_j(U^j(b))$, with $n_j = \lceil T/K \rceil$. For any U which is not in $\mathcal{A}(b)$, then there exists k such that $n_k \geq \lceil T/K \rceil$.

Now, $I(U) \geq M_k(U^k) \geq M_j(U^j(b)) = I(U(b))$. This finishes the proof □

14.3.7 Computational Problems

For the constrained problem, we can compute the best sequence using a procedure similar to the one used in [36] with quadratic programming techniques.

As for the unconstrained case, from Lemma 88, it is easy to see that when $T/K \in \mathbb{N}$, then the best routing policy (which minimizes the function I) is the round robin policy.

However, when $T/K \notin \mathbb{N}$, then little was know before. Here, we can come up with the optimal routing sequence using the following procedure.

First, we compute the optimal sequence $b = (n_1, \cdots, n_k)$ using Lemma 88. This sequence is unique up to a permutation. Note that by symmetry of the cost function, all those permutation will perform the same. Moreover, since all balanced sequences with the same slope in the routing sequence are shifts of each other, then for all j such that $n_j = \lceil T/K \rceil$, all the waiting times $M_j(U^j(b))$ are equal.

The construction of $U^j(b)$ for any j can be done by any known method to construct balanced sequence, as for example, by using the bracket sequence formula given in Theorem 101.

14.3.8 A routing example

Let us assume that we want to route customers into three parallel event graphs E_1, E_2, E_3 with different service times. For some reason, the proportions of customers sent to E_1, E_2, E_3 must be $1/2, 1/3$ and $1/6$ respectively. This is a case where the routing policy is constrained to stay in the set $\mathcal{A}(3, 2, 1)$ with a period equal to $T = 6$.

The proportions $1/2, 1/3, 1/6$ are not balanceable. Therefore, it is not possible to find a routing sequence which is balanced for each event graph. However, we can use the regular ordering to compare several policies.

Let $U = (1, 1, 1, 2, 3, 2)$ and $U' = (1, 1, 2, 1, 3, 2)$.
If we consider the sequence of customers sent in E_1 we get $U_1 = (1, 1, 1, 0, 0, 0)$ and $U_1' = (1, 1, 0, 1, 0, 0)$. In Example 12, we showed that $U_1' \trianglelefteq U_1$.
If we consider the sequence of customers sent in E_2 we get $U_2 = (0, 0, 0, 2, 0, 2)$ and $U_2' = (0, 0, 2, 0, 0, 2)$. The sequence U_2' is balanced. Therefore, $U_2' \trianglelefteq U_2$.
Finally, if we consider the sequence of customers sent in E_2 we get $U_3 = (0, 0, 0, 3, 0)$ and $U_3' = (0, 0, 0, 0, 3, 0)$, which are equal.

Since the sequence U' is more regular than U for all the event graphs, then the maximal waiting time under U' is smaller than the maximal waiting time under U (in the stochastic sense). This is true for all the distributions of the service times in the three event graphs.

However, if we consider $U'' = (1, 3, 1, 2, 1, 2)$ then $U_1'' \trianglelefteq U_1'$ but $U_2' \trianglelefteq U_2''$. The routing sequences U' and U'' are not comparable for the regularity ordering. For some distributions of the service times in the three event graphs then U' will be better than U'' and for some other distributions, U'' will be better than U'.

14.4 Application 2: Assignment to queues with no buffer with redundancy

In this section, we present an example of optimal control of admission into a single buffer queue with redundancy.

– Transmission opportunities occur at time $T_0, T_1, T_2, \ldots$. At each transmission opportunity, a controller can decide to actually transmit a packet, or not. If transmission occurs, we assume that it is instantaneous. Define $\zeta_n = T_n - T_{n-1}$.

– At a down link node in the network a packet that has been transmitted is buffered in a single buffer till its service there is completed. The duration of the service is exponentially distributed and it has expectation μ^{-1}. We assume that the delay between the transmitter and the down link node is zero (the results of this section do not change if we take this time to be any other constant period).

– If a packet is transmitted before the previous packet has completed its service time then it is lost.

A model that can be used to solve this problem was introduced and studied in Chapter 8. Here, we further focus on the following problem arising from higher layer network considerations. We consider the problem of redundancy of packet transmission: each packet is retransmitted at k consecutive transmission opportunities, which we call a frame, so as to decrease the probability of losses. Thus, only if all the k packets in a frame are lost then the frame, or equivalently the original information packet, is considered to be lost.

Our goal, roughly speaking, is to obtain a control with two objectives: on one hand it maximizes the average throughput, and on the other it minimizes the loss probabilities of frames. Since the throughput is the acceptance rate minus the losses, the problem can be formulated as maximization of the average number of accepted packets, on one hand, and minimizing the loss probabilities of frames, on the other.

We now formulate more precisely our problem. We define a control policy as a sequence $\{u_i\}$, $i \in \mathbb{N}$, where $u_i = 1$ if a transmission is scheduled at the ith transmission opportunity, and it is 0 otherwise. We consider here the case where the controller has no information on the state of the buffer (nor on which packet is the beginning of a frame).

Define $D(n) = \sum_{i=0}^{n} u_i$. The actual nth transmission occurs at time $p_u(n)$ (see Definition 14.1). We define the process X_n to be the number of packets in the buffer just prior to time $p_u(n)$. Define $G(u)$ as

$$G(u) = \limsup_{N \to \infty} \frac{1}{N} \left(\lambda D(N) + \sum_{j=1}^{N} \mathbb{E}^u \min_{i=0,\dots,k-1} X_{j-i} \right)$$

where λ is a negative constant. We consider the problem of minimizing G over all sequences u.

Note that $D(N)/N$ is the acceptance rate, whereas the second term is the average number of frames lost among the first N ones. The cost thus contains a term responsible for maximizing the throughput and a term for minimizing the loss rates. The parameter λ can be viewed as a Lagrange multiplier related to a constrained problem; indeed, in Chapter 8 a similar problem (for the case $k = 1$) is studied where the loss rate is to be minimized subject to a constraint on the rate of acceptance of packets.

Next, we focus on the cost $G(u)$. We shall restrict to policies u that are periodic (see Chapter 8 for a justification for doing so). Our main result, stated below, is that $G(u)$ is regular-preserving. (This allows us to conclude, in particular, that there exists an optimal sequence which is balanced.)

Theorem 107. *Assume that ζ_i are stationary and that the service times are i.i.d. exponentially distributed, and independent of the sequence ζ_k. Let $u \in P(T, n)^\infty$. Then*
(i) G can be rewritten as

$$G(u) = \frac{\lambda n}{T} + \frac{1}{n} \sum_{i=1}^{n} \mathbb{E} e^{-\mu d_i^k}.$$

where d_i^k is defined in (14.1).
(ii) $G(u)$ is regular-preserving for $n \geq K$.

Proof. (i) Let Σ_n denote the length of the time interval that started when the $(n - K + 1)$th packet was transmitted, and ended when the nth packet is transmitted. $f_n(u) := \min_{k=0,\ldots,K-1} X_{n-k}$ equals 0 if and only if there has been no service completion during Σ_n, and is otherwise 1. Thus, the expectation of $f_n(u)$ equals the probability that an exponential random variable with parameter μ is greater than or equal to Σ_n. Using the periodicity of u, this yields (i).

(ii) This follows directly from Theorem 108. Note that since $e^{(-\mu x)}$ is convex in x, $g(x) = \mathbb{E} e^{(-\mu x)}$ is convex too. □

14.5 Appendix: properties of the gap sequences

This appendix gives several properties of the gap sequences which have some interest by their own.

The two following lemmas are straightforward consequences of the definition of the gaps.

Lemma 89. *The gap sequences satisfy the following properties:*

$$d_j^1 + d_{j+1}^1 + \cdots + d_{j+k}^1 = d_j^{k+1}, \tag{14.11}$$
$$d_1^i + d_2^i + \cdots + d_n^i = iT, \tag{14.12}$$
$$d_j^i + d_{j+i}^{n-i} = T. \tag{14.13}$$

Proof. The proof is a straightforward consequence of the definition of the gaps. □

Lemma 90. *Let u and u' be two sequences in $P(T, n)$. If u' is a shift of u or a mirror of u (i.e. $\forall j, u_j = u'_{j+k}$ or $u_j = u'_{-j+k}$), then $\forall i, \{d^i(u)\} = \{d^i(u')\}$.*

Proof. The proof also follows directly from the definition of the gaps. □

Remark 40. The converse of Lemma 90 is false in general, as shown by the following counter example[1]; choose u and u' such that $d^1(u) = (1, 1, 2, 0, 2, 1, 1, 0)$

---

[1] This counter example was provided to the authors by Jérôme Galtier

and $d^1(u') = (1,1,0,2,0,1,1,2)$. u' is not a mirror nor a shift of u. However, for all order i, we have $\{d^i(u)\} = \{d^i(u')\}$. We will prove in the following (see Remark 41) that the converse of Lemma 90 is true in the special case of balanced sequences.

Lemma 91. *If $u \in P(T,n)$, then $d^1(u) \in P(n,T)$ and $d^1(d^1(u))$ is a shift of u.*

Proof. By using Equation (14.12), then $d^1(u) \in P(n,T)$. Now, let u' be a shift of u such that $p_{u'}(1) = 0$ and such that $d^1(u') = d^1(u)$. We will show that $d^1(d^1(u')) = u'$. Since $p_{u'}(i+1) = \inf\{j|\sum_{k=0}^j u'_k \geq i+1\}$, then $p_{u'}(i+1) - p_{u'}(i) = \#\{j|\sum_{k=0}^j u'_k = i\}$. Now we compute

$$d^1_i(d^1(u')) = \#\{j| \sum_{k=0}^j d^1_k(u') = i\}$$
$$= \#\{j|p_{u'}(j) - p_{u'}(0) = i\}$$
$$= \#\{j|p_{u'}(j) = i\} = u'_i.$$

$\square$

We can construct the different gaps by computing in the set $\mathbb{Z}[X]/(X^n - 1)$. Indeed, from the set $d^i(u)$, we can define the polynomial

$$P^i_u(X) \stackrel{\text{def}}{=} \sum_{j=1}^n d^i_j(u)X^{j-1}.$$

Note that by definition, we have $P^i_u(X) = (X^{i-1}+\cdots+X+1)P^1_u(X) \bmod (X^n - 1)$.

Lemma 92. *Let $i > 1$. The original sequence u can be retrieved from $d^i(u)$ if and only if i and n are relatively prime.*

Proof. There is a one to one correspondence between u and $d^1(u)$. Therefore, we rather show that the knowledge of $d^i(u)$ enables one to compute $d^1(u)$ if and only if i and n are relatively prime. If i and n are relatively prime, then the polynomials $X^n - 1$ and $X^{i-1} + \cdots + X + 1$ are relatively prime. Using the Bezout equality, there are two polynomials $u(X)$ and $v(X)$ in $\mathbb{Z}[X]$ and a non-zero integer k such that $k = (X^{i-1} +\cdots+ X + 1)u(X) + v(X)(X^n - 1)$. Therefore, by multiplying by $P^1_u(X)$, we get $kP^1_u(X) = P^i_u(X)u(X) \bmod (X^n - 1)$. The knowledge of $P^1_u(X)$ induces the knowledge of $d^1(u)$, which in turn gives all gaps by using Equation 14.11.

On the other hand, if i and n are not relatively prime, then $\gcd((X^{i-1} + \cdots + X + 1), X^n - 1) = k(X)$, and if we denote by $q(X) \stackrel{\text{def}}{=} (X^n - 1)/k(X)$, then $q(X) \in \mathbb{Z}[X]$ and $q(X)(X^{i-1} + \cdots + X + 1) = 0 \bmod (X^n - 1)$. We have $P^i_u(X) = P^1_u(X)(X^{i-1}+\cdots+X+1) \bmod (X^n - 1)$ as well as $P^i_u(X) =$

$(P_u^1(X) + q(X))(X^{i-1} + \cdots + X + 1) \bmod (X^n - 1)$. Therefore, two different gap sequences of order one are compatible with the gap sequence of order i.
□

Remark 41. Note that if for a sequence u, all the gaps of all order i differ by at most one and if $\{d^i(u)\} = \{d^i(u')\}$ for all i, then u and u' are balanced by lemma 81. On the other hand, all balanced sequences are shift of each other (see Theorem 101) (It is interesting to see that in this case, all mirror transformations are also shifts). This is an answer to the question asked in Remark 40 in the case where the gap sequences are all formed by two consecutive numbers.

14.6 Appendix: relations between regularity and multimodularity

It is interesting to see if regular preserving functions have some kind of discrete convexity properties. In this appendix we will show that all regular preserving functions are multimodular in some sense whereas the reverse is not true.

Let $e_i \in \mathbb{N}^T$ denote the vector with all components equal to 0 except the *i-th* component which is equal to 1. Define $b_i = e_{i-1} - e_i$ for $1 \leq i < T$, $b_T = e_{T-1} - e_0$ and $\mathcal{F} = \{b_1, b_2, \cdots, b_T\}$.

Since we deal with sequences which sums are fixed to n, we use an adapted version of the definition of multimodularity (see Chapter 12).

Definition 25. *A function f on $P(T, n)$ is multimodular with respect to $\mathcal{F}$ if for all $u \in P(T, n)$, $v, w \in \mathcal{F}$, $v \neq w$,*

$$f(u + v) + f(u + w) \geq f(u) + f(u + v + w), \tag{14.14}$$

whenever $u + w, s + v, s + v + w \in \mathbb{N}^T$.

This is not the classical definition given by Hajek in [59], which considers arbitrary sequences in $\mathbb{N}^T$. However, both definitions are closely related. One can show that the projection of a multimodular function (in the sense of Hajek) on $P(T, n)$ is multimodular according to Definition 25. (see Chapter 12 for a deeper insight on the definition given here.)

Theorem 108. *Let $f(u)$ be a function that only depends on $d^i(u)$, which is also denoted by $H_i(d^i(u))$. If $H_i(d^i(u))$ can be put under the from $\sum_{j=1}^n g(d_j^i(u))$, where g is a convex function, then it is is multimodular on $P(T, n)$.*

Proof. According to the definition of multimodularity, we need to check the inequality (14.14) for all $u \in P(T, n)$.

First note that for all $1 \leq k \leq T$, there exists j_k such that:

$$d_j^i(u + b_k) = \begin{cases} d_j^i(u) - 1 & \text{if } j = j_k, \\ d_j^i(u) + 1 & \text{if } j = j_k + i, \\ d_j^i(u) & \text{otherwise.} \end{cases} \qquad (14.15)$$

Now, for any $1 \leq k, l \leq p_{n+i}$ such that $j_k \neq j_l \pm i$, then b_l and b_k modify different gaps. Therefore,

$$f(u + b_k) + f(u + b_l) - f(u) - f(u + b_l + b_k) = 0.$$

For all $1 \leq k, l \leq T$ such that $j_k = j_l - i$, then,

$$\begin{aligned}
&f(u + b_k) + f(u + b_l) - f(u) - f(u + b_l + b_k) \\
&= \left(g(d_{j_k + 2i}^i(u)) + g(d_{j_k}^i(u)) + g(d_{j_k}^i(u) - 1) + \right. \\
&\quad \left. g(d_{j_k + i}^i(u) + 1) + g(d_{j_k + i}^i(u) - 1) + g(d_{j_k + 2i}^i(u) + 1) \right) - \\
&\quad \left(g(d_{j_k}^i(u)) + g(d_{j_k + 2i}^i(u)) + g(d_{j_k + i}^i(u)) + \right. \\
&\quad \left. g(d_{j_k}^i(u) - 1) + g(d_{j_k + 2i}^i(u) + 1) + g(d_{j_k + i}^i(u)) \right) \\
&= -2g(d_{j_k + i}^i(u)) + g(d_{j_k + i}^i(u) + 1) + g(d_{j_k + i}^i(u) - 1)
\end{aligned}$$

This is positive since g is convex. □

The following corollary covers the case when several gap orders are used.

Corollary 13. *If for all i, $H_i(d^i(u)) = \sum_{j=1}^n g_i(d_j^i(u))$, where g_i are convex functions, then define the function*

$$f(u) \stackrel{def}{=} \sum_{i=1}^n c_i H_i(d^i(u)), \qquad (14.16)$$

where c_i are non negative constants. Then f is multimodular.

Proof. For all i, the function $f_i(u) \stackrel{def}{=} H_i(d^i(u))$ is multimodular. A positive linear combination of multimodular functions is multimodular. □

Remark 42. Note that the functions $f_i(u)$ also form a "characteristic set" of regular preserving in the same way sums of convex functions play an important role among Schur convex functions ([88]).

Remark 43. Functions of the form 14.16 have been used in [36] as a criterion to test regularity of a sequence. In that paper, the minimization problem of functions of this form is written as under quadratic assignment formulation. This makes computations possible for small n and T and can be used for optimization purposes.

Remark 44. Finally, one may remark that regular preserving functions constitute an essential class among multimodular functions. Indeed, as shown in Section 14.3.2, if a given sequence in $P(T, n)$ minimizes all possible regular preserving functions, then this sequence in balanced.

References

1. E. Altman, S. Bhulai, B. Gaujal, and A. Hordijk. Optimal routing to m parallel queues with no buffers. In *Alerton Conference*, Monticello, Il, September 1999.
2. E. Altman, S. Bhulai, B. Gaujal, and A. Hordijk. Open-loop routing to m parallel servers with no buffers. *Journal of Applied Probability*, 37(3), September 2000.
3. E. Altman, B. Gaujal, and A. Hordijk. Optimal open-loop control of vacations, polling and service assignment. Technical Report 3261, INRIA, 1998.
4. E. Altman, B. Gaujal, and A. Hordijk. Admission control in stochastic event graphs. *IEEE Transaction on Automatic Control*, 45(5):854–868, 2000.
5. E. Altman, B. Gaujal, and A. Hordijk. Balanced sequences and optimal routing. *Journal of the ACM*, 47(4):752–775, 2000.
6. E. Altman, B. Gaujal, and A. Hordijk. Multimodularity, convexity and optimization properties. *Mathematics of Operations Research*, 25(2):324–347, 2000.
7. E. Altman, B. Gaujal, and A. Hordijk. Optimal open-loop control of vacations, polling and service assignment. *Queueing Systems*, 36:303–325, 2000.
8. E. Altman, B. Gaujal, and A. Hordijk. Simplex convexity with application to open-loop stochastic control in networks. In *39th Conf. on Decision and Control*. IEEE, 2000.
9. E. Altman, B. Gaujal, and A. Hordijk. Regular ordering and applications in control policies. *Journal of Discrete Event Dynamic Systems*, 12:187–210, 2002.
10. E. Altman, B. Gaujal, A. Hordijk, and G. Koole. Optimal admission, routing and service assignment control: the case of single buffer queues. In *CDC*, Tampa Bay, Fl., dec 1998. IEEE.
11. E. Altman and G. Koole. Control of a random walk with noisy delayed information. *Systems and Control Letters*, 24:207–213, 1995.
12. E. Altman and G. Koole. On submodular value functions of dynamic programming. *Stochastic Models*, 14(5), 1998.
13. E. Altman, Z. Liu, and R. Righter. Scheduling of an input-queued switch to achieve maximal throughput. Technical Report 96-020, Leavey School of Business and Administration, Santa Clara,, 1996.
14. E. Altman and S. Stidham. Optimality of monotonic policies for two-action markovian decision processes with applications to control of queues with delayed information. *Queuing Systems: Theory and Applications*, 21:267–291, 1995.
15. E. Altman and S. Stidham. Optimality of monotonic policies for two-action markovian decision processes, with applications to control of queues with de-

layed information. *QUESTA*, 21:267–291, 1995. special issue on optimization of queueing systems.

16. P. Arnoux, C. Mauduit, I. Shiokawa, and J. Tamura. Complexity of sequences defined by billiards in the cube. *BSMF (Bulletin de la Societé Mathématique de France)*, 122:1–12, 1994.

17. S. Asmussen, A. Frey, T. Rolski, and V. Schmidt. Does markov-modulation increase the risk? *ASTIN Bull*, 25:49–66, 1995.

18. .S. Asmussen and G. Koole. Marked point processes as limits of Markovian arrival streams. *Journal of Applied Probability*, 30:365–372, 1993.

19. S. Asmussen and C.A. O'Cinneide. On the tail of the waiting time in a markov-modulated M/G/1 queue. Technical report, Lund University, 1998.

20. F. Baccelli. Ergodic theory of stochastic Petri networks. *Annals of Probability*, 20(1):375–396, 1992.

21. F. Baccelli and P. Brémaud. *Elements of queueing theory*. Springer-Verlag, 1994.

22. F. Baccelli, S. Foss, and B. Gaujal. Free choice nets, an algebraic approach. *IEEE Transaction on Automatic Control*, 41(12):1751–1778, 1996.

23. F. Baccelli, G. Gohen, G. J. Olsder, and J.-P. Quadrat. *Synchronization and Linearity*. Wiley, 1992.

24. F. Baccelli and Z. Liu. On a class of stochastic recursive equations arising in queueing theory. *Annals of Probability*, 21(1):350–374, 1992.

25. F. Baccelli and V. Schmidt. Taylor series expansions for Poisson driven (max,+) linear systems. *The annals of Applied Probability*, 6(1):138–185, 1996.

26. A. Bar-Noy, R. Bhatia, J. Naor, and B. Schieber. Minimizing service and operation costs of periodic scheduling. In *Ninth Annual ACM-SIAM Symposium on Discrete Algorithms SODA*, 1998.

27. M. Bartroli and S. Stidham. Towards a unified theory of structure of optimal policies for control of network of queues. Technical report, Department of Operations Research, University of North Carolina, Chapel Hill, 1987.

28. Y. Baryshnikov. Complexity of trajectories in rectangular billiards. *Communications in Mathematical Physics.*, 175:43–56, 1995.

29. Y. Baryshnikov. Complexity of trajectories in rectangular billiards. *Commun. Math. Phys.*, 174:43–56, 1995.

30. N. Bäuerle and T. Rolski. A monotonicity result for the work-load in markov-modulated queues. *J. Applied Probability*, 35:741–747, 1998.

31. F. Bause and P. Kritzinger. *Stochastic Petri nets - An introduction to the theory*. Advanced Studies in Computer Science - Vieweg Verlagsgesellschaft, 1996.

32. J. Berstel and P. Séébold. Morphismes de Sturm. *Bulletin Belgian Mathematics Society*, 1:175–189, 1994.

33. R.K. Boel and J.H. Van Schuppen. Distributed routing for load balancing. In *Proceedings of the IEEE*, pages 210–221, 1989.

34. A. A. Borovkov. *Asymptotic methods in queueing theory*. Wiley, New York, 1984.

35. A. A. Borovkov and S. Foss. Stochastically recursive sequences and their generalizations. *Siberian Advances in Mathematics*, 2(1):16–81, 1992.

36. S. Borst and K.G. Kamakrishnan. Optimization of template-driven scheduling mechanisms. *J. of Discrete Event Dynamics Systems*, 1998.

37. J.Y. Le Boudec and P. Thiran. *Network calculus, A theorey of deterministic queing systems for the internet*. Number LNCS 2050. Springer Verlag, 2000.

38. M. Canales and B. Gaujal. Marking optimization and parallelism in marked graphs. Technical Report 2049, INRIA, 1993.
39. C.S. Chang. *Performance Garanties in communication networks*. Springer, 2000.
40. C.S. Chang, X. Chao, and M. Pinedo. A note on queues with Bernoulli routing. In *29th Conference on Decision and Control*, 1990.
41. G. Ciardo and C. Lindemann. Analysis of deterministic and stochastic petri nets. In *5th International Workshop on Petri Nets and Performance Models,*, pages 34–43, 1993.
42. E.G. Coffman, Z. Liu, and R.R. Weber. Optimal robot scheduling for web search engines. *Journal of Scheduling*, 1, 1999.
43. M. B. Combé and O. Boxma. Optimzation of static traffic allocation policies. *Theoretical Computer Science*, 125:17–43, 1994.
44. J. Desel and J. Esparza. *Free Choice Petri Nets*. Cambridge Tracts in Theorical Computer Science, 1995.
45. W. Fisher and K. S. Meier-Hellstern. The markov-modulated Poisson process (MMPP) cookbook. *Performance Evaluation*, 18:149–171, 1992.
46. A. Fraenkel. Complementary and exactly covering sequences. *Journal of Combinatorial Theory*, 14:8–20, 1973.
47. B. Gaujal. Optimal allocation sequences of two processes sharing a resource. *Journal of Discrete Event Dynamic Systems*, 7:327–354, 1997.
48. B. Gaujal and S. Haar. A limit semantics for timed petri nets. In R. Boel and G. Stremersch, editors, *Discrete Event Systems: Analysis and Control. Proceedings of WODES*, pages 219–226. Kluwer, 2000.
49. B. Gaujal, A. Hordijk, and D. van der Laan. On orders and bounds for multimodular functions. Technical Report MI 2001-29, Leiden University, 2001.
50. B. Gaujal and E. Hyon. Optimal routing policy in two deterministic queues. *Calculateurs Parallèles*, 2001. accepted for publication, also available as INRIA RR-3997.
51. B. Gaujal and E. Hyon. Routage optimal dans des réseaux de files d'attentes déterministes. In *MSR'2001*, Toulouse, 2001. in french.
52. P. Glasserman and D. D. Yao. Monotone optimal control of permutable gsmps. *Mathematics of Operation Research*, 19:449–476, 1994.
53. P. Glasserman and D. D. Yao. *Monotone Structure in Discrete Event Systems*. Wiley, New York, 1994.
54. M. Gondran and M. Minoux. *Graphs and algorithms*. Eyrolles, 1983.
55. R. Graham. Covering the positive integers by disjoint sets of the form $\{[n\alpha + \beta] : n = 1, 2, \cdots\}$. *Journal of Combinatorial Theory*, 15:354–358, 1973.
56. D. Gross and C. Harris. *Fundamentals of queueing theory*. Series in Probabilty and mathematical statistics. Wiley, 1974.
57. J. Gunawardena, editor. *Idempotency*. Publications of the Newton Institute. Cambridge University Press, 1996. to appear.
58. B. Hajek. Optimal control of two interacting service stations. *IEEE Trans. Aut. Cont.*, 29:491–499, 1984.
59. B. Hajek. Extremal splittings of point processes. *Mathematics of Operation Research*, 10(4):543–556, 1985.
60. R. Hariharan, V. G. Hulharni, and S. Stidham. A survey of research relevant to virtual circuit routing in telecommunication networks. Technical Report WC/OR/TR 90-13, University of North Carolina at Chapel Hill, 1990.

61. A. Hordijk. Comparison of queues with different discrete-time arrival processes. *Probability in Engineering and Information Sciences*, 15:1–14, 2001.

62. A. Hordijk and G. Koole. On the assignment of customers to parallel queues. *Probability in the Engineering and Informational Sciences*, 6:495–511, 1992.

63. A. Hordijk and G. Koole. On the optimality of LEPT and μc rules for parallel processors and dependent arrival processes. *Adv. Applied Probability*, 25:979–996, 1993.

64. A. Hordijk, G. Koole, and J. A. Loeve. Analysis of a customer assignment model with no state information. *Probability in the Engineering and Informational Sciences*, 8:419–429, 1994.

65. A. Hordijk, J. A. Loeve, and J. Tiggelman. Analysis of a finite-source customer assignment model with no state information. *Mathematical Methods of Operation Research*, 47:317–336, 1998.

66. A. Hordijk and J.A. Loeve. Optimal noncyclic server allocation in a polling model. In *Proceedings of the 36th IEEE Conference on Decision and Control*, San-Diego, California, December 1997.

67. A. Hordijk and R. Schassberger. Weak convergence for generalized semi-markov processes. *Stoch. Proc. Appl.*, 12:271–291, 1982.

68. A. Hordijk and D. van der Laan. Periodic routing to parallel queeus with bounds on the average waiting time. Technical Report MI 2000-44, Leiden University, 2000.

69. A. Hordijk and D. van der Laan. Bounds for deterministic periodic routing sequences. In K. Aardal and B. Gerards, editors, *8th international IPCO Conference*, pages 236–250. Springer, 2001.

70. A. Hordijk and D. van der Laan. Note on the convexity of the stationary waiting time as a function of the density. *Prob. Engin. Inform. Sci.*, 17:503–508, 2003.

71. A. Hordijk and D. van der Laan. Periodic routing to parallel queues and billiard sequences. *Mathematical Methods of Operations Research*, 2003. To appear.

72. A. Hordijk and D. van der Laan. The unbalance and bounds of the average waiting time for periodic routing to one queue. *Mathematical Methods of Operations Research*, 2003. To appear.

73. A. Hordijk and D. A. van der Laan. Comments on waiting times of determisnitic queues and Farey's intervals. private communication.

74. A. Hordijk and D. A. van der Laan. Periodic routing to parallel queues with bounds on the average waiting time. Technical Report MI N. 2000-44, Leiden University, 2000.

75. P. Hubert. Propriétés combinatoires des suites définies par le billard dans les triangles pavants. *heoretical Computer Science*, 164(1-2):165–183, 1996.

76. A. Itai and Z. Rosberg. A golden ratio control policy for a multiple-access channel. *IEEE Trans. on Automatic Control*, 29:712–718, 1984.

77. Traffic control and congestion control in b-isdn, 1995. Perth.

78. A. Jean-Marie. The waiting time distribution in Poisson-driven deterministic systems. Technical Report RR3088, INRIA, 1999.

79. G. Koole. On the pathwise optimal Bernoulli routing policy for homogeneous parallel servers. *Mathematics of Operations Research*, 21:469–476, 1996.

80. G. Koole. Structural results for the control of queueing systems using event-based dynamic programming. Technical Report WS-461, Vrije Universiteit Amsterdam, 1996.

81. G. Koole. On the static assignment to parallel servers. *IEEE Trans. on Automatic Control*, 44:1588–1592, 1999.

82. Z. Liu and R. Righter. Optimal load balancing on distrinuted homogeneous unreliable processors. *Journal of Operations Research*, 46:563–573, 1998.

83. J. A. Loeve. *Markov Desicion Chains With Partial Information*. PhD thesis, Leiden University, 1995. Chap 8: Regularity of words and periodic policies.

84. M. Lothaire. *Algebraic Combinatorics on Words*, chapter Sturmian Words. To appear, available on www-igm.univ-mlv.fr/ berstel/Lothaire/.

85. M. Lothaire. *Mots*, chapter Tracé de droites, fractions continues et morphismes itérés (J. Berstel). Hermes, 1991.

86. M. Lothaire. *Mots*, chapter Tracé de droites, fractions continues et morphismes itérés. Hermes, 1991.

87. D. M. Lucantoni, K. S. Meier-Hellstern, and M. F. Neuts. A single-server queue with server vacations and a class of non-renewal arrival processes. *Advances in Applied Probability*, 22:275–705, 1990.

88. A. W. Marshall and I. Olkin. *Inequalities: Theory of Majorization and its Applications*, volume 143 of *Mathematics in Science and Engineering*. Academic Press, 1979.

89. R. Morikawa. Disjoint sequences generated by the bracket function i-vi. Technical Report 26(1985), 28(1988), 30(1989), 32(1992), 34(1993), Bulletin of the Faculty of Liberal Arts, Nagasaki University, 1993.

90. R. Morikawa. On eventually covering families generated by the bracket function i-v. Technical Report 23(1982), 24(1983), 25(1984), 26(1985), 25(1985), 36(1995), Bulletin of the Faculty of Liberal Arts, Nagasaki University, 1995.

91. M. Morse and G. A. Hedlund. Symbolic dynamics II- Sturmian trajectories. *American Journal of Mathematics*, 62:287–306, 1940.

92. M.F. Neuts. *Structured Stochastic Matrices of $M|G|1$ Type and their Applications*. Marcel Dekker, 1989.

93. M. Newman. Roots of unity and covering sets. *Mathematics Annals*, 191:279–282, 1971.

94. B. Parvaix. Propriétés d'invariance des mots sturmiens. *Journal de Théorie des Nombres de Bordeaux*, 9:351–369, 1997.

95. M. L. Puterman. *Markov decision processes*. Wiley Series in Probabilities and Mathematical Statistics. Wiley, 1994.

96. R. T. Rockafellar. *Convex analysis*. Princeton mathematical series. Princeton University press, 1972.

97. T. Rolski. Comparison theorems for queues with dependent interarrival times. In *lecture Notes in Control and Information Sciences*, volume 60, pages 42–71. Springer Verlag, 1983.

98. S.M. Ross. Average delay in queues with non-stationary poisson arrivals. *Journal of Applied Probability*, 15:602–609, 1978.

99. M. Senechal. *Quasicrystals and geometry*. Cambridge University Press, 1995.

100. L. I. Sennott. *Stochastic Dynamic Programming and the Control of Queueing Systems*. Wiley-Interscience Series in Probability and Statistics. John Wiley & Sons, 1998.

101. H. Shirakawa, M. Mori, and M. Kijima. Evaluation of regular splitting queues. *Comm. Statist. Stochastic Models*, 5:219–234, 1989.

102. Y. G. Sinai. *Introduction to Ergodic Theory*. Princeton University Press, Princeton, N.J., 1976.

103. P.D. Sparaggis, C.G. Cassandras, and D.F. Towsley. On the duality between routing and scheduling systems with finite buffer spaces. *IEEE Trans. on Automatic Control*, 38:1440–1446, 1993.

104. S. Stidham. Optimal control of admission to a queueing system. *IEEE Transaction on Automatic Control*, AC-30(8):705–713, 1985.

105. R. Sznajder and J. A. Filar. Some comments on a theorem of hardy and littlewood. *Journal of Optimization Theory and Application*, 75(1):201 – 208, 1992.

106. R. Tijdeman. On disjoint pairs of sturmian bisequences. Technical Report W96-02, Leiden University, The Netherlands, 1995.

107. R. Tijdeman. On complementary triples of Sturmian sequences. *Indagationes Mathematicae*, pages 419–424, 1996.

108. R. Tijdeman. Fraenkel's conjecture for six sequences. Technical Report W98-24, Leiden University, The Netherlands, 1998.

109. R. Tijdeman. Intertwined periodic sequences, Sturmian sequences and Beatty sequence. *Indagationes Mathematicae (N.S.)*, 9:113–122, 1998.

110. G. Urvoy, Y. Dallery, and G. Hebuterne. Cac procedures for leaky bucket-constrained sources. *Performance Evaluation*, 41(2-3):117–132, 2000.

111. S. Vamvakos and V. Anantharam. On the departure process of a leaky bucket system with long-range dependent input traffic. *Queueing Systems*, 23(1-3):191–214, 1998.

112. D. A. van der Laan. Routing jobs to servers with deterministic service times. Technical Report MI N. 2000-20, Leiden University, 2000.

113. L. Vuillon. Combinatoire des motifs d'une suite sturmienne bidimensionnelle. *Theoritical Computer Science*, 209:261–285, 1998.

114. R.R. Weber and S. Stidham. Optimal control of service rates in networks of queues. *Advances in Applied Probability*, 19:202–218, 1987.

115. H. S. Wilf. *Generatingfunctionology*. Academic Press, 2nd edition, 1994.

Index

Printing and Binding: Strauss GmbH, Mörlenbach

4. For evaluation purposes, manuscripts may be submitted in print or electronic form (print form is still preferred by most referees), in the latter case preferably as pdf- or zipped ps-files. Lecture Notes volumes are, as a rule, printed digitally from the authors' files. To ensure best results, authors are asked to use the LaTeX2e style files available from Springer's web-pages at

http://www.springer.de/math/authors/index.html

Macros in LaTeX2.09 and TeX are available on request from: lnm@springer.de Careful preparation of the manuscripts will help keep production time short besides ensuring satisfactory appearance of the finished book in print and online. After acceptance of the manuscript authors will be asked to prepare the final LaTeX source files (and also the corresponding dvi-, pdf- or zipped ps-files) together with the final printout made from these files. The LaTeX source files are essential for producing the full-text online version of the book

(http://www.springerlink.com/link/service/series/0304/tocs.htm)..

The actual production of a Lecture Notes volume takes approximately 8 weeks.

5. Authors receive a total of 50 free copies of their volume, but no royalties. They are entitled to a discount of 33.3 % on the price of Springer books purchased for their personal use, if ordering directly from Springer.

6. Commitment to publish is made by letter of intent rather than by signing a formal contract. Springer-Verlag secures the copyright for each volume. Authors are free to reuse material contained in their LNM volumes in later publications: A brief written (or e-mail) request for formal permission is sufficient.

Addresses:
Professor J.-M. Morel, CMLA,
École Normale Supérieure de Cachan,
61 Avenue du Président Wilson, 94235 Cachan Cedex, France
E-mail: Jean-Michel.Morel@cmla.ens-cachan.fr

Professor F. Takens, Mathematisch Instituut,
Rijksuniversiteit Groningen, Postbus 800,
9700 AV Groningen, The Netherlands
E-mail: F.Takens@math.rug.nl

Professor B. Teissier, Université Paris 7
Institut Mathématique de Jussieu, UMR 7586 du CNRS
Équipe "Géométrie et Dynamique", 175 rue du Chevaleret
75013 Paris, France
E-mail: teissier@math.jussieu.fr

Springer-Verlag, Mathematics Editorial, Tiergartenstr. 17,
69121 Heidelberg, Germany,
Tel.: +49 (6221) 487-8410
Fax: +49 (6221) 487-8355
E-mail: lnm@springer.de